Colin Powell

MATHEMATICS AND STATISTICS
FOR USE IN THE BIOLOGICAL AND PHARMACEUTICAL SCIENCES

MATHEMATICS AND STATISTICS

FOR USE IN THE BIOLOGICAL
AND PHARMACEUTICAL SCIENCES

LEONARD SAUNDERS
*Professor of Pharmaceutical Chemistry
University of London*

AND

ROBERT FLEMING
*Lecturer in Pharmaceutical Chemistry
School of Pharmacy, University of London*

SECOND EDITION
Revised and enlarged
to incorporate
an introduction to computer techniques

LONDON
THE PHARMACEUTICAL PRESS
17 Bloomsbury Square WC1A 2NN

First published 1957
Second impression (revised) 1966
Second edition 1971

Copyright © 1971 by the Pharmaceutical Society of Great Britain

No part of this book may be reproduced in any form or by any means, including photocopying, without written permission from the publisher

ISBN 0 85369 077 4

Made and printed in Great Britain by William Clowes & Sons, Limited, London, Beccles and Colchester

CONTENTS

		Page
Preface to the Second Edition		vii
Preface to the First Edition		ix
Chapter 1.	Arithmetic and computing	1
Chapter 2.	Algebra	27
Chapter 3.	Graphs	51
Chapter 4.	Series, e and natural logarithms	66
Chapter 5.	Differential calculus	77
Chapter 6.	Higher derivatives and partial differentiation	93
Chapter 7.	Integration	106
Chapter 8.	Trigonometry	121
Chapter 9.	Differential equations	136
Chapter 10.	Equations and series for describing experimental measurements	148
Chapter 11.	Probability	159
Chapter 12.	Statistical analysis of repeated measurements	172
Chapter 13.	Comparison of data by statistical methods. Tests of significance	190
Chapter 14.	Some applications of statistics to biological assay and bacteriology	210
Chapter 15.	Some applications of statistics in pharmacy	236
Appendix I.	Fundamental constants, approximations and conversion factors	257
Appendix II.	Triangles, lengths, areas, volumes and analytical geometry	259
Appendix III.	Standard integrals	265
Appendix IV.	Stirling's approximation	266
Appendix V.	Mean and variance of the binomial and Poisson distributions	268
Appendix VI.	The normal distribution	270
Appendix VII.	Variance of a function of variates	273
Appendix VIII.	Some theorems in statistics	275

CONTENTS

	Page
APPENDIX IX. Regression variance	277
APPENDIX X. Solution of the diffusion equation	279
APPENDIX XI. Binomial theorem for any index	282
APPENDIX XII. Statistical tables	284
FOUR-FIGURE LOGARITHMS	288
FOUR-FIGURE ANTILOGARITHMS	290
ANSWERS TO PROBLEMS	293
INDEX	301

COMPUTER PROGRAMMES

COMP 1.	Fortran programme to give a table of values from an algebraic equation	36
COMP 2.	Use of a library programme to solve simultaneous equations	47
COMP 3.	Fortran programme for Newton solution of a bi-exponential equation	90
COMP 4.	Programme to integrate a function by Simpson's rule	117
COMP 5.	Programme to determine the area under a curve by Simpson's rule	119
COMP 6.	Minimum by parabola	157
COMP 7.	Computer programme for mean, variance, standard deviation and error, and limits of error	181
COMP 8.	Computer programme for t	194
COMP 9.	Linear regression	206
COMP 10.	Fortran programme for bioassay	218

PREFACE TO THE SECOND EDITION

THE growing availability of digital computers is changing the approach of the quantitative scientist to mathematics. In this change, some of the elegance of the subject is being replaced by the cruder methods of numerical analysis. Rather than wrestling with algebra and calculus to solve equations, the method of finding rough numerical solutions and then refining them by successive approximations is both quicker and easier when a computer is available.

Some knowledge of the elements of computing is necessary for graduates in the science-based subjects of pharmacy and medicine if they are going to be in a position to make use of the further developments in computing which are certain to come in the next decade. I have therefore added an introductory course in computing to this book. This course is spread throughout the chapters as additions to the previous text, and so, for a course in mathematics and statistics only, these sections may be omitted and perhaps used in a subsequent course in computing.

In my experience, the best way to learn the elements of programming for the solution of mathematical and statistical problems is to study programmes which have been proved to be successful and which solve problems with which the student is already familiar. I have adopted this approach in the computing sections.

Explaining a computer language is a dull operation which has been undertaken in a number of manuals. I have therefore set out to explain the language of Fortran IV, one of the most widely used programming languages for mathematical work, in terms of its use to solve problems, most of which were in the text of the previous edition.

When the reader has fully understood the ten programmes in this book he should be in a position to write programmes to solve his own problems. However, he will need to have available a reasonably complete manual of the language which he is using and for Fortran IV I would strongly recommend *A Guide to Fortran IV Programming* by Daniel D. McCracken, J. Wiley & Sons Inc., New York. Another book recommended for further reading is *The Statistical Analysis of Experimental Data* by J. Mandel, Interscience, New York.

LONDON   L. SAUNDERS
July, 1971

PREFACE TO THE FIRST EDITION

WE hope that this course will prove helpful to students in schools, technical colleges and universities, and also to practitioners in pharmacy, medicine, biology and chemistry who may wish to brush-up their mathematical knowledge.

Unfortunately, it is almost impossible to write a book on mathematics in the form of an easy narrative, and a certain amount of time and concentration will be required in reading a book of this type. We suggest that readers should go through each chapter twice, the first time to get a general idea of the subject matter, and the second time to follow the arguments in detail, preferably working out the proofs and examples on paper for themselves. Finally, an attempt should be made to solve the problems at the end of the chapter, comparing the answers with those given at the end of the book. The problems in statistics in Chapters 11, 12 and 13 can be worked out most readily with the help of an accurate set of square and square root tables, such as Barlow's Tables (see below), although it is possible to solve them without this aid. One of the problems of Chapter 10 requires log-log graph paper, and Problems 2 and 3 of Chapter 12 require arithmetic probability graph paper.

We have not attempted to give a comprehensive set of references, but suggest that the following books, which contain extensive bibliographies, should be consulted for further reading:

> G. Stephenson, *Mathematical Methods for Science Students*. Longmans, Green and Co. Ltd., London.
> G. J. Kynch, *Mathematics for Chemists*. Butterworths Scientific Publications, London.
> O. L. Davies, *Statistical Methods in Research and Production*. Oliver and Boyd, Ltd., Edinburgh.
> J. H. Burn, D. J. Finney and L. G. Goodwin, *Biological Standardisation*. Oxford University Press, London.

A valuable set of square and square root tables for use in statistical calculation is included in

> L. J. Comrie (Ed.), *Barlow's Tables*. Spon Ltd., London.

When accurate values of statistical quantities are required reference should be made to

> R. A. Fisher and F. Yates, *Statistical Tables for Biological, Agricultural and Medical Research*. Oliver and Boyd Ltd., Edinburgh.

PREFACE TO THE FIRST EDITION

We are indebted to Professor Sir Ronald A. Fisher, Cambridge, Dr. Frank Yates, Rothamsted, and to Oliver and Boyd Limited, Edinburgh, for permission to include in Appendix XII abridged versions of tables from their book.

We thank our colleagues and friends who have helped us by reading the script and by giving us useful suggestions. We particularly thank Dr. J. W. Fairbairn, Dr. J. R. Hodges and Dr. B. A. Wills for helping with Chapter 14 and for supplying us with numerical data from their own experimental work. Finally, we acknowledge the important contribution of Mr. S. C. Jolly, of the Scientific Publications Department of The Pharmaceutical Society of Great Britain, who has arranged the text in a more presentable form than our original script, and to the help and advice received from the staff of the printers.

L. SAUNDERS
R. FLEMING

LONDON
February, 1957

CHAPTER 1

Arithmetic and Computing

TRADE is said to be responsible for the development of numbering systems. The earliest numerals were represented by notches on a stick or by scratches on a piece of pottery, but these methods of numeration and the earliest numbering systems, including that used by the Romans, were inadequate in that they did not contain the number zero. The earliest records of arithmetic calculations are said to date from about 3400 B.C., and from the beginning the successful application of mathematics to practical problems has been based on accurate arithmetic. The development of slide rules, calculating machines and computers has reduced much of the labour of arithmetic, but there still remains the necessity for checking results, an operation that should never be omitted.

1. Numbers, factors and primes

The 'alphabet' of the decimal system consists of the ten familiar symbols 0, 1, 2, 3, 4, 5, 6, 7, 8 and 9. All numbers, however large or small, can be expressed in terms of these symbols, except irrational numbers (see § 8).

Most whole numbers, called *integers*, can be expressed as the product of several numbers, called *factors* of the original number. For instance, the integer 12 can be written as 2×6, or $2 \times 2 \times 3$, or 4×3, the sign '$\times$' between the numbers indicating multiplication.

Certain integers cannot be split into factors, and these are called *primes*. The smaller primes are 2, 3, 5, 7, 11, 13, 17, 19, 23, 29, 31, 37, 41 and 43.

2. Zero and infinity

When zero and a number are added together, the answer is equal to the number: thus $4+0 = 4$. Multiplication of a number by zero gives the answer zero: thus $4 \times 0 = 0$.

If a finite number is divided by a very small number, the *quotient*, i.e. the result of a division, is a very large number, e.g. $1/0 \cdot 000001 = 1$ million. If the very small number is made smaller still so that its value approaches zero, the quotient becomes an extremely large number and approaches the value 'infinity', which is denoted by the symbol '∞'. Thus

$$1/n \to \infty \quad \text{as} \quad n \to 0.$$

The symbol '$\to$' means 'approaches'; in words, the relation just quoted is 'one divided by n approaches infinity as n approaches zero'.

CHAPTER I

3. Fractions, decimals and negative numbers

A fraction consists of two integers written one above the other, the upper figure being the *numerator* and the lower the *denominator*. Division of the numerator by the denominator results in a decimal, and if the quotient is less than 1, the symbol zero should always be written in front of the decimal point, e.g. $\frac{1}{4} = 0\cdot 25$.

The minus sign '−' is always written in front of a negative number, but the positive sign '+' is often omitted. The multiplication of two positive numbers or two negative numbers always results in a positive product: thus $2 \times 2 = 4$ and $(-2) \times (-2) = 4$. The multiplication of a positive number and a negative number always results in a negative product: thus $2 \times (-2) = -4$. It follows, therefore, that the square of any number, either positive or negative, is always positive.

If several factors are multiplied together and there is an even number of negative factors, the product will be positive. If there is an odd number of negative factors, the product will be negative: thus

$$(-2) \times 4 \times 3 \times (-4) \times (-1) = -96,$$

since there are three negative factors.

4. Roots

The square root of a number is the quantity which when multiplied by itself is equal to the original number. For example, the square root of 4 is 2, or it may be −2, since either of these numbers multiplied by itself gives 4. Square roots are denoted by the symbol '$\sqrt{}$'; thus $\sqrt{4}$ equals ± 2, the symbol '$\pm$' meaning plus or minus.

A cube root of a number is the quantity which when multiplied by itself to three factors gives the original number. Cube roots are denoted by the symbol '$\sqrt[3]{}$'; for example $\sqrt[3]{8}$ equals 2, since $2 \times 2 \times 2 = 8$.

The nth root of a number is the quantity which when multiplied by itself to n factors gives the original number. It is denoted by the symbol '$\sqrt[n]{}$'. Thus if a equals $\sqrt[n]{x}$, then

$$a \times a \times a \ldots \text{ to } n \text{ factors} = x.$$

5. The metric system

One of the reforms that emerged from the French Revolution was a revision of the French measuring system. In 1799 the *metre* became the standard unit of length; it was defined as one ten-millionth part of the meridian passing through Paris at sea-level. The practical working standard for the metre consists of a platinum–iridium bar with two fine marks engraved on it and kept at 0°C. In 1964, the Conférence Générale des Poids et Mesures adopted a newly rationalised set of metric units, the Système International d'Unités (SI Units). In this system, the basic units are the *metre*, the *kilogramme*, the *second*, the *ampere*, the *kelvin* or degree absolute of temperature, and the *candela* which is the unit of luminous intensity. The *litre* is defined as the cubic decimetre. In the publication of the Symbols Committee of the Royal Society, 'Symbols, Signs, and Abbreviations', London 1969, it is recommended that this set of units based on the mks (metre-kilogramme-second) should be adopted in all scientific work.

The use of some familiar units which are not directly compatible with SI units should be, according to this publication, progressively abandoned. Units in this category are the pressure units of the *atmosphere* and the *millimetre of mercury*; the energy unit, the *calorie*; the temperature unit of the *degree Celsius* or *centigrade*; *electrostatic* and *electromagnetic* units.

Some other units whose use it is recognised will continue for some time and which are compatible with SI units and may be directly converted to corresponding SI units by multiplying by 10 to a positive or negative integral power, are the ångström unit of length; the litre; the pressure unit, the bar; the erg and the dyne. Two of these units, the litre and the bar, are certain to be used extensively in many contexts and so it is difficult to understand why this Royal Society publication recommends that the

TABLE 1

Prefixes for SI units

Factor by which basic unit is multiplied	Prefix name	Prefix abbreviation
10^{12}	tera	T
10^{9}	giga	G
10^{6}	mega	M
10^{3}	kilo	k
10^{2}	hecto	h
10	deka	da
10^{-1}	deci	d
10^{-2}	centi	c
10^{-3}	milli	m
10^{-6}	micro	μ
10^{-9}	nano	n
10^{-12}	pico	p
10^{-15}	femto	f
10^{-18}	atto	a

The prefix should be written immediately in front of the unit it qualifies so as to avoid any ambiguity, as for example in the use of m for milli and for metre; if the unit qualified by the prefix is raised to a power so is the prefix, for example mm^2 is $(mm)^2$ or $10^{-6} m^2$ *not* $m(m^2)$ *or* $10^{-3} m^2$.

use of these names should be progressively abandoned. In commerce generally it seems likely that the litre will be used as a working unit of volume—the name 'litre' is much more attractive than the awkward 'cubic decimetre'. Similarly, a pressure unit near to normal atmospheric pressure is certain to be widely used. The bar, which is equal to 10^5 SI units (newton per square metre) and corresponds to approximately 750 mm of mercury (1 atmosphere pressure equals 1·01325 bar), is ideally suited for this purpose.

TABLE 2

Dimensions of derived SI units

Derived unit	Definition	Dimensional formula	Unit in metre-kilogramme-second system (SI or mks)
Area	length squared	L^2	m^2
Volume	length cubed	L^3	$m^3 = 1000$ litres
Velocity	length per unit time	LT^{-1}	$m\ s^{-1}$
Density	mass per unit volume	ML^{-3}	$kg\ m^{-3}$
Flow (volume)	volume per unit time	L^3T^{-1}	$m^3\ s^{-1}$
Acceleration	velocity change per unit time	LT^{-2}	$m\ s^{-2}$
Force	product of mass and acceleration	MLT^{-2}	newton, N
Pressure	force per unit area	$ML^{-1}T^{-2}$	$N\ m^{-2} = 10^{-5}$ bar
Work (or energy)	product of force and length	ML^2T^{-2}	joule, J
Power (or rate of working)	work per unit time	ML^2T^{-3}	$J\ s^{-1}$
Viscosity	force per unit area per unit velocity gradient*	$ML^{-1}T^{-1}$	$N\ s\ m^{-2}$
Kinematic viscosity	ratio of viscosity to density	L^2T^{-1}	$m^2\ s^{-1}$
Diffusion coefficient	rate of movement of solute per unit area per unit concentration gradient†	L^2T^{-1}	$m^2\ s^{-1}$
Surface tension	force per unit length (or energy per unit area)	MT^{-2}	$N\ m^{-1}$ ($J\ m^{-2}$)

* Velocity gradient is velocity per unit length, i.e. T^{-1}, so that the units of viscosity are $MLT^{-2}/(L^2 \times T^{-1})$, i.e. $ML^{-1}T^{-1}$.

† Concentration is mass per unit volume, i.e. ML^{-3}. Concentration gradient is concentration per unit length, i.e. ML^{-4}. Hence the units of diffusion coefficient are $MT^{-1}/(ML^{-4} \times L^2)$, i.e. L^2T^{-1}. *Note :* There is no term connected with concentration in the final unit.

A notable inconsistency in the publication cited above is that whereas the basic unit of mass is the kilogramme, the proposed unit of quantity of a substance is the mole (mol) or gramme-molecule. If a consistent set of mks units is to be used it is much more straightforward to adopt the kilogramme-molecule, kmol, as the basic unit of quantity. Molar concentrations in moles per litre are then exactly equal to the mks concentrations in kilomoles per cubic metre.

A list of prefixes to scale the SI units up or down by factors of 10 is given in Table 1, the names of and abbreviations for units derived from the basic set of SI units is given in Table 2.

The unit of time, the *second*, is based upon astronomical observations, and is an exception to the general rule of the metric system, because in the larger units, the minute and the hour, a factor of 60 is used instead of 10. The introduction of the factor 60 is attributed to the astronomers of Babylon.

The basic units of the metric system are not always convenient for practical use, and multiples or submultiples of these basic units, denoted by the prefixes listed in Table 1, may be used.

Very small units of length are required when considering, for example, particles of atomic size. In SI units, the nanometre (nm) and picometre (pm) are used.

$$1 \text{ nanometre} = 10^{-9} \text{ metre} = 10 \text{ ångström (Å)}$$

$$1 \text{ picometre} = 10^{-12} \text{ metre} = 0.01 \text{ Å}.$$

6. Dimensions

In mensuration, three units are regarded as fundamental: mass (M), length (L) and time (T). Other units such as force and velocity are derived from these. A list of dimensions of a number of derived units is given in Table 2.

For expressing heat, luminous intensity, and electrical quantities, a new dimension, in addition to M, L and T, is introduced in each case. For heat, the new dimension is temperature K, the SI symbol for degrees kelvin; for luminous intensity it is the candela (cd); for electricity, it is the ampere (A), defined in terms of forces between conductors carrying this current.

7. Dimensional analysis

Dimensional analysis is a useful method for determining the general form of equations relating several measurable quantities. The basic rule is that the net dimensions on the two sides of the equation should be identical.

EXAMPLE. *Given that the rate of volume flow ($\dot{v}$) of a liquid through a capillary is inversely proportional to the length of the tube (l) and is also a function of the pressure drop across the tube (p), the radius of the tube (r) and the viscosity of the liquid (η), use dimensional analysis to establish the form of the function.*

From the information given relating volume-flow rate, tube dimensions and viscosity of the liquid, it follows that

$$\dot{v} = \frac{1}{l} \cdot f(p, r, \eta),$$

where f( ) means 'function of'. It is assumed that the unknown function can be expressed in the form $kp^x r^y \eta^z$, where k is a dimensionless constant, and x, y and z are unknown indices. Therefore

$$\dot{v} = \frac{k}{l} \cdot p^x r^y \eta^z.$$

Consider the left-hand side of the equation; the dimensions of $\dot{v}$ are volume per unit time, i.e. $L^3 T^{-1}$. There is no mass term and this is indicated by M^0. The dimensions of the left-hand side, therefore, are $M^0 L^3 T^{-1}$.

The dimensions of the right-hand side are

$$L^{-1}(\text{pressure})^x(\text{length})^y(\text{viscosity})^z = L^{-1}(ML^{-1}T^{-2})^x(L)^y(ML^{-1}T^{-1})^z.$$

Collecting terms gives

$$= M^{(x+z)} L^{(-1-x+y-z)} T^{(-2x-z)}.$$

According to the principles of dimensional analysis, the dimensions on the right-hand side of the equation are equal to those on the left-hand side. Hence, equating the indices of M gives

$$0 = x+z. \qquad \text{(i)}$$

Equating the indices of L gives

$$3 = (-1-x+y-z). \qquad \text{(ii)}$$

Equating the indices of T gives

$$-1 = -2x-z. \qquad \text{(iii)}$$

Adding equations (i) and (iii) gives, after multiplying throughout by -1,

$$1 = x, \quad \text{and hence} \quad z = -1.$$

Substituting these values in equation (ii) gives

$$3 = -1-1+y+1, \quad \text{i.e.} \quad y = 4.$$

The form of the equation is therefore

$$\dot{v} = \frac{1}{l} \cdot k p r^4 \eta^{-1} = \frac{k p r^4}{l \eta}.$$

The numerical value of the dimensionless constant k cannot be derived from dimensional analysis, and it must be found either by experiment or by more detailed theoretical treatment of the flow of a liquid through a capillary.

8. Indices

The limitations of the ten basic symbols of arithmetic become immediately apparent when the number of molecules in 1 kilogramme-molecule (Avogadro number) is written down; it is

$$602,000,000,000,000,000,000,000,000.$$

Such numbers are cumbersome and are better expressed by using indices. As has been explained in § 1, numbers other than primes can

ARITHMETIC AND COMPUTING 7

be expressed as the products of smaller numbers called factors. When these factors are identical, the number of such factors can be indicated by an index number written as a superscript to the right of the factor. For example

$$32 = 2 \times 2 \times 2 \times 2 \times 2 = 2^5.$$

Expressed in words, 2^5 is 'two raised to the power of five', or simply as 'two to the fifth'. Similarly, the Avogadro number can be written as 602×10^{24}, or more usually as $6 \cdot 02 \times 10^{26}$ kmol^{-1}. It should be noted that a number raised to the power of one is equal to the number; thus $10^1 = 10$.

A number raised to a negative integral index, i.e. a negative whole number, is equal to the reciprocal of the number raised to an equal positive index: thus

$$2^{-5} = \frac{1}{2^5} = \frac{1}{32}.$$

A fractional index indicates the root of the number, e.g. $8^{1/3}$ means the cube root of 8.

Decimal indices also indicate that a root is taken: thus

$$32^{0 \cdot 2} = 32^{1/5} = \sqrt[5]{32} = 2.$$
$$2^{0 \cdot 6} = 2^{3/5} = (2^{1/5})^3 = (\sqrt[5]{2})^3.$$

Indices in multiplication and division. The product of a number raised to a power and the same number raised to another power is the number raised to the sum of the two powers: thus

$$2^3 \times 2^5 = (2 \times 2 \times 2)(2 \times 2 \times 2 \times 2 \times 2) = 256 = 2^8, \text{ i.e. } 2^{(3+5)}.$$

The quotient of a number raised to a power and the same number raised to another power is the number raised to the difference of the two powers: thus

$$\frac{2^5}{2^3} = \frac{2 \times 2 \times 2 \times 2 \times 2}{2 \times 2 \times 2} = 2 \times 2 = 2^2, \text{ i.e. } 2^{(5-3)}.$$

If the denominator has a higher index than has the numerator, the quotient has a negative index: thus

$$\frac{2^3}{2^5} = \frac{2 \times 2 \times 2}{2 \times 2 \times 2 \times 2 \times 2} = \frac{1}{2 \times 2} = \frac{1}{2^2} = 2^{-2}, \text{ i.e. } 2^{(3-5)}.$$

Any number raised to zero power is equal to 1, for

$$\frac{2^3}{2^3} = \frac{8}{8} = 1, \text{ or } \frac{2^3}{2^3} = 2^{(3-3)} = 2^0.$$

A positive number raised to any power (positive or negative) gives a positive answer: thus

$$2^4 = 16; \quad 2^{-4} = \frac{1}{2^4} = \frac{1}{16}.$$

A negative number raised to an even integral power (positive or negative) gives a positive answer: thus

$$(-2)^4 = 16; \quad (-2)^{-4} = \frac{1}{(-2)^4} = \frac{1}{16}.$$

A negative number raised to an odd integral power (positive or negative) gives a negative answer: thus

$$(-2)^3 = -8; \quad (-2)^{-3} = \frac{1}{(-2)^3} = -\frac{1}{8}.$$

It is convenient at this point to digress from the properties of indices to define irrational and rational numbers. An *irrational number* is a number whose value cannot be found exactly, e.g. $\sqrt{2}$, $\sqrt{3}$, $\sqrt{5}$, e, π, etc. A *rational number* is a number whose value can be expressed exactly.

Use of indices for expressing large and small numbers. It has already been shown that a very large number like the Avogadro number can be expressed in a compact way by using indices. Similarly, very small numbers can be expressed as a number between 1 and 10 multiplied by 10 raised to a negative integral power. For example, the wavelength of the green line in the mercury spectrum is 0·00005461 cm, and using indices this can be conveniently expressed as $5·461 \times 10^{-5}$ cm.

In expressing very large or very small numbers in the form $a \times 10^n$, where n is a positive or negative integer and a is a number between 1 and 10, the value of n can be found by applying the following rules:

(i) For very large numbers, n is a positive integer one less in value than the number of figures before the decimal point: thus

$$602{,}300 = 6·023 \times 10^5, \quad \text{i.e.} \quad n = 6-1.$$

(ii) For very small numbers, n is a negative integer one greater in value than the number of successive noughts immediately after the decimal point in the number: thus

$$0·0073423 = 7·3423 \times 10^{-3}, \quad \text{i.e.} \quad n = -(2+1).$$

Conversely, to write in full numbers expressed in the form $a \times 10^n$, it follows that if n is positive the number of figures before the decimal point will be $n+1$, and if n is negative the number of successive noughts immediately following the decimal point will be $-n-1$.

Calculations involving very large numbers or very small numbers are simplified if the numbers are expressed as a product of a number between 1 and 10 and 10 raised to the appropriate power.

EXAMPLE 1. 1 *kilogramme-molecule of a gas at* $0°C$ *and* 1 *bar* (= 750 *mmHg* = 10^5 Nm^{-2}) *pressure occupies a volume of* $22·7$ m^3. *How many molecules of the gas will there be in a 1-litre flask evacuated to a pressure of* $0·1$ μbar *at* $0°C$?

$22·7$ m^3 of gas at $0°C$ and 1 bar pressure contains

$$6 \times 10^{26} \text{ molecules.}$$

ARITHMETIC AND COMPUTING

Hence, 1 m³ of gas at 0°C and 1 bar pressure contains

$$\frac{6 \times 10^{26}}{22 \cdot 7} \text{ molecules,}$$

and 1 litre of gas at 0°C and 0·1 μbar pressure contains

$$\frac{10^{-3} \times 6 \times 10^{26} \times 0 \cdot 1 \times 10^{-6}}{22 \cdot 7} \text{ molecules.}$$

The most reliable method for evaluating an expression of this type is to write each quantity as a number between 1 and 10 multiplied by 10 raised to the appropriate power, and then to rearrange the expression so that the numbers between 1 and 10 are in one group and the tens raised to the various powers are in another group: thus

$$\frac{10^{-3} \times 6 \times 10^{26} \times 10^{-7}}{22 \cdot 7} = \frac{6}{2 \cdot 27} \times \frac{10^{26} \times 10^{-10}}{10}$$

$$= 2 \cdot 6 \times 10^{(26-10-1)}$$

$$= 2 \cdot 6 \times 10^{15}$$

EXAMPLE 2. *A stock solution contains 5 g of a drug in 100 ml. What volume of this solution should be diluted with water to give 20 ml of a solution containing 10 microgrammes of the drug in 1 ml?*

Since 0·05 g of the drug is contained in 1 ml of stock solution, the volume of this solution containing 1 microgramme (i.e. 1×10^{-6} g) will be

$$1 \times 10^{-6}/0 \cdot 05 \text{ ml.}$$

To make 20 ml of the required solution, 200 microgrammes of the drug are required. The volume of the stock solution containing 200 microgrammes of drug will be

$$\frac{200 \times 10^{-6}}{0 \cdot 05} = \frac{2 \times 10^2 \times 10^{-6}}{5 \times 10^{-2}} = \frac{2}{5} \times 10^{(2-6+2)}$$

$$= 0 \cdot 4 \times 10^{-2} = 4 \times 10^{-3} \text{ or } 0 \cdot 004 \text{ ml.}$$

EXAMPLE 3. *Calculate the charge in coulombs on a single ion of a monovalent metal, given that 96,500 coulombs are required to deposit 1 gramme-equivalent of the metal from solution and that 1 gramme-equivalent contains 6×10^{23} ions.*

The quantity of electricity associated with 1 gramme-equivalent of the metal is 96,500 coulombs. Therefore, the quantity associated with 1 ion is $96{,}500/6 \times 10^{23}$ coulombs. Therefore, the charge on a single ion is

$$\frac{96{,}500}{6 \times 10^{23}} \cdot 1 = \frac{9 \cdot 65}{6} \cdot \frac{10^4}{10^{23}}$$

$$= 1 \cdot 61 \times 10^{-19}.$$

9. Logarithms

The logarithm of a number is the power to which another number, called the base of the logarithm, is raised to become equal to the original number. Thus $10^2 = 100$, and, therefore, according to the above definition, the logarithm of 100 to the base 10 is 2. This is written $\log_{10} 100 = 2$, the base of the logarithm being shown as a subscript; expressed algebraically, if $M = a^x$, then $\log_a M = x$.

Systems of logarithms. Two systems of logarithms are in general use: common logarithms and natural logarithms. They differ in the number used as the base. Common logarithms based on the number 10 are used for all arithmetic calculations and are usually signified by the symbol 'log', no subscript being used. Natural, or Naperian, logarithms are based on a quantity denoted by the letter 'e' and are signified by the symbol 'ln' or '$\log_e$'. The quantity e is an irrational number which can be expressed as a sum of a series. The particular properties of e and natural logarithms are described in Chap. 4.

Common logarithms. Ordinary mathematical tables give the common logarithms of numbers between 1 and 10 to four places of decimals. For more accurate work logarithms to five and seven places of decimals have been published.

The logarithm of a number consists of two parts: an integral part called the *characteristic* and a decimal part called the *mantissa*. The mantissa, which is the quantity given by the logarithm tables, is always positive. The characteristic may be positive or negative, depending upon whether the number is greater or less than unity, and is written down by inspecting the number and applying the following rules:

(i) If the number is greater than 1, the characteristic is a positive integer one less in value than the number of figures before the decimal point. For example, the number 2371 has four figures in front of the decimal point and its logarithm has, therefore, a characteristic of 3. From logarithm tables the mantissa is found to be 0·3749, and so log 2371 = 3·3749. This is the power to which the base 10 must be raised to equal 2371, i.e. $10^{3 \cdot 3749} = 2371$.

(ii) If the number is less than 1, the characteristic is a negative integer, one greater in value than the number of successive noughts immediately after the decimal point. For example, the number 0·00271 has two successive noughts immediately after the decimal point, and log 0·00271 has, therefore, a characteristic of -3, written $\bar{3}$ by convention and called 'bar three'. The mantissa of 2·71, obtained from the logarithm tables, is 0·4330, so that log 0·00271 = $\bar{3}$·4330.

The device of separating the logarithm into a characteristic and mantissa is equivalent to dividing the number into two factors, one of which is 10 raised to an integral power and the other is a number between 1 and 10 whose logarithm can be found directly from logarithm tables. Thus

$$2371 = 10^3 \times 2 \cdot 371 = 10^3 \times 10^{0 \cdot 3749} = 10^{(3 + 0 \cdot 3749)};$$
$$0 \cdot 00271 = 10^{-3} \times 2 \cdot 71 = 10^{-3} \times 10^{0 \cdot 4330} = 10^{(-3 + 0 \cdot 4330)}.$$

To convert a logarithm back into an ordinary number, i.e. to find the antilogarithm of a logarithm, the reverse procedure to that described above is used. The number corresponding to the mantissa is found from the antilogarithm tables and then multiplied by 10 raised to the integral power indicated by the characteristic. For example, antilog $1 \cdot 1057$ is found by looking up $0 \cdot 1057$ in the antilogarithm tables and multiplying the figure $1 \cdot 276$ thus obtained by the figure indicated by the characteristic, i.e. 10^1. Thus antilog $1 \cdot 1057 = 10^1 \times 1 \cdot 276 = 12 \cdot 76$. Similarly, antilog $\bar{2} \cdot 6136 = 10^{-2} \times 4 \cdot 108 = 0 \cdot 04108$. Since the mantissa is always a positive decimal, i.e. greater than 0 and less than 1, and since $0 = \log 1$ and $1 = \log 10$, the antilogarithm of the mantissa is always a number between 1 and 10.

Logarithms in multiplication and division. From the rules of indices already outlined in § 8, it follows that

(i) the logarithm of the *product* of two numbers is equal to the *sum* of the logarithms of the numbers, and

(ii) the logarithm of the *quotient* of two numbers is equal to the *difference* of the logarithms of the numbers.

Logarithms simplify the processes of multiplication and division by converting these processes into addition and subtraction respectively.

EXAMPLE. *Evaluate by logarithms* $217 \cdot 2 \times 0 \cdot 0075 / 33 \cdot 71$.

Inspection of logarithm tables gives the logarithms: thus

$$\log 217 \cdot 2 = 2 \cdot 3369, \log 0 \cdot 0075 = \bar{3} \cdot 8751 \text{ and } \log 33 \cdot 71 = 1 \cdot 5277.$$

Therefore

$$\log \frac{217 \cdot 2 \times 0 \cdot 0075}{33 \cdot 71} = \frac{\log 10^{2 \cdot 3369} \times 10^{\bar{3} \cdot 8751}}{10^{1 \cdot 5277}}$$

$$= \log 10^{(2 \cdot 3369 - 3 + 0 \cdot 8751 - 1 \cdot 5277)}$$

$$= \log 10^{\bar{2} \cdot 6843}$$

$$= \bar{2} \cdot 6843.$$

Therefore

$$217 \cdot 2 \times 0 \cdot 0075 / 33 \cdot 71 = \text{antilog } \bar{2} \cdot 6843 = 0 \cdot 04834.$$

Logarithms in the evaluation of roots and powers. The use of logarithms to find the powers and roots of a number is based on the rules of indices already outlined in § 8. These rules are illustrated algebraically in Chap. 2, but their practical utilisation in arithmetic is discussed here.

The logarithm of the nth power of a number is equal to n times the logarithm of the number.

EXAMPLE. *Evaluate* $(2 \cdot 31)^5$.

$$\log 2 \cdot 31 = 0 \cdot 3636$$
$$\log (2 \cdot 31)^5 = 5 \times 0 \cdot 3636 = 1 \cdot 8180$$
$$\text{antilog } 0 \cdot 8180 = 6 \cdot 577$$
$$\text{antilog } 1 \cdot 8180 = 65 \cdot 77.$$

Therefore

$$(2 \cdot 31)^5 = 65 \cdot 77.$$

Similarly, the logarithm of the nth root of a number is equal to $1/n$ times the logarithm of the number. In calculating the roots of numbers less than 1, it is convenient to obtain the negative characteristic of the answer directly as an integer, and this can be done by rewriting the logarithm so that its negative characteristic is exactly divisible by n. This procedure is illustrated in the following example.

EXAMPLE. *Evaluate* $\sqrt[4]{0 \cdot 00217}$.

$$\log \sqrt[4]{0 \cdot 00217} = \tfrac{1}{4} \log 0 \cdot 00217 = \tfrac{1}{4} \times \bar{3} \cdot 3365$$
$$= \tfrac{1}{4}(\bar{4} + 1 \cdot 3365) = \bar{1} + 0 \cdot 3341.$$
$$\text{antilog } 0 \cdot 3341 = 2 \cdot 158$$
$$\text{antilog } \bar{1} \cdot 3341 = 0 \cdot 2158.$$

Therefore

$$\sqrt[4]{0 \cdot 00217} = 0 \cdot 2158.$$

10. Logarithmic scales, pH and pK

The concentration of hydrogen ions in very pure water is $0 \cdot 0000001$, or 10^{-7}, gramme-equivalent per litre. The hydrogen-ion concentration of a solution is of great importance in, for instance, biochemistry, since quite small changes in this quantity have large effects on the net electrical charge carried by a biocolloid. A change of hydrogen-ion concentration from 10^{-6} to 10^{-4} gramme-equivalent per litre will result in a complete reversal of the electrical charge on an albumin molecule in solution. Sørensen, a Danish scientist, developed a pH scale in order to express these small concentrations in a convenient way. The symbol 'p' simply means $-\log$, and 'H' the hydrogen-ion concentration: thus

$$\text{pH} = -\log [\text{H}^+],$$

where $[\text{H}^+]$ is the hydrogen-ion concentration in gramme-equivalents per litre. If $[\text{H}^+] = 0 \cdot 0000001$, i.e. 10^{-7}, then

$$\text{pH} = -\log 10^{-7} = -(-7 \log 10) = -(-7 \times 1) = 7.$$

The pH scale provides a system whereby the hydrogen-ion concentration of a liquid can be conveniently expressed in terms of small numbers. The usual pH range lies between 0 and 14, since the hydrogen-ion concentration in a normal solution of a strong acid is approximately 1 gramme-equivalent per litre, i.e. $\text{pH} = 0$, and that of a normal solution of sodium hydroxide is approximately 10^{-14} gramme-equivalent per litre, i.e. $\text{pH} = 14$. The logarithmic nature of the pH scale means that if the pH of a solution is increased by one unit the hydrogen-ion concentration is diminished to one-tenth of its original value.

A similar scale, the pK scale, is used to express the ionic dissociation constants (K values) of weak electrolytes: thus

$$\text{pK} = -\log \text{K}.$$

pH and pK values are usually non-integral and some care is necessary in calculating them from [H⁺] and K values.

As an example, the dissociation constant K of acetic acid is 1.75×10^{-5} at 20°. Therefore

$$pK = -\log(1.75 \times 10^{-5}) = -(\log 1.75 + \log 10^{-5}) = -(\log 1.75 - 5)$$
$$= 5 - \log 1.75 = 5 - 0.2430 = 4.76.$$

The pK value is less in magnitude than the negative index of 10 in the constant K.

Non-integral values of pH are calculated in a similar way. Suppose the hydrogen-ion concentration of a solution is 2×10^{-4} gramme-equivalent per litre. Then

$$pH = -\log(2 \times 10^{-4}) = -(\log 2 - 4) = 4 - \log 2 = 4 - 0.3010 = 3.70.$$

In mks (SI) units, concentrations in kilomole or kilogramme equivalents per cubic metre (kmol m⁻³ or keq m⁻³) are exactly equal to the centimetre–gramme–second (cgs) values in moles or gramme-equivalents per litre.

11. Slide rules

The processes of multiplication, division, extraction of square roots, obtaining the powers of numbers, etc., are very easily carried out mechanically by using the slide rule. This instrument consists of several scales, the

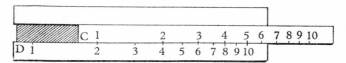

Fig. 1a. Multiplication by the slide rule

most important of which are the two adjacent identical scales, each graduated with numbers from 1 to 10. The numbers occur in ascending order from left to right and are spaced logarithmically, for example, the space between 2 and 4 is the same as that between 4 and 8, since

$$\log 4 - \log 2 = \log 8 - \log 4.$$

Fig. 1a shows these two scales. The lower scale D is fixed and the upper scale C is engraved on a central sliding section so that it moves parallel to D.

To multiply 2 by 4, the number 1 on the sliding scale C is set opposite to the 2 on the lower fixed scale, and the answer to this multiplication is the figure on scale D opposite to the number 4 on scale C, i.e. 8. A sliding cursor, consisting of a transparent piece of plastic bearing an engraved hair line, facilitates the reading of corresponding numbers on the scales.

If the scales were not logarithmic, the process would have added 2 and 4, but as in fact it adds log 2 and log 4, the numbers have been, in effect, multiplied together, and by reading off the answer from scale D the antilogarithm is obtained directly.

Division, involving the subtraction of logarithms, is accomplished similarly. For example, to evaluate 6/4 the denominator 4 on the sliding scale C is set opposite the numerator 6 on the lower scale D. The answer (see Fig. 1b) is the reading 1·5 on scale D opposite the mark 1 (or 10 in

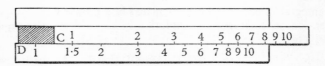

FIG. 1b. Division by the slide rule

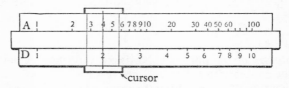

FIG. 1c. Squares and square roots by the slide rule

other cases when the mark 1 is off scale D) on scale C. The accuracy obtainable with a 10-in. slide rule is about 1 in 500, and these rules should not be used to work out calculations requiring an accuracy greater than this, e.g. gravimetric assays. Greatly improved accuracy is achieved by printing the scales as a helix or spiral on the surface of a cylinder so that a scale considerably longer than 10 in. is obtained without having an inconveniently long rule to work with. A reasonably small calculator of this type, e.g. Otis-King's, will give an accuracy of 1 in 3000, which approaches the accuracy of four-figure logarithm tables. It must be noted that slide rules, like logarithm tables, deal only with mantissae, and the position of the decimal point in the final answer must be found by rough calculation.

For finding roots and powers of a number by means of the slide rule, other scales are used and the root or power is read off directly as the number. The sliding scale is not used for these operations. Many slide rules have an upper fixed scale A showing numbers from 1 to 100, while the lower fixed scale D shows numbers from 1 to 10 (see Fig. 1c). Numbers on scale A are the squares of their opposite numbers on scale D. To find the square root of a number N, the cursor is set to N on scale A and $\sqrt{N}$ is then read off directly as the number indicated by the cursor on scale D.

To find squares, the reverse procedure is used: N is set on scale D and N^2 is read off directly on scale A.

Mechanical and electronic calculating machines are very much more accurate than slide rules, but are correspondingly more expensive.

12. Energy

Since energy appears in many forms, for instance as mechanical, electrical and heat energy, the interconversion of units is of particular importance.

Mechanical energy (or work) is equal to the magnitude of the product of a constant force F and the distance d through which the force acts: thus $W = Fd$.

When a force of 1 newton is displaced through 1 metre the work done is 1 joule.

Work of expansion is frequently encountered in the theory of gases. If a gas is expanded in a cylinder of uniform cross-sectional area A against a constant external pressure p, and if the piston moves through d m, then since pressure is force per unit area and the total force is $p \times A$, the work done (total force × distance) will be $(p \times A)d$. As $A \times d$ is the change in volume Δv, the work done is $p \Delta v$.

The work of expansion of a gas has been expressed in litre-atmospheres (l-atm). This is the work done by a gas in expanding by 1 litre against a constant external pressure of 1 atmosphere.

$$1 \text{ l-atm} = 101 \cdot 33 \text{ joules.}$$

Heat energy has been expressed as calories (cal) or kilocalories (kcal). The calorie is the amount of heat required to raise the temperature of 1 gramme of water from $14 \cdot 5°$ to $15 \cdot 5°C$. The kilocalorie is equal to 10^3 calories,

$$1 \text{ cal} = 4 \cdot 184 \text{ joules}; \quad 1 \text{ kcal} = 4184 \text{ joules.}$$

In SI the use of the calorie will be discontinued and heat energies will be expressed in joules.

Electrical energy is the product of current, time and potential difference. The electrical energy dissipated in 1 second when a current of 1 ampere flows through a potential drop of 1 volt is 1 joule. The industrial unit of electrical energy is the kilowatt-hour (kWh), which is 1000 times the energy dissipated in 1 hour when a current of 1 ampere flows through a potential drop of 1 volt.

$$1 \text{ kWh} = 3 \cdot 6 \times 10^6 \text{ joules} = 860 \text{ kcal.}$$

The gas constant R appears in the equation relating pressure, volume and absolute temperature for an ideal gas, i.e. $PV = nRT$. In the theory of physical chemistry, the ideal gas has been taken as a model system, and other systems that are not ideal are compared with it. As a direct result of this, R occurs in equations that are not concerned with gases, for instance in the equation for the electromotive force of an electrical cell.

For 1 kilogramme-molecule of an ideal gas, $R = PV/T$. Thus R is the product of pressure and volume divided by the temperature, and has the dimensions ML^2T^{-2}/K. The dimensions of R are energy per degree of temperature per kilogramme-molecule. The numerical value of R depends upon the energy units chosen. In mks units, $R = PV/nT$, n being the number of kilogramme-molecules (kmol) of gas; R is 8314·3 J kmol^{-1} K^{-1}. The symbol K is used to denote temperature in degrees kelvin (K = °C+273·15).

13. Digital computers

Digital computers consist of banks of electronic and magnetic units in which information can be stored in binary, that is, ON/OFF form, together with circuits for scanning and combining the information. Each

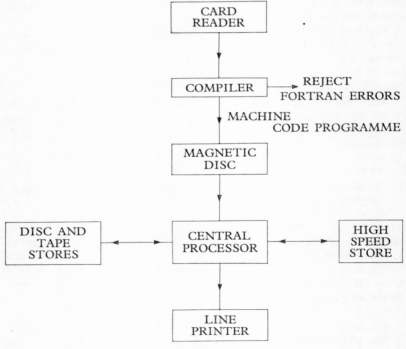

FIG. 1d. Computer units

ON/OFF item is called a *byte* and the bytes are combined together in *words*. The computer *word* occupies one *cell* and the word length varies from one computer to another.

All information, instructions and numbers put into the computer have to be coded into this binary form so that they can be stored and handled. It is no longer necessary for the computer user to do this coding; information can be submitted in ordinary letters and numbers and these are converted by a programme called a *compiler*, which is part of the computer installation, into the machine language. Translation back is effected before the results are given in the output. In general, the user of a computer need not therefore be greatly concerned with machine language; high level languages have been developed so that programmes may be written in a reasonably intelligible form. The language used in this book is the widely used Fortran IV; nearly all computer installations have a compiler which will translate Fortran IV into machine code.

All computer languages are very demanding in terms of accuracy and, for example, a mistake such as putting a comma in a programme where there should have been a full stop may cause the compiler to reject a whole programme. An accurate programme which works is a valuable commodity and it may take quite a long time to remove all the faults from a new programme.

In this introductory course to computing we have adopted the approach that it is more profitable for a beginner to study programmes which are known to work so that some useful output is obtained, rather than starting with the frustrating task of writing his own programmes which are unlikely, in the first attempts, to give any worthwhile output. Most types of problem have at some stage been solved by computing methods and it is more important for the general user to know where to find and how to use established (library) programmes or subprogrammes than to acquire great facility in writing his own. Sufficient knowledge of computer language is however required in order to be able to put data into established programmes, to print out the results in a desired form, and to modify library programmes.

Numbers are put into the computer in their ordinary form and in the compiler they are converted to binary ON/OFF form as in the following example. The decimal number 435 is in fact $4 \times 10^2 + 3 \times 10^1 + 5 \times 10^0$. Binary numbers are similarly expressed with 2 in place of 10, the figures in the number are then 0 or 1 corresponding to OFF and ON for the byte in the computer: $2^2 = 4$, $2^3 = 8$, $2^4 = 16$, $2^5 = 32$, $2^6 = 64$, $2^7 = 128$, $2^8 = 256$, $2^9 = 512$; the number 435 is less than 512 so there will be zero in the 2^9 position, 1 in 2^8, leaving $435 - 256 = 179$, 1 in 2^7 leaves $179 - 128 = 51$, there will be zero for 2^6 since it is greater than 51, 1 for 2^5 leaves $51 - 32 = 19$, 1 for 2^4 leaves $19 - 16 = 3$, 0 for 2^3, 0 for 2^2, 1 for 2 and 1 for 2^0 gives 435 in binary as $110110011 = 2^8 + 2^7 + 2^5 + 2^4 + 2^1 + 2^0 = 256 + 128 + 32 + 16 + 2 + 1 = 435$.

For the programmes considered in this book, the input to the computer consists of punched cards which contain the programme of instructions for the operations in Fortran IV language, together with the numerical data on which the operations are to be carried out. These punched cards are read at high speed by light beams which pass through the holes in the cards and record pulses in photocells. The information from the cards passes to the compiler where programme and numbers are converted to binary form. In a multi-use computer, the compiled programme is then stored on magnetic discs or tapes to await its turn for passing into the central processor. Included in the compiler is a programme for detecting errors in the input which gives the user an error message in his output showing where the errors are which need to be corrected before the programme can be compiled.

14. Input to the computer

The pack (deck) of punched cards which is put into the computer normally consists of three sections, the first of which is the job control card section. This gives the user's job number and states which compiler is required, and gives information about the length of the programme; the exact form is specified by the computer centre. The next group of cards is the source programme deck consisting of the programme cards in Fortran which are converted into an object programme in binary code by the compiler. A third group of cards is the data section, giving the numerical data for the programme which is read from these cards in a format which is precisely defined in the Fortran section. In some short programmes all the data required is included in the Fortran section and so no data cards are required. The three groups of cards, job control, Fortran and data, are usually separated by special cards specified by the computer centre. For the CDC 6600, these are called end-of-record (E.O.R.) cards. After the last card in the deck, another special card, the end-of-file (E.O.F.) card, is required.

The cards have 80 columns and the holes are punched out by a card punch which has a keyboard similar to that of a typewriter but with capital letters only together with decimal numbers and a set of other characters such as brackets, a comma, full stop, asterisk, solidus (slash), plus, minus, etc. Each character occupies one vertical column of the card and is printed out above the set of holes representing it and in Fortran each statement requires a separate card. Trivial errors such as a missed comma or full stop may cause a programme to fail; to the compiler a comma is just as important a pattern of holes as any other character.

A Fortran card is shown in Fig. 1e. It is subdivided into four fields. Columns 1 to 5 are reserved for statement numbers and column 1 is used additionally for comment cards. If a C is punched in column 1, this card is ignored by the compiler but its contents appear in the print-out of the programme; comments such as descriptions of each section of the

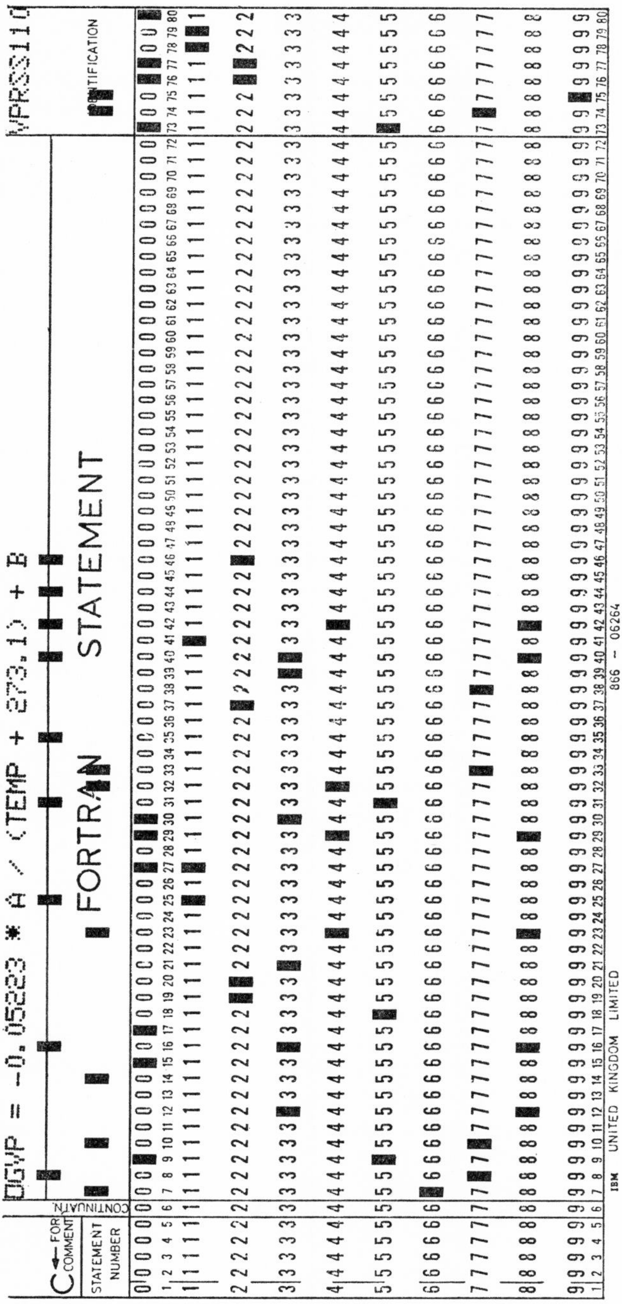

FIG. 1e. Fortran card showing a statement from the programme COMP 1 in Chapter 2

programme may be put in by this means. It is best to start a programme with a set of C cards explaining its purpose.

Column 6 is a field reserved for indicating continuation; if a Fortran statement exceeds the columns available on one card any number punched into column 6 of the next card indicates that it is to be read as a continuation of the previous card. The main field, from column 7 to column 72, is the region in which the statement is punched. There are no rigid rules about spacing in this field and blanks are ignored by the compiler so that statements may be spaced out so that they can be read intelligibly.

The final field, columns 73 to 80, is reserved for the programmers identification of his cards, usually by a group of letters appearing on each card followed by consecutive numbers. It is useful to have a zero as the last number so that if the programme is revised additional cards may be added without altering the numbering of the existing cards. These columns are ignored by the compiler but appear in the print-out. This numbering is helpful if the card deck is dropped or otherwise becomes mixed up.

In the Fortran statements, usually at the end, precise instructions about the way in which the results are to be printed out are given in a format statement; titles in words may also be included.

The data cards consist of 80 columns which are divided into fields by the read format statement in the programme. When punching the data this format, which is specified by the programmer, must be rigidly observed.

In writing programmes, forms are used which correspond to the Fortran and data cards with one space for each character. Card punching is often carried out by professionals and the programmer passes the forms with Fortran and data to the puncher who prepares the deck of cards including job cards and other specialised cards required by a particular centre. The programmer should distinguish clearly between the letter O and zero, the letter I and one and the letter Z and two. Various conventions are used to make these distinctions, one is to put a bar over or under the letter O, $\bar{O}$ or $\underline{O}$, to put clear horizontal lines at the top and bottom of letter I and to put a horizontal line through the middle of Z, $\not{Z}$. In the programmes in this book, the letter O has been underlined to distinguish it from zero.

15. Numbers, indices, logarithms and square roots in Fortran

Ordinary numbers in Fortran are called *real* numbers and should always be shown with a full stop as a decimal point even when there is nothing after it; this is to distinguish them from integers, discussed more fully in § 18 (page 25). Integers are used for counting and as subscripts and many computers will not perform arithmetic directly with mixtures of real numbers and integers.

Real numbers may also be expressed in floating point form—this is similar to the notation used in § 8 (page 8) of this chapter. The number is given as a real number multiplied by a power of ten; 220·7 would be written 2·207E2 or 0·2207E3, the number after E being the power of 10 by which the real number in front of E, which must have a decimal point, is raised. The Avogadro number in this notation would be 6·023E26 while Planck's constant $4·135 \times 10^{-15}$ electron-volt second would be 4·135E-15, the number after E being a positive or negative integer without a decimal point.

In Fortran, logarithms may be obtained simply by writing ALOG( ) for natural logarithms and ALOG10( ) for common logarithms, the number whose logarithm is required being shown in the brackets. When the compiler comes to these combinations of characters with brackets after them (the brackets are an essential part of the function) it works out the logarithm from a programme supplied through the compiler.

There is no antilog to base 10 supplied in the basic Fortran functions; however, antilogs to base e may be found by the function EXP( ). Logarithms to base 10 are converted to natural logarithms by multiplying by ln(10) = 2·3026. To obtain the antilogarithm of say 3·406 to base 10 write EXP(2·3026*3·406), the asterisk being the multiplication sign in Fortran.

The situation is a little more complicated with a negative logarithm. If it is in the form $\bar{6}$·2173, the characteristic is negative but the decimal part, the mantissa, is positive. It should therefore be converted to a completely negative number $-6+0·2173 = -5·7827$ and the antilog is EXP(2·3026*−5·7827).

16. Input/Output (I/O) formats for real numbers

The input format for numbers is usually placed near the beginning of a programme and it gives precise instructions as to the way in which numbers are to be read from a data card. The usual format is the F type for real numbers, a real number being an ordinary positive or negative number with a decimal point. In the simplest form for a single real number, the input statements are

    READ(5,1) X
1   FORMAT (F 8·0)

The READ statement has in brackets the code number 5 which is used in some computers to indicate the input tape or disc, the second number which must be separated from the 5 by a comma is the format statement number; outside the brackets a name, in this case X, is given to indicate the name by which the cell containing this number will be identified in the Fortran programme. The format must always be numbered with the number given in the READ statement. FORMAT also has brackets which contain the details; F stands for a real number, 8 for the number of columns of the data card in which the number is contained and it is a rigid requirement that the punching of the data card must agree exactly with the input format. The decimal point is essential for an F format and after the decimal point a number can be put to indicate the number of decimal places in the input but if this is done. the placing of the numbers within the 8 columns of the data card is rigidly fixed; the last column (the eighth in this case) of the data card field will be read as the last figure after the decimal point indicated in the F specification. A more flexible arrangement is to put zero after the decimal point and then put a decimal point in the input number on the data card; the latter then overrides the zero in the format and the number may be placed anywhere in the field of eight columns of the data card.

When several numbers are to be read in, their names are stated in an input list, thus labelling the cells in which they are stored; these names are then used in the programme. Each number may be given a separate format, in which case the details are all enclosed in the same FORMAT statement brackets, separated by commas.

```
      READ(5,8) X,Y,Z
   8  FORMAT(F 12·0, F 8·0, F 10·0)
```

This means that three real numbers with decimal points are to be read; the first one is labelled X and is to be found in the first twelve columns of the data card, the second is labelled Y and is to be found in the next 8 columns, and the last one, labelled Z, is to be found in the next ten columns.

Alternatively, if the numbers are all given the same number of columns in the data card, the format could be modified to

```
   8  FORMAT(3F 12·0)
```

The 3 in front of F means that three numbers are to be read, each of which is contained in a twelve-column field of the data card.

When a large amount of numerical data is to be read, it is convenient to use subscripted variables rather than inventing a name for each number. Suppose we have thirty values of a quantity to read in, then if we call it X we can distinguish the values and put them in separate cells of the computer by using numerical subscripts to X written X(1), X(2), X(3), X(4), etc.

If this is done, the first statement in the programme must tell the compiler how many values of X there are so that the correct number of cells is reserved for them. This is done by a DIMENSION statement.

```
      DIMENSION X(30)
```

reserves thirty cells for the storage of X. If several quantities are to be subscripted they may all be included in the one DIMENSION statement, separated by commas; for example

```
      DIMENSION X(30), Y(24), Z(35).
```

Rejection will occur if DIMENSION is not placed as the first compiled statement in the programme.

A READ statement after DIMENSION need only list the overall name of the whole set of numbers. The format must however be arranged so that instructions are given for finding each number on the data cards. If they are more than can be contained on one card, separate specifications for each card may be given within the same FORMAT brackets, separated by the solidus (slash), /, which in an input format statement means skip to the next card. For example

>   DIMENSION X(30)
>   READ(5,3) X
> 3 FORMAT(10F 8·0/10F 8·0/10F 8·0)

would mean that each of the thirty values of X is contained with a decimal point in a field of eight columns on the data cards. As each card has eighty columns, ten values are put on each of three cards. Alternatively, FORMAT(10F 8·0) could be used. This FORMAT would then be applied to each data card until all the values indicated in the READ list coupled with the DIMENSION statement are obtained.

Double subscripts, separated by a comma within the brackets, Y(3,4), or treble subscripts, Z(6,3,9), may also be used. In such cases the DIMENSION statement declares the largest value for each subscript.

The subscripts themselves are positive integers; zero and negative numbers are not acceptable.

For output, the F-type format is also used for real numbers but in this case the number after the decimal point is important and will give the number of figures after the point which will be printed. The number before the decimal point in the format states the number of printing spaces to be given to the number. An output statement would have the form

>   WRITE(6,2) X,Y,Z
> 2 FORMAT(1H0, 3F 20·4)

The 6 in the WRITE brackets is a code for the output tape and 2 is the format statement number. In the FORMAT the H is a symbol used to print titles. The line printer will print as written in the FORMAT any set of letters after the H except the first, the number of spaces being exactly indicated by a number in front of H. The first space after H is reserved for a carriage control symbol which is not printed out; a blank immediately after H means ordinary spacing between lines in the print-out, zero means double spacing between lines, and one means start a new page.

The format for the numbers is separated from the H format by a comma and as given above means that the three numbers will be printed out on one line each occupying 20 spaces and having four figures after the decimal point. By specifying more positions than are necessary for the numbers, they are spread out in a readable form.

To print out values of a subscripted variable it is only necessary to give the variable name in the WRITE statement unless some special arrangement is required; the numbers are then printed out successively in order of increasing subscript according to the output format. For example, with X dimensioned to 30 values

  WRITE(6,25) X
25 FORMAT(1H0, 5F 20·5)

would give the first five values of X in one line with 20 spaces for each number and five figures after the decimal point; the H format gives double spacing with the next set of five values on the next line and so on until all thirty values have been printed. As in the READ format, the arrangement indicated in the brackets will be used repeatedly taking a new line for each repetition until all the values in the WRITE list have been printed. If the FORMAT for each line is identical it is therefore unnecessary to use slashes with repetitive statements for each.

17. Floating point numbers

When large or small numbers are required or when the magnitude of the answers to a calculation is uncertain, it is useful to use the floating point E type format mentioned in § 15 (page 20). For input, E is followed by a number which gives the number of columns on the data card in which the number, including sign, decimal point, and exponent, is to be found; this is followed by a full stop with a number to indicate the number of decimal places; however, as with F formats a zero may be put here and the decimal point put on the data card. The exponent must be punched at the right-hand side of the card field specification and is preceded either by + or − or by E on the card. For example, to read in the Avogadro number $6·02 \times 10^{26}$ kmol^{-1}

  READ(5,3) AN
3 FORMAT(E10·0)

The card would be punched as below

Column number 1 2 3 4 5 6 7 8 9 10
          6 . 0 2 E 2 6

In place of E in column 8, + could be used.

For output the E specification gives a real number with a zero before the decimal point which is followed by the first number, and then E followed by the exponent. In order to obtain the output in the more convenient form of a number between one and ten multiplied by ten to a power, a scale factor 1P is used in front of E which multiplies the real number by 10 and reduces the exponent by 1. For example, to write the Avogadro number

  WRITE(6,8) AN
8 FORMAT(1H1, 1PE 20·2)

would give, if AN was used to represent this number in the programme, a new page with a print-out of 6·02E 26.

As with F, a number in front of E indicates that number of quantities in the specification; e.g. (1P4E 20·4) in an output format would print four numbers of this type.

18. Integers

In Fortran, numbers used for counting or for subscripts which are always integers are treated differently from real numbers. They are distinguished by the fact that a real number in E or F form always has a decimal point even when zero or when there is nothing after the decimal point; for example, 0· and 26· and 46·87 are all real numbers, whereas 26 without a decimal point is an integer. The names used for real numbers in programmes start with any letter except those from I to N; names starting with the letters I, J, K, L, M, and N are reserved for integers.

Many computers will not perform arithmetical operations on mixtures of integers and real numbers. The former should therefore be reserved for counting and for subscripts.

There is an I/O specification, I, which is used to read in integers and to write them out. It is used exactly as in F except that there is no decimal point, and integers punched on cards to be read in I format are read with the right-hand column in the field on the card representing the digit, the next to the left, the tens, the next the hundreds, and so exact placing on the card is important.

```
       READ(5,2) L,M
   2   FORMAT(2I4)
```

means that two integers are to be put in cells labelled L and M (both integer names) being read from a card in which each of them has four spaces. If the integers were 16 and 107 they would be punched as shown below.

Column number 1 2 3 4 5 6 7 8
                        1 6   1 0 7

Blank spaces are read as zeros and so if 16 had been punched in columns 1 and 2 it would have been read as 1600.

Problems

1. Express as powers of 2
   - (i) $2^7 \times 2^5$
   - (ii) $2^{18} \div 2^{12}$
   - (iii) $(2^3)^4$
   - (iv) $8 \times \sqrt{2}$
   - (v) $2^{-4} \times 2^{0.3}$
   - (vi) $2^3 \times 2^{1/5} \div \sqrt{2}$
2. Express as numbers between 1 and 10 multiplied by an integral power of 10
   - (i) 0·00373
   - (ii) 2,756,000
   - (iii) 0·00102
   - (iv) 1324
   - (v) 0·00000107
   - (vi) 15·7394

3. A fatty acid molecule in a film on the surface of water occupies an area of 0.25 nm². If the molecular weight of the acid is 130, how many milligrammes of the acid will there be in such a film 1 m² in area?
Number of molecules in 1 kmol $= 6 \times 10^{26}$.

4. How many molecules of gas will there be in a 10-litre flask at 0°C evacuated to a pressure of 0.1 Nm⁻²? 1 kmol of gas occupies 22.7 m³ at 0°C and 1 bar ($= 10^5$ Nm⁻²) pressure.

5. Given that $\log 2 = 0.3010$, write down, without using logarithm tables, the values of the *logarithms* of

   (i) $2^3 \times 2^7$ (ii) $2^{0.6}$ (iii) 0.032 (iv) 1.28
   (v) $4 \times 8 \times 2$ (vi) 80   (vii) 50,000   (viii) 625

6. Evaluate by means of logarithm tables and a slide rule

   (i) $2.303 \times 7.251 \div 27.32$ (ii) $\sqrt{726.3}$ (iii) $854.3 \times 7.31^2 \div \sqrt{871}$
   (iv) $\sqrt{0.3721}$ (v) 531^3 (vi) $0.01742 \times 0.1618 \div 2274$
   (vii) $\sqrt[3]{728}$ (viii) $(0.0765)^{1/4} \times \sqrt{87.32 \div 0.875^2}$

7. Work out the pK values of the following weak acids, given their dissociation constants (K) at 25°C

   (i) HCN 7.24×10^{-10} (ii) $H.CO_2H$ 1.77×10^{-4}
   (iii) $C_2H_5.CO_2H$ 1.34×10^{-5} (iv) $CH_2Cl.CO_2H$ 1.35×10^{-3}
   (v) C_6H_5OH 1.20×10^{-10}

8. The critical linear velocity (LT^{-1}) of a liquid flowing through a tube is a function of the radius of the tube and of the density and viscosity of the liquid. Use dimensional analysis to deduce the form of the relation.

9. Write input and output formats for twenty-four values of a quantity X, each value having four significant figures, three after the decimal point, the output to be printed in four rows of six values each, with double spacing.

CHAPTER 2
Algebra

THE ORIGIN and early development of algebra is credited to the ancient Egyptians. Algebra is the branch of mathematics in which the rules of calculation expressed arithmetically in the previous chapter are described in a more general way by using letters to represent numbers. For instance, the rules of multiplication enunciated in Chap. 1, §3 can be summarised as

$$a \times b = ab; \quad a \times -b = -ab; \quad -a \times b = -ab; \quad -a \times -b = ab.$$

Each of these rules is an algebraic statement called an *identity*, which is defined as a statement which is true for all numerical values of the letters a and b.

1. Algebraic definitions

Two types of symbols are used in algebra; firstly, letters are used to represent numbers, and, secondly, combinations of letters and symbols are used to describe operations performed on the letters. A number of familiar symbols (*operators*) are given in Table 1.

TABLE 1
Symbols for mathematical operations and their Fortran equivalents

Term	Symbol	Fortran equivalent	Term	Symbol	Fortran equivalent
addition	+	+	common logarithm	log	ALOG10( )
subtraction	−	−	natural logarithm	ln	ALOG( )
equality	=	= (see § 6)	differentiation	$\frac{d( )}{dx}$	none
approximately	≃	none			
multiplication	× or ·	*	integration	∫	none
division	/ or ÷	/	trigonometric functions	sin	SIN( )
square root	√	SQRT( )		cos	COS( )
				tan	none

Argument. An operator cannot stand alone. It must have some quantity on which to perform the operation. This quantity, which may be a letter, number, or group of letters and numbers, is called the *argument* of the operator. For example, the argument in log (x^2+4) is (x^2+4).

Expression. An algebraic expression is a group of letters, numbers and operators, e.g. $7x^3+2x-3$.

Function. In the expression $7x^3+2x-3$, there is only one symbol whose value can be varied and that is the letter x. The numbers 7, 3 and 2 are constants, and since 7 and 3 are associated with the letter x they are known as the *coefficients* of x^3 and x respectively. As there is only one variable, the expression is said to be a function of that variable, and this is indicated in a general sense by using the operator f( ). Expressions such as $3xy+7y^2$, in which there is more than one variable, will be a function of all the variables, denoted in this instance by f(x, y).

Statement. An algebraic statement can be an equation, identity or inequality. An *equation* (in which the symbol '=' occurs) involving one variable states that two expressions are equal for one value, or for a limited number of values, of the variable; thus the equation $x^2 = 4$ is true for two values of x only, and these are $+2$ and -2. An *identity* (in which the symbol '≡' occurs) with one variable states that two expressions are equal for all values of the variable, e.g. $(x+2)^2 \equiv x^2+4x+4$. *Inequalities* are indicated by the following symbols: $>$ means 'greater than', $<$ means 'less than', and $\gg$ and $\ll$ mean 'much greater than' and 'much less than', respectively. Two further symbols are sometimes used: $\geqslant$ means 'greater than or equal to' and $\leqslant$ means 'less than or equal to'. Thus

$x > y$ means that x is greater than y,
$x \gg y$ means that x is much greater than y,
$x \geqslant y$ means that x is greater than or equal to y,
$x \neq y$ means that x is not equal to y.

2. Indices

Indices can be algebraically expressed as letters, and by this means the general rules of behaviour of indices enunciated in Chap. 1, § 8 can be summarised. The following statements describe some of these rules.

(i) $a^n = a \times a \times a \times a \times \ldots$ to n factors.
(ii) $a^{-n} = 1/a^n = 1/(a \times a \times a \times \ldots$ to n factors).
(iii) $\sqrt[n]{a} = a^{1/n}$. (iv) $a^m \times a^n = a^{m+n}$. (v) $a^m/a^n = a^{m-n}$.
(vi) $(a^m)^n = a^m \times a^m \times a^m \times \ldots$ to n factors $= a^{mn}$.
(vii) $\sqrt[n]{a^m} = a^{m/n}$.

The statements (iii) to (vii) can be shown to be true for all values of a, m and n, i.e. they are identities.

3. Logarithms in algebra

The algebraic definition of a logarithm is
$$N = a^{\log_a N}.$$
The logarithm of a number is an index, and so logarithms obey the rules of indices given in § 2. Using this fact, some rules will now be developed.

If N, M and a are positive integers (logarithms of negative numbers have no simple meaning), then by definition
$$M = a^{\log_a M} \quad \text{and} \quad N = a^{\log_a N}.$$
Therefore
$$MN = a^{\log_a M} \times a^{\log_a N} = a^{(\log_a M + \log_a N)}.$$
But, from the definition of a logarithm,
$$MN = a^{\log_a (MN)}.$$
Hence, equating the indices of a, it follows that
$$\log_a MN = \log_a M + \log_a N. \tag{2.1}$$
Similarly,
$$M/N = a^{\log_a M}/a^{\log_a N} = a^{(\log_a M - \log_a N)}.$$
Hence
$$\log_a (M/N) = \log_a M - \log_a N. \tag{2.2}$$

Equations (2.1) and (2.2) describe algebraically the use of logarithms to simplify the processes of multiplication and division. The use of logarithms to find the powers and roots of a number is now illustrated: thus
$$M^n = (a^{\log_a M})^n = a^{n \log_a M},$$
and hence
$$\log_a (M^n) = n \log_a M. \tag{2.3}$$
Similarly,
$$\sqrt[n]{M} = (a^{\log_a M})^{1/n} = a^{(1/n) \log_a M},$$
and hence
$$\log_a \sqrt[n]{M} = (1/n) \log_a M. \tag{2.4}$$

A useful logarithmic relation is
$$\log_a (M/N) = -\log_a (N/M), \tag{2.5}$$
which follows from the relation
$$\log_a M - \log_a N = -(\log_a N - \log_a M).$$

The logarithms of 1 and 0 are unchanged for any value of the positive base a; for, by definition,
$$1 = a^{\log_a 1},$$
and it has been shown already in Chap. 1, § 8 that any number raised to the power 0 is equal to 1, i.e. $a^0 = 1$. Hence $\log_a 1 = 0$, i.e. the logarithm of 1 to any base a is 0.

The logarithm of 0 is the power to which a must be raised to equal 0.

This power is minus infinity, since if a is greater than 1, then
$$a^{-\infty} = 1/a^{\infty} = 1/(a \times a \times a \times \ldots \text{ to an infinite number of factors})$$
$$= 0.$$
That is
$$\log_a 0 = -\infty.$$

Changing the base of logarithms. It is often necessary to be able to convert a logarithm with one base to a logarithm with another base. The systems of logarithms most frequently used have the bases 10 and e, and they can be easily converted from one into the other by multiplying by a suitable factor.

If M, a and b are positive numbers, then, by definition,
$$M = a^{\log_a M}, \quad M = b^{\log_b M} \quad \text{and} \quad b = a^{\log_a b}.$$
Substituting for b in the second expression gives
$$M = a^{(\log_a b)(\log_b M)}.$$
Hence
$$\log_a M = (\log_a b)(\log_b M), \tag{2.6}$$
or
$$\log_b M = \log_a M / \log_a b.$$

The conversion factor for converting $\log_b M$ to $\log_a M$ is therefore $\log_a b$.

The usual conversion required is from common logarithms to natural logarithms, i.e. from base 10 to base e. Substituting in (2.6) gives the expression
$$\log_e M = (\log_e 10)(\log_{10} M).$$
The conversion factor $\log_e 10$ has the numerical value of 2·3026, and this value is used to convert common logarithms to natural logarithms. The reciprocal of this factor, 1/2·3026, is used to convert natural logarithms to common logarithms. For convenience, common logarithms are written 'log', and natural logarithms as 'ln'. Thus
$$\ln M = (\ln 10)(\log M) = 2·3026 \log M.$$
It should be noted that the natural logarithm is greater than the common logarithm because e is smaller than 10.

4. Identities

If two functions of one variable x are identical, i.e. equal for all values of x, then the coefficients of like powers of x are the same on both sides of the identity. This can be illustrated for a quadratic function: thus if
$$ax^2 + bx + c \equiv dx^2 + fx + g,$$
then taking all the terms over to the left-hand side gives
$$(a-d)x^2 + (b-f)x + (c-g) \equiv 0.$$

ALGEBRA

This statement can only be true for all values of x if the coefficients $(a-d)$, $(b-f)$ and $(c-g)$ are each equal to 0, i.e. if $a = d$, $b = f$ and $c = g$.

5. Equations

If two functions of a variable x are equal for a limited number of values of x, the statement of equality is an *equation*, and the values of x which satisfy the equation are called the *solutions* or *roots* of the equation. The number of solutions depends on the degree of the equation, i.e. the highest power of x in the equation.

An equation of the first degree in x is satisfied by a single value of x; for example, the equation

$$a(x-a) = ab - 2b(x+b)$$

is satisfied by a single value of x. To find this solution, the terms are multiplied out and rearranged to bring all terms in x on to one side and other terms on to the opposite side. When a term is transferred from one side of an equation to the other, its sign is changed. Thus, multiplying out the example gives

$$ax - a^2 = ab - 2bx - 2b^2,$$

which on rearranging and factorising becomes

$$(a+2b)x = (a+2b)(a-b).$$

Hence, $x = (a-b)$, providing that $(a+2b) \neq 0$. When $a + 2b = 0$, the rearranged equation becomes $0 = 0$, which is true for all values of x, and the original equation is an indentity in x.

Quadratic equations. Equations of the second degree in x can have two solutions. This may be illustrated by the very simple quadratic equation $x^2 = 9$, which can have the solutions $x = +3$ and $x = -3$; for either of these values of x, the value of x^2 is 9.

A solution to a quadratic equation of the general form

$$ax^2 + bx + c = 0$$

can be found in terms of a, b and c. Any quadratic equation can be reduced to the general form by rearranging it so that all terms are on the left-hand side: thus

$$5x^2 + 5x + 2 = 2x^2 + 7$$

can be arranged to give

$$3x^2 + 5x - 5 = 0.$$

The general solution is obtained by rearranging the equation to the form

$$ax^2 + bx = -c,$$

and dividing by the coefficient of x^2 to give

$$x^2 + \frac{b}{a}x = -\frac{c}{a}.$$

If the square of half the coefficient of x is now added to both sides of the equation, the left-hand side becomes a perfect square: thus

$$x^2 + \frac{b}{a}x + \left(\frac{b}{2a}\right)^2 = -\frac{c}{a} + \left(\frac{b}{2a}\right)^2.$$

Rewriting the equation gives

$$\left(x + \frac{b}{2a}\right)^2 = \frac{-4ac + b^2}{4a^2}.$$

When the square root of both sides of the equation is taken, an uncertainty about the sign of the right-hand side is introduced, and the equation becomes

$$x + \frac{b}{2a} = \pm \frac{\sqrt{(b^2 - 4ac)}}{2a}.$$

where '$\pm$' means either the positive or negative value (just as $y^2 = 9$ can be satisfied by $y = \pm 3$). Hence

$$x = \frac{-b \pm \sqrt{(b^2 - 4ac)}}{2a}. \tag{2.7}$$

This general formula gives the two values of x which satisfy the original quadratic equation. If the latter can be factorised, the solutions can be found directly without using the formula; thus $x^2 = 3x - 2$ can be rearranged to give

$$x^2 - 3x + 2 = 0,$$

and the left-hand side can now be factorised into $(x-2)$ and $(x-1)$; hence

$$(x-2)(x-1) = 0.$$

This equation is true only if either of the factors is equal to 0, i.e. if $x = 2$ or $x = 1$. These values of x are therefore the two solutions to the original quadratic equation.

Complex numbers. The solutions to quadratic equations are of three types:

(i) If $b^2 > 4ac$, the two solutions obtained are simple numbers.
(ii) If $b^2 = 4ac$, the term under the root sign in (2.7) is equal to 0, and there is, in fact, only one solution (or two equal solutions) of the equation, namely $x = -b/2a$.
(iii) If $b^2 < 4ac$, the quantity under the root sign is negative, and the solutions are not simple numbers, since there is no number in the ordinary arithmetic counting system which when squared gives a negative number (as stated in Chap. 1, § 3, all positive and negative numbers have positive squares).

Quadratic equations in which $b^2 < 4ac$ occur in the theoretical treatment of a number of physical problems, and it is necessary therefore to be able

ALGEBRA

to solve them. To do this a new quantity 'i' is introduced. This is not a simple number, but has the property $i^2 = -1$, or $i = \sqrt{(-1)}$. All square roots of negative numbers can then be written in terms of i: thus

$$\sqrt{(-7)} = \sqrt{7} \times \sqrt{(-1)} = i\sqrt{7}; \quad \sqrt{(-16)} = \sqrt{16} \times \sqrt{(-1)} = 4i.$$

Numbers that contain i are called complex numbers. The solutions to quadratic equations in which $b^2 < 4ac$ are always complex numbers, i.e. they always contain i and cannot be further simplified.

EXAMPLE 1. *Solve the quadratic equation* $3x^2 - 7x + 6 = 2x^2 - 2x$.

Rearranging the equation gives $x^2 - 5x + 6 = 0$, which can be solved by (2.7), or more easily by direct factorisation, since

$$x^2 - 5x + 6 = (x-3)(x-2).$$

Hence $(x-3)(x-2) = 0$, and therefore $x = 3$ or 2.

EXAMPLE 2. *Solve the quadratic equation* $2x^2 - 7x + 3 = 0$.

Since in this case the left-hand side cannot be factorised easily, (2.7) is applied directly: thus

$$x = \frac{7 \pm \sqrt{(7^2 - 4.2.3)}}{2.2} = \frac{7 \pm \sqrt{25}}{4} = \frac{7 \pm 5}{4} = 3 \text{ or } \tfrac{1}{2}.$$

EXAMPLE 3. *Solve the quadratic equation* $4x^2 - 12x + 9 = 0$.

Applying (2.7) gives

$$x = \frac{12 \pm \sqrt{(12^2 - 4.4.9)}}{2.4} = \frac{12 \pm \sqrt{(144 - 144)}}{8} = \frac{12 \pm 0}{8} = 1\tfrac{1}{2}.$$

This is a case in which $b^2 = 4ac$. When this occurs, the left-hand side of the original equation always forms a perfect square: thus

$$4x^2 - 12x + 9 = 4(x^2 - 3x + 9/4) = 4(x - 3/2)^2,$$

which is the square of $2(x - 3/2)$.

EXAMPLE 4. *Solve the quadratic equation* $x^2 + x + 1 = 0$.

This equation is of the type in which $b^2 < 4ac$, and so the solutions are complex. Applying (2.7) gives

$$x = \frac{-1 \pm \sqrt{(1-4)}}{2} = \frac{-1 \pm \sqrt{(-3)}}{2} = \frac{-1 \pm \sqrt{3} \cdot \sqrt{-1}}{2} = -\tfrac{1}{2} \pm \frac{\sqrt{3}}{2}i.$$

Such solutions cannot be further simplified.

Equations of third and higher degrees. An equation of the third degree (or cubic equation) contains a term in x^3, and has the general form

$$ax^3 + bx^2 + cx + d = 0.$$

If the left-hand side can be factorised, the solutions are easily obtained: thus

$$x^3 - 2x^2 - x + 2 = (x-2)(x-1)(x+1),$$

so that $x = 2, 1$ and -1 are the solutions of the equation

$$x^3 - 2x^2 - x + 2 = 0.$$

If only a single factor can be found, the equation can be reduced to a quadratic equation, which can then be solved by application of (2.7). For example, in the cubic equation

$$2x^3-7x^2+7x-2 = 0,$$

the left-hand side has the factor $x-1$. Taking this factor out, the equation becomes

$$(x-1)(2x^2-5x+2) = 0.$$

Hence one solution is $x = 1$. The other solutions are obtained by applying (2.7) to the expression in the second bracket: thus

$$x = \frac{5\pm\sqrt{(25-16)}}{4} = \frac{5\pm 3}{4} = 2 \text{ or } \tfrac{1}{2}.$$

The three solutions of the cubic equation are therefore $x = 2, 1$ or $\tfrac{1}{2}$.

Solutions to a cubic equation may be three different simple numbers, or two of the solutions may be equal to one another with the third different, or all three solutions may be equal. Another possibility is that one solution may be simple while the other two are complex numbers.

Equations of higher degree than the third cannot be solved by any general algebraic rule. If they can be factorised when the equation is written in the form $f(x) = 0$, then some or all of the solutions can be found. These higher-degree equations can have up to n solutions, where n is the degree of the equation. The simple solutions, i.e. those that are not complex numbers, of any equation can always be found by graphical methods (see Chap. 3).

6. Algebra in Fortran

Variables. In order to make programmes as general as possible so that they can be used to solve problems involving different sets of numerical data, numbers in programmes which vary from problem to problem are expressed in terms of lettered names called variables which are similar to algebraic quantities. The rules for these Fortran variables are that they may consist of any groups of letters or numbers up to a total of six characters, the first of which must be a letter. As already stated in Chap. 1, § 18, names beginning with letters from I to N, inclusive, are reserved for integers (positive whole numbers used for counting or as subscripts); any other initial letters indicate variables corresponding to real numbers with a decimal point.

Simple algebraic letters like X,Y,Z may be used for Fortran variables but it is often helpful to use fuller names which give some indication of the nature of the quantity; for example, TEMP for temperature, TIME for time and PRESS for pressure.

Operators. Some Fortran operators are listed in Table 1 (page 27). The simple ones have uses similar to those of ordinary algebra except that the equality operator, =, has a special meaning in Fortran. The quantity on the right-hand side of the equality is put into a computer cell whose name is given on the left-hand side and in performing this operation the previous contents of this cell are eliminated. As a result a Fortran statement such as

$$\text{TEMP} = \text{TEMP} + 10\cdot$$

which would be nonsense in ordinary algebra means that the contents of a cell called TEMP are increased by 10· (the decimal point is required since TEMP represents a real number) and the increased value is put into the cell labelled TEMP, the previous value being destroyed.

One additional operator which is useful is the exponentiation operator **. This raises a real variable or number to a power shown on the right-hand side. The power need not be an integer, but if it is not a positive or negative integer, the number operated upon *must be positive* to avoid ambiguities with the roots of negative numbers.

Expressions. A number of operations may be combined together in Fortran as in ordinary algebra to give an expression and, by means of the equality operator, the value of the expression may be put into a cell with a name shown on the left-hand side of the equals sign; this name is then used to print out the answer.

The order of operations is that exponentiation is carried out first, then multiplication and division, then addition and subtraction, with a general sequence from left to right. The sequence of multiple operations is best controlled by the use of brackets which are used to avoid ambiguity in the same sense as in algebra.

$$X + Y/A - B$$

This expression would first divide Y by A and then X would be added to the result and B subtracted from it. If it is required to divide X plus Y by A minus B the above sequence would be altered by using two sets of brackets

$$(X+Y)/(A-B)$$

which would of course give a different answer.

Brackets. A major cause of rejection of learners' programmes is the failure to close brackets. Multiple bracketing is often used in complicated expressions and if they also involve functions and subscripted variables the numbers of pairs of brackets may be quite large. A good general rule to make sure that such an expression will be accepted by the compiler is to count the number of brackets facing each way; if they are not equal the programme will be rejected. A good example of a complicated set of brackets is given in the *t* test by the Fortran programme COMP 8, in Chap. 13, § 2.

Brackets form an essential part of a dimensioned, subscripted variable and of supplied functions such as EXP( ) and SQRT( ). The quantity inside the subscripted variable brackets must be an integer but it may be an integer variable or an expression, providing that on working out the answer is an integer. The supplied function brackets may contain real numbers, variables, or expressions.

7. COMP 1: Fortran programme to give a table of values from an algebraic equation

The variation of the physical properties of a substance with temperature may be compactly summarised by means of an algebraic equation. This equation may then be used to give a table of values of the property at round-figure temperatures.

The vapour pressure p of a liquid in terms of absolute temperature T may be summarised over a limited range by an equation of type

$$\log p = -A/T + B$$

where A and B are constants for a given liquid.

With p in millibars (1 mbar = 100 N m^{-2} = 0·75 mmHg) the values of A and B for n-hexane are A = 1655· (real number), B = 7·849.

The following Fortran programme will calculate the vapour pressure at each 10-degree interval between 0°C and 70°C and will print out the results in a table with headings for each column.

```
         PROGRAM VPRS(INPUT,OUTPUT,TAPE 5=INPUT, TAPE 6=OUTPUT)
C        Programmer's name
C        COMP 1
C        VAPOUR PRESSURE - TEMPERATURE TABLE
         READ(5,1) A,B
1        FORMAT(2F10.0)
         WRITE (6,2)
2        FORMAT(37H1TEMPERATURE,C    VAPOUR PRESSURE,MBAR)
         TEMP = -10.
         DO 3 I = 1,8
         TEMP = TEMP + 10.
         OGVP = -A/(TEMP + 273.1) + B
         VP = EXP(2.3026 * OGVP)
3        WRITE(6,4) TEMP,VP
4        FORMAT(1H0,F3.0,11X,F10.3)
         STOP
         END
```

This is the Fortran part of the programme. The first card is a header card required by the University of London Computer Centre for their CDC 6600 computer and it requires a programme name of up to five letters which is at the choice of the programmer, VPRS in this example;

ALGEBRA

it also gives the 5/6 input/output codes. Other computer centres might require different header cards or none at all. The second card is a comment card with the programmer's name; in subsequent programmes in this book these first two cards are omitted. The next two cards are comment cards giving the code title for the programme and a description of its purpose.

The READ card enables the programme to be used for any liquid for which A and B are known, the data card will have these two quantities contained in fields of ten columns each as given in the FORMAT card and both must be given a decimal point; if the full stop is omitted they will be misread.

The first WRITE gives headings to the columns of figures; H1 gives a new page and then the other 36 spaces after H will be printed out at the top of this new page with the exact characters and spacings punched on the format card after H1.

The next statement puts the value $-10\cdot$ into the real number cell called TEMP, which represents temperature in °C.

The DO statement on the next card is a very powerful one in Fortran; it permits the calculation of a series of values of one quantity from regularly incremented values of another. DO is followed by a statement number which must be attached to an executable statement and which marks the end of the DO loop. This is followed by an integer variable name which is altered in successive DO operations from the first value shown in the group of three integers separated by commas to the central value by increments shown in the third number. If the third number is 1, it may be omitted. In this programme, I, the integer variable, is simply a counter and does not appear elsewhere in the DO loop. DO 3 I = 1,8 simply means that the DO operation down to statement number three will be carried out successively 8 times.

The next card means that in each DO cycle TEMP will be increased by $10\cdot$, giving a first value of $0\cdot$; $\log p$ is then calculated for this value of TEMP but as a name beginning with L would be an integer variable it is called OGVP; the calculation of $\log p$ is made on the right-hand side of equality. In this expression, 273·1 is added to TEMP to give degrees kelvin.

To obtain the vapour pressure itself the antilogarithm of OGVP is evaluated by conversion to a natural logarithm and by use of the supplied function EXP( ) as described in § 15 of Chap. 1.

The final statement of the DO loop gives a WRITE instruction for the two variables TEMP and VP whose values will be lost in the next stage of the DO and will be replaced by new values. The FORMAT is outside the loop but the positioning of FORMAT in a programme is unimportant so long as it is numbered to correspond with the READ or WRITE. The DO loop could not end on a FORMAT since it is a non-executable statement; a non-executable statement such as DIMENSION or FORMAT is defined as one which does not cause any computation but only gives information to the compiler. When a DO is likely to end on such a statement a CONTINUE card is used. This does nothing except to mark the end of a DO loop; it must be numbered with the statement number shown after DO.

In the FORMAT, double spacing is called by the 1H0. As TEMP has two-digit round-figure values there is no point in specifying figures after the decimal point and three printing spaces are sufficient. The X specification gives a number of blank spaces in the print-out equal to the number in front of X and this spaces out the figures and brings them under the headings printed by the first WRITE statement.

The last two cards appear to be repetitive but both are required, the STOP is an instruction to stop executing the programme and may in some cases appear in other positions than the end of the set of Fortran cards. END must appear at the end of this set of cards and is a message to the compiler that the source programme which is to be compiled is complete.

The print-out from this programme is shown below.

```
TEMPERATURE,C     VAPOUR PRESSURE,MBAR
0                      61.512
10                    100.698
20                    159.396
30                    244.777
40                    365.734
50                    533.046
60                    759.523
70                   1060.115
```

8. Simultaneous equations

Two equations, each containing the variables x and y, are called simultaneous equations. There are a limited number of pairs of values of the variables that satisfy both equations, and these are called the simultaneous solutions of the equations.

Simple simultaneous equations are easily solved by eliminating one variable. Thus, the simultaneous solutions of the equations

$$x+y = 4 \quad \text{and} \quad x^2-xy-y^2 = 0$$

are found by substituting $y = 4-x$, obtained from the first equation, in the second one to give

$$x^2-x(4-x)-(4-x)^2 = 0.$$

Multiplying out and rearranging gives

$$x^2+4x-16 = 0.$$

The solutions for x are obtained by applying (2.7): thus

$$x = -2\pm\sqrt{20} = 2\cdot472 \text{ or } -6\cdot472.$$

ALGEBRA

As $y = 4-x$, the corresponding values of y are 1·528 and 10·472. The final results are given in pairs: thus

$$x = 2·472, \quad y = 1·528, \quad \text{or} \quad x = -6·472, \quad y = 10·472.$$

If the equations are of the first degree with respect to both variables, they are called linear equations, and there is only one pair of values of the variables which satisfies both equations. To find the simultaneous solutions of n linear equations, each containing n variables, matrices are used.

9. Matrices

Groups of simultaneous equations arise in the important computer technique of *optimisation* and in many other contexts. Since the input to a computer is preferably in the form of numbers, it is advantageous to separate the numerical coefficients of a set of equations from the algebraic unknowns and this is done in a systematic way by matrix representation.

A matrix is simply a table or array of numbers. The matrix itself does not call on any operation to be performed on the numbers but rules for combining matrices with one another by addition, subtraction, and multiplication have been developed mainly for the purpose of solving simultaneous equations.

The dimensions of a rectangular matrix are expressed as $M \times N$ where M is the number of rows of numbers in the tables and N is the number of columns. A 3×2 matrix is

$$A = \begin{pmatrix} 4 & 1 \\ 2 & 6 \\ 3 & 5 \end{pmatrix}$$

The numbers are enclosed in rounded brackets as shown. Each number is called an element and the elements are denoted by subscripts; A_{31} would be the element in row 3, column 1, that is, 3 in the above matrix. In computer form the 'subscripts' are put in brackets corresponding to a doubly subscripted variable, $A(3,1)$.

Some of the rules for combining matrices are as follows.

(*a*) *Equality.* Two matrices can only be equal when they are of the same dimensions and the statement

$$A = B$$

means that all the elements of A are equal to the corresponding elements of B.

(*b*) *Addition and subtraction.* Both matrices must have the same dimensions $M \times N$ in order to conform for these operations and the resultant is an $M \times N$ matrix whose elements are formed by adding or subtracting corresponding elements of the two matrices.

$$\begin{pmatrix} 4 & 1 \\ 2 & 6 \\ 3 & 5 \end{pmatrix} - \begin{pmatrix} 6 & 4 \\ 2 & 7 \\ -1 & 3 \end{pmatrix} = \begin{pmatrix} 4-6 & 1-4 \\ 2-2 & 6-7 \\ 3+1 & 5-3 \end{pmatrix} = \begin{pmatrix} -2 & -3 \\ 0 & -1 \\ 4 & 2 \end{pmatrix}$$

(c) *Multiplication.* In order to conform for the multiplication $A \times B$ two matrices A and B must be such that the number of columns in A equals the number of rows in B; if A has dimensions $L \times M$, then B will be $M \times N$ and the product matrix will have dimensions $L \times N$.

The rule for multiplication seems somewhat involved but it is designed so as to facilitate the handling of simultaneous equations.

The elements of the product matrix are formed by multiplying successively elements of a row of A by corresponding elements of a column of B and adding the results. The process is best illustrated by an example.

$$\begin{array}{ccc} A & \times & B & = & C \end{array}$$

$$\begin{pmatrix} 1 & 3 \\ 5 & 2 \\ 0 & 3 \end{pmatrix} \times \begin{pmatrix} 2 & 5 \\ 4 & 7 \end{pmatrix} = \begin{pmatrix} 1\times 2+3\times 4 & 1\times 5+3\times 7 \\ 5\times 2+2\times 4 & 5\times 5+2\times 7 \\ 0\times 2+3\times 4 & 0\times 5+3\times 7 \end{pmatrix} = \begin{pmatrix} 14 & 26 \\ 18 & 39 \\ 12 & 21 \end{pmatrix}$$

The elements of C are formed by the rule. $C(1,1)$ is obtained by multiplying elements of the first row of A by corresponding elements of the first column of B and adding the results. Similarly $C(3,2)$ is obtained from the third row of A and the second column of B. The rule means going along a row of A and down a column of B as indicated by the arrows in the example above.

The dimensions of $A(3\times 2)$ and $B(2\times 2)$ conform for this multiplication since the number of columns of A equals the number of rows of B; they would not however conform for the multiplication $B \times A$. In general the order of a matrix multiplication affects the result. When two matrices conform for both $A \times B$ and $B \times A$, as for example when they are both square matrices ($M = N$), the two products generally differ.

(d) *Inversion.* A square matrix A of dimensions $M \times M$ can have an inverse or reciprocal matrix written as A^{-1}, which is a square matrix of the same dimensions such that the product of A and A^{-1} is equal to the unit matrix I of these same dimensions. A unit matrix is a square matrix of any dimension in which all the elements are zero except those along the leading diagonal (i.e. the diagonal from the top left to bottom right corner), which are all 1. In the product of A and A^{-1}, the order of multiplication is irrelevant and

$$A \times A^{-1} = A^{-1} \times A = I.$$

ALGEBRA

The rules for finding the elements of A^{-1} from those of A, a process called the inversion of A, are somewhat complicated. They are illustrated below by finding the reciprocal of

$$A = \begin{pmatrix} 1 & 3 \\ 2 & -1 \end{pmatrix}$$

An extended matrix is formed by writing the 2×2 unit matrix alongside A:

$$AI = \begin{pmatrix} 1 & 3 & 1 & 0 \\ 2 & -1 & 0 & 1 \end{pmatrix}$$

Rows are then multiplied throughout by any factor and added and subtracted, again throughout the row, so as to give the unit matrix on the left-hand side; the right-hand side is then A^{-1}.

In this case, subtract twice row 1 from row 2 to give the zero in the first column, required for I:

$$\begin{pmatrix} 1 & 3 & 1 & 0 \\ 0 & -7 & -2 & 1 \end{pmatrix}$$

Row 1 is then multiplied by 7 and three times row 2 is added to it to give the zero in the second column required for the unit matrix:

$$\begin{pmatrix} 7 & 0 & 1 & 3 \\ 0 & -7 & -2 & 1 \end{pmatrix}$$

Row 1 is multiplied by $1/7$ and row 2 is multiplied by $-1/7$ to give the unit matrix

$$\begin{pmatrix} 1 & 0 & 1/7 & 3/7 \\ 0 & 1 & 2/7 & -1/7 \end{pmatrix}$$

so that

$$A^{-1} = \begin{pmatrix} 1/7 & 3/7 \\ 2/7 & -1/7 \end{pmatrix}$$

To prove that the result is correct carry out the multiplication $A \times A^{-1}$:

$$\begin{pmatrix} 1 & 3 \\ 2 & -1 \end{pmatrix} \times \begin{pmatrix} 1/7 & 3/7 \\ 2/7 & -1/7 \end{pmatrix} = \begin{pmatrix} 1/7+6/7 & 3/7-3/7 \\ 2/7-2/7 & 6/7+1/7 \end{pmatrix} = \begin{pmatrix} 1 & 0 \\ 0 & 1 \end{pmatrix}$$

For larger matrices finding elements of the reciprocals becomes more complicated but a computer programme can be written to invert any size of matrix up to the limit of the computer storage capacity. A library programme for this operation is included in the Scientific Subroutine Package (SSP)—see COMP 2, § 12 (page 47).

The unit matrix has a property similar to the number 1 in ordinary algebra and arithmetic, which is readily proved from the rule of matrix multiplication that when it multiplies another matrix which conforms with it, the other matrix is unchanged. If A is a square matrix and I is the unit matrix of the same dimensions

$$I \times A = A \times I = A$$

10. Matrices and simultaneous equations

As a result of the rules of matrix equality and multiplication, linear simultaneous equations can be expressed as a matrix of coefficients times a matrix of variables, which is set equal to a matrix of the constants. For example the two equations

$$x+3y = 11$$
$$2x-y-1 = 0$$

are rearranged so that the constants are on the right-hand side

$$x+3y = 11$$
$$2x-y = 1.$$

These equations can then be expressed in matrix form as

$$\begin{pmatrix} 1 & 3 \\ 2 & -1 \end{pmatrix} \times \begin{pmatrix} x \\ y \end{pmatrix} = \begin{pmatrix} 11 \\ 1 \end{pmatrix}$$

If the left-hand side is multiplied out we get

$$\begin{pmatrix} x+3y \\ 2x-y \end{pmatrix} = \begin{pmatrix} 11 \\ 1 \end{pmatrix}$$

The elements of the left-hand matrix are then equal to those of the right-hand one, giving the two equations.

In general a set of N simultaneous linear equations with N variables can be expressed in the matrix form

$$A \times X = B.$$

A is a square matrix of coefficients of the variables, the first row of A being the coefficients of the first equation and so on. The dimensions of A are $N \times N$. X is a single column matrix of the variables. A single column matrix is called a column vector; similarly, a single row matrix is called a row vector. B is a column vector of the constant terms.

The importance of this transformation is that multiplying through by the inverse of A solves the whole set of equations.

$$A^{-1} \times A \times X = A^{-1} \times B$$

but $A^{-1} \times A = I$ and $I \times X = X$

therefore $X = A^{-1} \times B.$

Returning to the two equations above the inverse of A has already been found in § 9,

$$X \text{ is } \begin{pmatrix} x \\ y \end{pmatrix} \text{ and B is } \begin{pmatrix} 11 \\ 1 \end{pmatrix}$$

$$A^{-1} = \begin{pmatrix} 1/7 & 3/7 \\ 2/7 & -1/7 \end{pmatrix}$$

therefore $\begin{pmatrix} x \\ y \end{pmatrix} = \begin{pmatrix} 1/7 & 3/7 \\ 2/7 & -1/7 \end{pmatrix} \times \begin{pmatrix} 11 \\ 1 \end{pmatrix} = \begin{pmatrix} 11/7+3/7 \\ 22/7-1/7 \end{pmatrix} = \begin{pmatrix} 2 \\ 3 \end{pmatrix}$

so that the solutions are $x = 2, y = 3$.

The value of this method is that by using a library programme based on matrix inversion any number of simultaneous linear equations up to the capacity of the computer may be solved together.

The method breaks down if the coefficients and constant term of one equation are equal to or are the same multiple of the coefficients and constant term of another. In this case the two equations give the same information and solution by any method is impossible without a further equation.

Another breakdown occurs with homogeneous linear equations in which all the constants are zero so that the vector B has zero elements. $A^{-1} \times B$ gives zeroes and the solutions are that all the variables are equal to zero. These are the trivial solutions to the equations. Non-trivial solutions may also exist and if so they are found by a different matrix method called diagonalisation. This gives the ratios of the solutions and to get separate values another relationship between the variables is required.

11. Determinants and homogeneous linear simultaneous equations

A square matrix has a determinant which has the same elements, but which does require operations to be performed on them. The determinant is always a square array of elements enclosed in vertical lines. The determinant of matrix

$$\begin{pmatrix} 2 & 5 \\ 4 & 7 \end{pmatrix} \text{ is written } \begin{vmatrix} 2 & 5 \\ 4 & 7 \end{vmatrix}$$

and calls for the four numbers to be multiplied across the diagonals; the first product across the leading diagonal is positive, the other is negative and the two are added together to give the value of the determinant

$$\begin{vmatrix} 2 & 5 \\ 4 & 7 \end{vmatrix} = 2 \times 7 - 5 \times 4 = 14 - 20 = -6.$$

A third order determinant is a 3×3 array which may be expanded along any row or down any column as follows. First of all there is the rule of alternating signs; the top left-hand corner element in the (1,1) position is given a positive sign, the elements next to it are taken as negative, the elements next to these as positive, and so on. Each element in the row or column by which the determinant is expanded is given the appropriate sign and is multiplied by a second order determinant formed from the original one by omitting the row and the column containing the multiplying element. For example,

$$\begin{matrix} + \\ - \\ + \end{matrix} \begin{vmatrix} 2 & 4 & -1 \\ 3 & 1 & 1 \\ -2 & 1 & 3 \end{vmatrix}$$

expanded down the first column gives

$$2\begin{vmatrix} 1 & 1 \\ 1 & 3 \end{vmatrix} - 3\begin{vmatrix} 4 & -1 \\ 1 & 3 \end{vmatrix} + (-2)\begin{vmatrix} 4 & -1 \\ 1 & 1 \end{vmatrix}$$

$$= 2(3-1) - 3(12+1) - 2(4+1) = 4 - 39 - 10 = -45.$$

The same value should be obtained by expansion down any other column or along any row.

Higher order determinants are reduced successively down to a set of second order determinants by repeated application of this rule. The arithmetic involved is tedious and for larger determinants expansion is best done by computer.

Determinants may be used to give solutions to linear simultaneous equations. If we take two such equations

$$a_1 x + b_1 y = c_1$$
$$a_2 x + b_2 y = c_2$$

then the standard method for solving for x would be to eliminate y by multiplying the first equation by b_2 and the second by b_1 and subtracting the second from the first, giving

$$x(a_1 b_2 - a_2 b_1) = b_2 c_1 - b_1 c_2.$$

From the rule for the expansion of a determinant it is seen that the quantity in the bracket on the left-hand side is

$$\begin{vmatrix} a_1 & b_1 \\ a_2 & b_2 \end{vmatrix} \quad \text{and the right-hand side is} \quad \begin{vmatrix} c_1 & b_1 \\ c_2 & b_2 \end{vmatrix}$$

and so x is given by a ratio of determinants

$$x = \begin{vmatrix} c_1 & b_1 \\ c_2 & b_2 \end{vmatrix} \div \begin{vmatrix} a_1 & b_1 \\ a_2 & b_2 \end{vmatrix}$$

ALGEBRA

The determinant in the denominator is the determinant of coefficients of the variables in the equations; the numerator determinant is the same with the column of coefficients of x replaced by a column of the constant terms. A similar result is found for y:

$$y = \begin{vmatrix} a_1 & c_1 \\ a_2 & c_2 \end{vmatrix} \div \begin{vmatrix} a_1 & b_1 \\ a_2 & b_2 \end{vmatrix}$$

The denominator is again the determinant of coefficients but the numerator has the column of y coefficients replaced by the column of constants.

This result is generalised in Cramer's rule, which applies to N linear simultaneous equations containing N variables. The value of any variable is equal to a ratio of determinants, the denominator being the determinant of coefficients of all the variables while the numerator is the same determinant with the column of coefficients of the variable being evaluated replaced by the column of constant values.

Consequently the solutions to a set of equations may be written straight out in determinant form:

$$3x+y-z = 12$$
$$x-4y+z = -3$$
$$2x+3y+2z = 5.$$

The determinant of coefficients is

$$\begin{vmatrix} 3 & 1 & -1 \\ 1 & -4 & 1 \\ 2 & 3 & 2 \end{vmatrix}$$

and the value of y, for example, would be given by replacing the centre column by the column of constants giving

$$\begin{vmatrix} 3 & 12 & -1 \\ 1 & -3 & 1 \\ 2 & 5 & 2 \end{vmatrix}$$

and dividing by the determinant of coefficients. These two determinants on expansion down the first column give respectively $3(-8-3)-1(2+3)+2(1-4) = -44$ and $3(-6-5)-1(24+5)+2(12-3) = -44$ and therefore y equals the second divided by the first determinant, $y = 1$.

Homogeneous linear equations. The determinant method is particularly useful for finding whether a set of linear homogeneous equations have non-trivial (i.e. non-zero) solutions. Homogeneous equations are those in which each term contains the variables to the same power (in linear equations this power is one) and there are no constant terms (since these would contain variables to the power zero).

With a set of equations of this type Cramer's rule would give the numerator determinant for each variable as zero, since each would contain a column of zeroes for the column of constants and expansion down this column would give zero. The only condition by which non-zero solutions could be obtained is by the denominator determinant also being zero; the ratio 0/0 can then have finite values.

To test whether a given set of linear homogeneous equations has non-trivial solutions the determinant of coefficients is evaluated; if it is zero then such solutions are possible. However, N homogeneous equations with N variables will only give ratios of the values of the variables; to evaluate them absolutely another relationship is required.

Consider the homogeneous equations

$$2x+4y-z = 0$$
$$5x+y+2z = 0$$
$$3x+y+z = 0$$

Clearly the trivial solutions $x = y = z = 0$, which would be given by Cramer's rule, satisfy these equations.

To find whether there is also a set of non-trivial solutions, evaluate the determinant of coefficients, expanding down the first column:

$$\begin{vmatrix} 2 & 4 & -1 \\ 5 & 1 & 2 \\ 3 & 1 & 1 \end{vmatrix} = 2(1-2)-5(4+1)+3(8+1) = 0$$

and so non-trivial solutions exist. To find their ratios to one of them, say z, divide through by z and write $X = x/z \quad Y = y/z$

$$2X+4Y-1 = 0 \qquad 2X+4Y = 1$$
$$5X+Y+2 = 0 \qquad 5X+Y = -2$$
$$3X+Y+1 = 0 \qquad 3X+Y = -1.$$

Only two of these equations give independent information—if the first two are solved then the third will be found to be satisfied. Solving the first two, which are no longer homogeneous in the new variables, by Cramer's rule gives

$$X = \begin{vmatrix} 1 & 4 \\ -2 & 1 \end{vmatrix} \div \begin{vmatrix} 2 & 4 \\ 5 & 1 \end{vmatrix} = (1+8)/(2-20) = -1/2$$

$$Y = \begin{vmatrix} 2 & 1 \\ 5 & -2 \end{vmatrix} \div \begin{vmatrix} 2 & 4 \\ 5 & 1 \end{vmatrix} = (-4-5)/(2-20) = 1/2.$$

The third equation $3X+Y = -3/2+1/2 = -1$ is automatically satisfied.

A further equation to give absolute values of the variables which is often used with homogeneous equations is the normalisation condition that the sum of the squares of the variables equals one:

$$x^2+y^2+z^2 = 1.$$

If this equation is divided by z^2 we get

$$X^2+Y^2+1 = 1/z^2$$

substituting the values of X and Y

$$1/4+1/4+1 = 1/z^2 = 6/4 \qquad z^2 = 4/6$$
$$z = \pm 2/\sqrt{6}$$

therefore, since

$$x/z = X = -1/2 \quad \text{and} \quad y/z = Y = 1/2$$
$$x = -1/\sqrt{6} \qquad y = 1/\sqrt{6} \qquad z = 2/\sqrt{6}$$

or

$$x = 1/\sqrt{6} \qquad y = -1/\sqrt{6} \qquad z = -2/\sqrt{6}.$$

12. COMP 2: Use of a library programme to solve simultaneous equations

It is not difficult to write a programme to solve a set of linear simultaneous equations by matrix or determinant methods, but there is no need to do this as a number of programmes for handling such equations are included in the Scientific Subroutine Package (SSP). This package, produced by IBM, is available at most computer centres on tape or magnetic disc and it is only necessary to call a subroutine by its name in a Fortran programme and supply the parameters required for its use. In addition, job control cards will be required to call the library subroutine from tape or disc.

SSP programmes do not contain input and output details and these have to be included in the Fortran programme in which the subroutine is called.

```
C       COMP 2
C       SOLUTION OF SIMULTANEOUS EQUATIONS
        DIMENSION COEFF(4,4),CONST(4)
        READ(5,1) COEFF,CONST
1       FORMAT(10F8.0)
        CALL SIMQ(COEFF,CONST,4,KS)
        WRITE(6,2) CONST,KS
2       FORMAT(10H1SOLUTIONS//5H C1 =,F20.10//5H C2 =,F20.10
       1 //5H C3 =,F20.10//5H C4 =,F20.10//5H KS =,I2)
        STOP
        END
```

COMP 2 solves four simultaneous equations with four unknowns. The matrix of coefficients is called COEFF and the vector of constants is CONST. COEFF is read in columns as a single string of numbers, those in the first column first, followed immediately by the second column, and so on. There is no need to distinguish the columns in the data input as the dimensioning breaks the sixteen values into columns of four. However, there must be sixteen values and where a term in one variable is missing in an equation it must be given a coefficient of 0·0. For example, the four equations

$$x_1 + 2x_2 - 3x_3 - x_4 = 8$$
$$2x_1 + x_4 = 7$$
$$2x_1 - 3x_2 + x_3 = -9$$
$$5x_1 + 2x_2 + x_3 + x_4 = 14$$

in matrix form would be

$$\begin{matrix} A & \times & X & = & B \end{matrix}$$
$$\begin{pmatrix} 1 & 2 & -3 & -1 \\ 2 & 0 & 0 & 1 \\ 2 & -3 & 1 & 0 \\ 5 & 2 & 1 & 1 \end{pmatrix} \begin{pmatrix} x_1 \\ x_2 \\ x_3 \\ x_4 \end{pmatrix} = \begin{pmatrix} 8 \\ 7 \\ -9 \\ 14 \end{pmatrix}$$

and COEFF would be entered as the string of sixteen real numbers.

1· 2· 2· 5· 2· 0· −3· 2· −3· 0· 1· 1· −1· 1· 0· 1·

and CONST would be

8· 7· −9· 14·

There is no need to distinguish between A and B on the data card; they are entered as a continuous set of 20 numbers; the dimensioning takes the first 16 as A and the last 4 as B. The read FORMAT gives identical specifications for the two data cards and so, as described in § 16 of Chap. 1 (page 24), a slash is unnecessary. FORMAT(10F8·0) is sufficient.

The details of the subroutine SIMQ are given in the IBM publication on the Scientific Subroutine Package. Operating instructions are also given on a set of C cards included in the programme and therefore printed in the computer output. The purpose of the subroutine is to obtain the solutions to a set of simultaneous equations A.X = B (A, X and B being matrices). The usage is

CALL SIMQ (A,B,N,KS)

This is the statement which is put into the Fortran programme. The four parameters are defined as follows:

A is the matrix of coefficients stored columnwise. These are destroyed in the computation. The size of matrix A is N×N.

B is the vector of constants, of length N. These are replaced by the final solution values X, and so in the computation, vector B becomes transformed into vector X. In order to write out the solutions B must be put in the WRITE list.

N is the number of equations and variables; N must be greater than one.

KS is an output digit which acts as an indicator of the validity of the calculation. KS = 0 for normal solutions but KS = 1 for singular solutions. These occur if two equations give almost the same information, that is, their coefficients are all equal or are in a constant ratio to one another; in such there are only N−1 equations for N variables and reliable results cannot be obtained. KS should be printed out and if it has the value 1, the solutions are not reliable and a further relationship between the variables should be sought.

In the brackets after CALL SIMQ in the Fortran programme, other names for A and B may be declared; the numerical value of N can be put in or can be put as a statement in the programme. If different names are used for A and B, the names A and B are used within the subroutine and are then translated back to the names given at the end of the subroutine.

In this programme A is called COEFF and B is called CONST; note that in the WRITE list, CONST (B in the subroutine) is used to obtain the solutions, called C1, C2, C3 and C4 in the FORMAT titles. The FORMAT requires a continuation number in column 6 of the Fortran card.

The output from this programme with the data given above is

```
SOLUTIONS
C1  =         1.0000000000
C2  =         3.0000000000
C3  =        -2.0000000000
C4  =         5.0000000000
KS = 0
```

Problems

1. Show that $\log_a b = 1/\log_b a$.
2. If $ax^3+7x^2+3x+b \equiv cx^2+dx+4$, write down the values of a, b, c and d.
3. Solve, by factorisation,
   (i) $x^2-7x+12 = 0$ (ii) $5x^2+7x-6 = 0$
4. Use equation (2.7) to solve the equations
   (i) $x^2+x+3 = 0$ (ii) $6x^2-4x+8 = 0$
   (iii) $-4x^2+3x+2 = 0$ (iv) $x^2+7x+2 = 0$

5. Solve the simultaneous equations
   (i) $x^2+3y = 7$ and $x+y = 3$
   (ii) $3x+2y = 0$ and $x+y = -4$
6. If $x-3$ is a factor of $x^3-2x^2-5x+6 = 0$, what are the three solutions to this equation?
7. Put the following simultaneous equations into matrix form

$$x+y+2z = 7$$
$$2x-y+3z = 16$$
$$4x+2y-z = -4$$

Solve them using determinants and Cramer's rule.

8. The costs of two purification processes per cubic metre of solution, in terms of parts per million (PPM) of impurity, are found to be

$$\text{PROCA} = 1 \cdot 062 + 0 \cdot 104 \times \text{PPM}/550$$
$$\text{PROCB} = 0 \cdot 009 + 0 \cdot 667 \times \text{PPM}/550$$

where PROCA and PROCB are the costs of the processes. Modify COMP 1 (page 36) so as to produce a table with headings showing PPM from 50 to 1500 in increments of 50 with corresponding values of PROCA and PROCB.

From this table estimate the value of PPM at which the costs become equal. What statement would you add to the programme to get an accurate value of this break-even value of PPM?

9. The absorbances of a solution containing four light absorbing solutes were measured at four wavelengths, the molar absorption coefficients of the four substances were known, and the results were used to construct the four simultaneous equations

$$1500C_1 + 51 \cdot 6C_2 + 41 \cdot 4C_4 = 1209$$
$$26 \cdot 2C_1 + 1152C_2 + 0 \cdot 4C_3 + 82 \cdot 2C_4 = 1176$$
$$34 \cdot 4C_1 + 35 \cdot 6C_2 + 2540C_3 + 294C_4 = 2953$$
$$33 \cdot 9C_1 + 68 \cdot 6C_2 + 343C_4 = 500 \cdot 8$$

Use COMP 2 (page 47) to estimate the values of the concentrations, C_1, C_2, C_3 and C_4.

CHAPTER 3

Graphs

IN ORDER to establish quickly the meaning of a set of experimental results, it is best to plot them as a graph. But this is not a suitable method of keeping them as a permanent record; it is much neater and less space-consuming to express them as an equation. The task of finding an equation to fit the experimental curve is not easy, unless it happens to be a straight line. Certain types of equations give characteristic curves, and it is the purpose of this chapter to illustrate this. The applications of these equations to practical problems are discussed in Chap. 10.

1. Rectangular co-ordinates

In graphs on rectangular co-ordinate graph paper, the magnitude of one variable is plotted along the horizontal axis, called the x-axis, and all values along this axis are known as *abscissae*. The magnitude of the other variable is plotted along the vertical axis, called the y-axis, and the values along this axis are known as *ordinates*. The position of a point on a graph is defined by its abscissa and ordinate, which together form its co-ordinates. The point of intersection of the two axes is called the *origin* (point O, see Fig. 3a). If the angle between the co-ordinates x and y is a right angle, the axes are said to be rectangular.

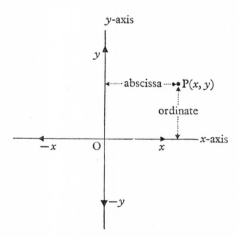

FIG. 3a. Sign convention for rectangular co-ordinates

A sign convention has been adopted with regard to x and y. To the right of the y-axis, x is positive and to the left, it is negative. Above the x-axis, y is positive and below, it is negative. This convention is illustrated in Fig. 3a.

The position of a point P is defined by its two co-ordinates (x, y), the value of x always being given first. Every point on a curve will have co-ordinates which satisfy the equation for that line.

2. Equations of the first degree in both variables

Equations of the first degree are equations which contain no exponent other than 1, and they are represented graphically by straight lines. The general form of the equation is

$$y = a + bx,$$

where y is the dependent variable, x is the independent variable and a and b are constants. The coefficient of x, i.e. b, defines the slope of the line

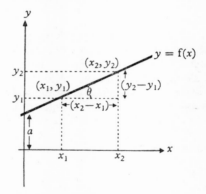

FIG. 3b. Slope and intercept of a line

representing the equation, i.e. the ratio of the increase in y corresponding to an increase in x. If y_1 is increased to y_2 and the corresponding increase in x is from x_1 to x_2, then

$$b = (y_2 - y_1)/(x_2 - x_1) = \Delta y / \Delta x = \text{slope of the line},$$

where Δx and Δy are the increases in x and y respectively. The ratio $\Delta y / \Delta x$ also measures the tangent of the angle θ which the line representing the equation makes with the x-axis (see Fig. 3b). It is seen from the figure that the distance between the origin and the intercept of the line $y = f(x)$ on the y-axis is equal to a. This follows quite simply from the general expression for the line $y = a + bx$. Anywhere along the y-axis, x is zero, and hence the term bx is also zero; it follows therefore that $y = a$ when $x = 0$.

Slope of the line. When y increases with x, b is positive; when y decreases as x increases, b is negative. The sign of the slope is the same as the sign of the tangent of the angle which the line makes with the x-axis.

It is seen from further consideration of the general equation

$$y = a + bx$$

that when $a = 0$ (as in $y = 3x$), the line passes through the origin, for when $x = 0$, $y = 0$ also. When $b = 1$ (as in $y = x+3$), the line is at an

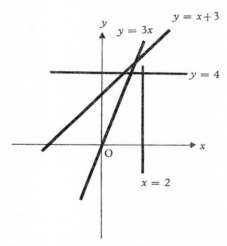

FIG. 3c. Graphs of some linear equations

angle of 45° to the x-axis, if the same scale is used for both axes. If the y term is missing (as in $x = 2$), the line is perpendicular to the x-axis, and if the x term is missing (as in $y = 4$), the line is parallel to the x-axis. These relations are illustrated in Fig. 3c.

Determination of the equation for a straight line that will pass through the point (x_1, y_1). Let the equation of the line be

$$y = a + bx.$$

Then if this line is to pass through the point (x_1, y_1), these co-ordinates must satisfy the equation of the line. Hence

$$y_1 = a + bx_1,$$

and by subtraction

$$y - y_1 = b(x - x_1). \tag{3.1}$$

This is the equation of the line passing through the point (x_1, y_1).

EXAMPLE 1. *Find the equation of the line having a slope of 3 which passes through the point* (5, 4).

Substituting in (3.1) gives
$$y-4 = 3(x-5).$$
Therefore $y = 3x-11$, the required equation, since when $x = 5$, $y = 4$.

Similarly, it is possible to find the equation of the line that passes through two points having the co-ordinates (x_1, y_1) and (x_2, y_2). At the point (x_2, y_2)
$$y_2 = a+bx_2.$$
At the point (x_1, y_1)
$$y_1 = a+bx_1.$$
Subtracting the second equation from the first gives
$$y_2-y_1 = b(x_2-x_1).$$
The slope of the line is b, which is equal to $(y_2-y_1)/(x_2-x_1)$, and substituting this value in (3.1) gives
$$y-y_1 = \frac{y_2-y_1}{x_2-x_1}(x-x_1). \tag{3.2}$$

EXAMPLE. *Find the equation of the straight line that passes through the points* (2, 3) *and* (8, 12).

Substituting in (3.2) gives
$$y-3 = \frac{12-3}{8-2}(x-2).$$
Hence
$$y-3 = 1\cdot5x-3,$$
and the required equation is $y = 1\cdot5x$.

Parallel lines. If two or more lines have the same slope, they will be parallel. For example, both the equations
$$2y = 6x+7 \quad \text{and} \quad y = 3x+6$$
have the same slope of 3, and their plots are parallel.

Points of intersection. Two straight lines will have only one pair of co-ordinates which satisfy the equations of both the lines. It is possible to find these co-ordinates by solving the equations simultaneously.

EXAMPLE. *Find the co-ordinates of the point of intersection of the two lines*
$$y = 4x+3 \quad \text{and} \quad y = 2x+7.$$
Subtracting the second equation from the first gives
$$0 = 2x-4,$$
and hence $x = 2$. Substituting this value of x in one of the equations gives
$$y = 4.2+3 = 11.$$
Therefore the two lines intersect at the point (2, 11).

Change of origin. The reference point for co-ordinates is the origin. If the axes are moved, the co-ordinates will alter in accordance with the change of origin. Thus, in Fig. 3e, if (h, k) are the co-ordinates of the new origin O′ with reference to the old origin O, then the new co-ordinates of any point (x_1, y_1) are related to the old co-ordinates (x, y) by the equations

$$x_1 = x-h \quad \text{and} \quad y_1 = y-k.$$

This relation holds only if the new axes are parallel to old axes.

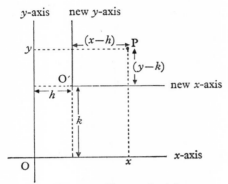

FIG. 3e. Change of origin

EXAMPLE. *Given that the co-ordinates of a point are (2, 3) and an equation $y = 2x+4$, what are the values of these co-ordinates and the form of the equation if the origin of the axes is moved to a new position (3, 5) and the new axes are parallel to the old axes?*

The new co-ordinates (x_1, y_1) are given by

$$x_1 = x-h = 2-3 = -1,$$

and

$$y_1 = y-k = 3-5 = -2.$$

The new co-ordinates of the point are therefore $(-1, -2)$.

The equation, when referred to the new axes, becomes

$$y_1+k = 2(x_1+h)+4.$$

Substituting the values of the new co-ordinates of the origin gives

$$y_1+5 = 2(x_1+3)+4.$$

Hence

$$y_1 = 2x_1+5.$$

3. Equations of the first degree in y and second degree in x

The general form of the equation is

$$y = ax^2+bx+c. \qquad (3.3)$$

If x and y are given a set of values and plotted on rectangular co-ordinate graph paper, a curve called a *parabola* is formed. Any point P (see

Fig. 3f) on the curve is equidistant from a fixed point on the axis of the parabola, called the *focus* F, and a straight line drawn perpendicular to the axis of the parabola, called the *directrix*. For example, the equation $y = ax^2$ represents a parabola with its *vertex*, i.e. the point where the curve of the parabola cuts the axis of the parabola, at the origin and with the axis of the parabola coinciding with the y-axis.

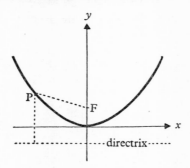

FIG. 3f. Parabola $y = ax^2$

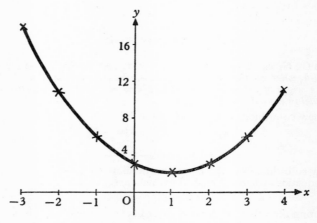

FIG. 3g. Graph of $y = x^2 - 2x + 3$

EXAMPLE. *Plot the graph representing the equation* $y = x^2 - 2x + 3$.

The corresponding values of x and y are obtained by substituting in the equation: thus when

$x =$	-3	-2	-1	0	1	2	3	4,
$y =$	18	11	6	3	2	3	6	11.

The curve representing these values is shown in Fig. 3g.

Most curves can be represented along limited sections by parabolic equations.

4. Equations of the first degree in y and third degree in x

The general form of the equation is

$$y = ax^3 + bx^2 + cx + d. \tag{3.4}$$

A graphical representation of this type of equation may take one of several forms, two of which are shown in Fig. 3h.

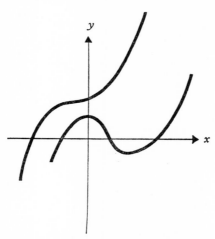

Fig. 3h. Forms of cubic equation

5. Equations of the second degree in both x and y

Some well-known curves occur in this group, which includes the circle, ellipse and rectangular hyperbola.

Circle. A circle with its centre at the origin of a rectangular co-ordinate system and radius r (see Fig. 3i) represents the algebraic equation

$$x^2 + y^2 = r^2.$$

Geometrically, a circle is a curve such that all points on it are equidistant from a fixed point, the centre.

The general form of the equation of a circle with any point as its centre is

$$x^2 + y^2 + ax + by + c = 0. \tag{3.5}$$

If the x and y terms are missing and c is negative, the centre of the circle is at the origin of the co-ordinate system and its radius is $\sqrt{(-c)}$.

Ellipse. An ellipse is a curve such that the sum of the distances of any point P on it from two fixed points F_1 and F_2, called the *foci*, is constant: thus in Fig. 3j

$$PF_1 + PF_2 = \text{constant}.$$

If the ellipse is symmetrical about the axes of a rectangular co-ordinate system, it is represented by the general equation

$$(x^2/a^2)+(y^2/b^2) = 1. \tag{3.6}$$

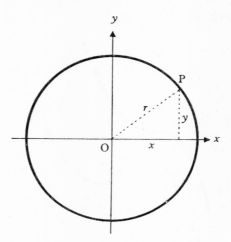

FIG. 3i. Circle with centre at origin

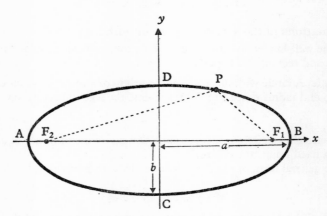

FIG. 3j. Ellipse $x^2/a^2+y^2/b^2 = 1$

The line AB is called the major axis, and the line CD the minor axis (or the maximum and minimum diameters, respectively). It can be shown that the length of these diameters are $2a$ and $2b$ respectively, since it follows from (3.6) that when $y = 0$, $x = \pm a$, and when $x = 0, y = \pm b$.

GRAPHS

Rectangular hyperbola. A hyperbola is a curve such that the difference of the distances of any point on it P from two fixed points F_1 and F_2, called the *foci*, is constant: thus in Fig. 3k

$$PF_1 - PF_2 = \text{constant}.$$

At their extremities, both branches of the curve merge into straight lines, called *asymptotes*. If the asymptotes are at right angles, the hyperbola is a

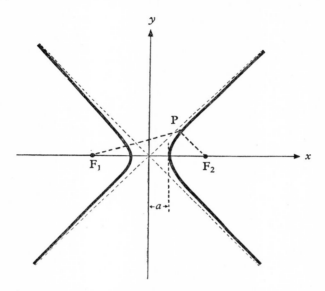

Fig. 3k. Rectangular hyperbola

rectangular hyperbola. A hyperbola that is symmetrical about both the x- and y-axis with its centre at the origin has the general equation

$$(x^2/a^2) - (y^2/b^2) = 1. \tag{3.7}$$

From this equation it can be seen that when $y = 0$, $x = \pm a$. But the meaning of y when $x = 0$ is complex, since y^2 is then equal to $-b^2$, i.e.

$$y = \pm\sqrt{(-b^2)} = \pm bi,$$

where $i = \sqrt{(-1)}$. This means that the curve does not cut the y-axis, as is apparent in Fig. 3k.

Reciprocal curves. A reciprocal curve is a type of hyperbola in which the asymptotes are the x- and y-axis. It is represented by the general equation $xy = a$, and the form of the curve can be seen by turning the axes in Fig. 3k through 45°.

One of the basic laws of physical chemistry, Boyle's law, relating the volume of a given mass of gas to its pressure at constant temperature, is summarised by a formula of this type, i.e. $pv =$ constant.

6. Equations of the form $y = ax^n$

When n, x and y are positive, curves corresponding to the equation $y = ax^n$ resemble parabolas. They are sometimes called parabolic curves, although this is not strictly accurate as only the curves representing $y = ax^2$ and $y = ax^{1/2}$ are true parabolas.

If n is negative, the curves are of a different type; they are now asymptotic to the axes and do not pass through the origin, since when $x = 0$, y becomes infinite, and vice versa. In the region where x and y are both positive, the curves resemble reciprocal curves and are sometimes called hyperbolic curves, although only that representing $y = ax^{-1}$ is a true hyperbola.

7. Exponential and logarithmic curves

'Exponent' is another name for index, and exponential equations are those in which a variable occurs as an index, e.g. $y = a^x$. In all the general

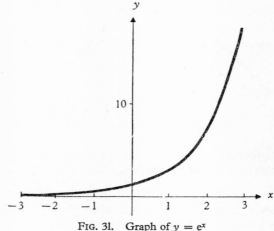

FIG. 31. Graph of $y = e^x$

equations so far discussed, the variable has been raised to a constant power; in exponential equations, a constant is raised to a variable power. An example of an exponential curve is shown in Fig. 31; this is the graph of $y = e^x$ (e $= 2 \cdot 72$). It should be noted that as x increases, y increases very rapidly and that all curves pass through the point (0, 1), since when $x = 0, y = a^0 = 1$ (see Chap. 1, § 8).

The general equation of a logarithmic curve is

$$y = b \log x + a,$$

and if y is plotted against log x, a straight line is obtained whose intercept on the y-axis at log $x = 0$ is a and whose slope is b. A logarithmic equation may be regarded as an exponential equation in which x and y have been interchanged. For example, $x = BA^y$ becomes on taking logarithms

$$\log x = \log B + y \log A.$$

Hence

$$y = \frac{1}{\log A} \cdot \log x - \frac{\log B}{\log A}.$$

As A and B are constants, log A and log B are also constants, and the equation has the same form as has the general equation with $b = 1/\log A$ and $a = -\log B/\log A$.

The main features of the curve representing a logarithmic equation are that (i) there are no simple values of y for negative values of x (see Chap. 2, § 3), (ii) y approaches minus infinity as x approaches 0, (iii) the curve flattens out as x increases, and (iv) a is the value of y when log $x = 0$, i.e. when $x = 1$.

Trigonometric curves are described in Chap. 8.

8. Graphical solution of equations

An equation of the form $f(x) = 0$ can always be solved by graphical methods, even in cases where algebraic solution is extremely difficult. It is important to note that the ordinary graphical methods give only the simple solution of the equation and give no information on the complex solutions, i.e. those involving i. The simple solutions are found by the following procedure. The graph of $y = f(x)$ is plotted in the usual way by giving x a series of values and calculating the corresponding values of y. The solutions of the equation $y = f(x)$ are those values of x at which $y = 0$, i.e. the values of x at which the curve cuts the x-axis. Solutions can be found with a high degree of accuracy by calculating values of y for a large number of values of x near the points at which the curve cuts the x-axis. If the curve does not cut the x-axis, there is no simple solution to the equation.

EXAMPLE. *Solve graphically the equation* $x^3 - 3x^2 - 13x + 15 = 0$.

If y is put equal to $x^3 - 3x^2 - 13x + 15$, then by giving x a series of numerical values the corresponding values of y may be found; thus

$x =$	-4	-3	-2	-1	0	1	2	3	4,
$y =$	-45	0	21	24	15	0	-15	-24	-21.

Without plotting the results, it can be seen that two solutions are $x = -3$ and $x = 1$, since these are the values of x when $y = 0$. As the equation is a cubic equation, there may be a third solution. Examination of the curve (see Fig. 3m) shows that after $x = 4$ the curve is beginning to rise again and so may cut the x-axis at higher values of x. Extending the above table, when $x = 5, y = 0$; when $x = 6, y = 45$. Therefore, $x = 5$ is the third solution to this cubic equation. This is a simple example in which all the solutions are integers; in equations in which the solutions are not integers, a good deal of arithmetic may be necessary in order to obtain accurate results.

The computer may be used as in COMP 1, Chap. 2, § 7 (page 36) to give a table of values of x and y from which the solutions may be estimated. More exact values are obtained by applying Newton's method, COMP 3, Chap. 5, § 9 (page 90), to improve the approximate results.

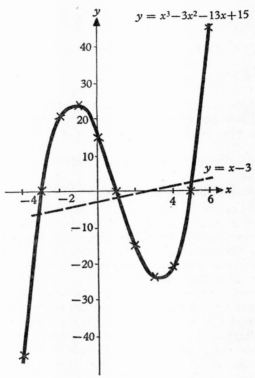

FIG. 3m. Graphs of $y = x^3 - 3x^2 - 13x + 15$ and $y = x - 3$.

9. Graphical solution of simultaneous equations

Any pair of simultaneous equations with two variables, e.g. $y = f_1(x)$ and $y = f_2(x)$, can be solved by plotting the curves representing the two functions and finding the values of x and y at the points at which the two curves intersect. If they do not intersect, there are no simple solutions. This graphical method is useful when algebraic solution is difficult or impossible.

EXAMPLE. *Solve graphically the simultaneous equations*
$$y = x^3 - 3x^2 - 13x + 15 \quad \text{and} \quad y = x - 3.$$

The first equation has already been plotted (see Fig. 3m). The second equation is represented by a straight line with the intercept at $x = 0$ of -3 and a slope of 1 (this line is the broken line in Fig. 3m). The simultaneous solutions are obtained by finding the values of x and y at the points at which the straight line cuts the curve.

10. Polar co-ordinates

Whereas rectangular co-ordinates (also called Cartesian co-ordinates) are extensively used in graphical work, it is sometimes more convenient to use polar co-ordinates, especially for expressing curves with a natural centre.

The rectangular co-ordinates of the point P in Fig. 3n are (x, y). In the polar co-ordinate system, the position of any point is defined by its distance r from a fixed point O, called the origin of the polar co-ordinate system,

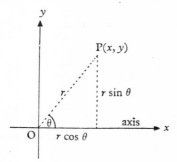

FIG. 3n. Polar co-ordinates

and the angle θ which the line joining the point to the origin makes with a fixed line (the x-axis in the figure) which also passes through the origin, called the axis of the system. The point P is then defined as (r, θ).

Using polar co-ordinates, the equation of a circle with its centre at the origin becomes $r = constant$, and the equation of a straight line passing through the origin is $\theta = constant$. To convert the equation of a curve expressed in rectangular co-ordinates to polar co-ordinates with the same origin and the x-axis as the axis of the polar system, use is made of the expressions $y = r \sin \theta$ and $x = r \cos \theta$.

11. Three-dimensional co-ordinates

For three-dimensional problems a third axis, z, is added to the x and y axes of a rectangular co-ordinate system, the z-axis being at right angles to the plane containing the other two.

The x value of a point P is the length of the perpendicular from the point P to the yOz plane, O being the origin of the three dimensional co-ordinate system; the y value is the length of the perpendicular from P to the xOz plane and the z value is the length of the perpendicular to the xOy plane—see Fig. 3o.

64 CHAPTER 3

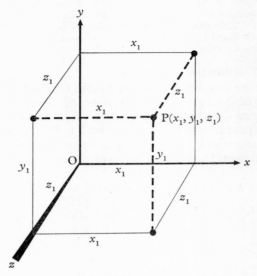

FIG. 3o. Three-dimensional co-ordinates

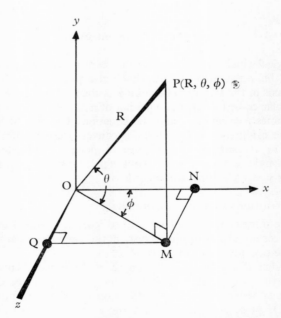

FIG. 3p. Polar three-dimensional co-ordinates

Problems in three dimensions always appear to be very difficult when they are expressed on a flat piece of paper, a problem is often greatly simplified by making a set of three-dimensional axes out of a piece of wire and examining the problem in the complete three dimensions.

When such a problem has a natural centre, as in considering electron distributions about the nucleus of an atom, it is better to use three-dimensional polar co-ordinates in place of rectangular co-ordinates. In this polar system an origin, an axis and a plane containing the axis are required to be defined. The co-ordinates of any point are then given by the distance R of the point P from the origin O; the second co-ordinate is an angle θ between OP and the defined plane and the third co-ordinate is an angle ϕ between the projection of OP on the defined plane and the axis, which is in this plane—see Fig. 3p. If the defined plane is thought of as an equatorial plane of a sphere of radius R, then θ is an angle of latitude and ϕ is an angle of longitude.

The relationship between the polar and rectangular co-ordinates of a point can be seen from Fig. 3p in which the polar axis is the x-axis and the defined plane is the xOz plane. Making the constructions shown with right angles indicated by ∟ then

in $\triangle$POM $y = PM = OP \sin \theta$, $OP = R$, therefore $y = R \sin \theta$
             $OM = OP \cos \theta = R \cos \theta$
in $\triangle$OMN $z = MN = OM \sin \phi = R \cos \theta \sin \phi$
in $\triangle$OMQ $x = MQ = OM \cos \phi = R \cos \theta \cos \phi$

note that the values of x and z for M are the same as those for P.

Problems

1. Solve graphically the equations
   (i) $x^2 - 3x + 2 = 0$ (ii) $x^3 - 2x^2 - x + 2 = 0$
   (iii) $x^3 - x - 1 = 0$
   (Use the following values of x: 3, 2, 1·5, 1, 0·5, 0, −0·5, −1, −2.)

2. Using the curve plotted in Problem 1, solve simultaneously
$$y = x^3 - x - 1 \quad \text{and} \quad y = x + 1$$

3. From the slope and intercept of the line obtained by plotting the following values of x and y, find the algebraic relation between these variables:

y	−12	−9	−6	−3	0	3	6	9
x	−2	−1	0	1	2	3	4	5

CHAPTER 4

Series, e and natural logarithms

AN ALGEBRAIC series is a set of terms, all of which conform to a definite pattern. Approximations based on the binomial, exponential and logarithmic series are frequently used to simplify equations which would otherwise be intractable.

The Greek letter Σ is used to indicate the sum of a number of terms of a series; a subscript and a superscript denote the limits of the terms summed. This symbol is followed by an expression showing the general term (the rth term) of the series. For example, $\sum_{1}^{4}(4+3r)$ means the sum of this series for positive integral values of r between 1 and 4 inclusive: thus

$$\sum_{1}^{4}(4+3r) = (4+3)+(4+6)+(4+9)+(4+12) = 46.$$

Two well-known series are the arithmetic progression (A.P.) and the geometric progression (G.P.).

1. Arithmetic progression

The pattern of an arithmetic progression is such that each term differs from the one preceding it by a constant quantity called the *difference*. The general form of an arithmetic progression is

$$a, (a+d), (a+2d), (a+3d), \ldots \{a+(n-1)d\},$$

where a is the first term, d is the difference and n is the number of terms in the series. A numerical example of an arithmetic progression is 2, 5, 8, 11, in which $a = 2$ and $d = 3$.

The sum of any number of terms of an arithmetic progression can be derived as follows. Let S denote the sum of n terms. The nth term will be, by analogy with the earlier terms, $a+(n-1)d$. Hence

$$S = a+(a+d)+(a+2d)+ \ldots +\{a+(n-2)d\}+\{a+(n-1)d\}.$$

If this series is now written in the reverse order, its sum will be unchanged: thus

$$S = \{a+(n-1)d\}+\{a+(n-2)d\}+ \ldots +(a+2d)+(a+d)+a.$$

Addition of the terms of these two equations in pairs gives the sum $2a+(n-1)d$ for each pair. Since there are n pairs altogether,

$$2S = n\{2a+(n-1)d\}.$$

SERIES, e AND NATURAL LOGARITHMS

Hence
$$S = \tfrac{1}{2}n\{2a+(n-1)d\}. \tag{4.1}$$
This formula can be used to find the sum of any number of terms of an arithmetic progression.

2. Geometric progression

The pattern of a geometric progression is such that the ratio of any term to the one preceding it is constant. The general form of a geometric progression is $a, ar, ar^2, ar^3, \ldots ar^{n-1}$, where a is the first term, r is the constant ratio and n is the number of terms in the series. A numerical example of a geometric progression is 2, 6, 18, 54, 162, in which $a = 2$ and $r = 3$.

The sum of any number of terms of a geometric progression can be derived as follows. Let S denote the sum of n terms. The nth term will be ar^{n-1}. Hence
$$S = a+ar+ar^2+ar^3+ar^4+ \ldots +ar^{n-3}+ar^{n-2}+ar^{n-1}.$$
Multiplication of each term of the series by r and moving one place to the right gives
$$rS = \quad ar+ar^2+ar^3+ar^4+ \ldots +ar^{n-3}+ar^{n-2}+ar^{n-1}+ar^n.$$
Subtracting the second equation from the first results in all terms excepting the first and the last on the right-hand side cancelling out, and thus
$$(1-r)S = a-ar^n,$$
or
$$S = \frac{a(1-r^n)}{1-r}. \tag{4.2}$$

This formula can be used to find the sum of any number of terms of a geometric progression. If the ratio r is greater than 1, it is more convenient to use the formula in the form obtained by multiplying the numerator and denominator by -1: thus
$$S = \frac{a(r^n-1)}{r-1}. \tag{4.3}$$

3. Infinite and convergent series

When a series is continued to an infinite number of terms, it is called an *infinite* series. The sum of such a series is often infinite, but when the magnitude of each successive term diminishes the sum may be finite, in which case the series is said to be *convergent*.

The sum of n terms of a convergent series in which all the terms are positive approaches a definite limiting value as n increases, the later terms of the series making a negligible contribution to the sum.

An example of a convergent series is a geometric progression with a ratio of less than 1. Consider the sum of the geometric progression 1, 0·2, 0·04, 0·008, 0·0016, 0·00032, ... Summing the first few terms arithmetically gives

$1+0·2$	$= 1·2$
$1+0·2+0·04$	$= 1·24$
$1+0·2+0·04+0·008$	$= 1·248$
$1+0·2+0·04+0·008+0·0016$	$= 1·2496$
$1+0·2+0·04+0·008+0·0016+0·00032$	$= 1·24992.$

However many terms are taken the sum will never exceed 1·25, and this is the limiting value of the sum of the series as $n \to \infty$.

This limiting sum can also be deduced from (4.2), which can be rewritten as

$$S = \frac{a}{1-r} - \frac{ar^n}{1-r}.$$

When r is positive and less than 1, the value of the second term on the right-hand side of the equation decreases as n (the number of terms summed) increases, becoming negligibly small when n becomes very large, since $r^n \to 0$ as $n \to \infty$. Hence $ar^n/(1-r) \to 0$ as $n \to \infty$.

The value of the first term on the right-hand side of the equation is independent of the value of n, and the limiting sum, as n is made infinite, is therefore equal to this term. Thus as $n \to \infty$,

$$S = \frac{a}{1-r} \quad (r < 1). \tag{4.4}$$

In the geometric progression above, $a = 1$ and $r = 0·2$, so that

$$S = \frac{1}{1-0·2} = \frac{1}{0·8} = 1·25.$$

Tests for convergency. It is not always easy to determine whether a given series is convergent. A necessary condition for convergency is that the later terms of the series should become successively smaller in magnitude. Some series which fulfil this condition do not converge properly, for example the series $1, \frac{1}{2}, \frac{1}{3}, \frac{1}{4}, \ldots$ does not have a finite sum for an infinite number of terms. However, complete tests for convergency are beyond the scope of this book.

4. Irrational numbers

Irrational numbers are numbers that cannot be expressed as integers, as finite fractions, or as decimals with a finite number of figures after the decimal point. Examples of irrational numbers are $\sqrt{7}, \sqrt{3}, \pi$ and e. These numbers can often be expressed as the limiting sum of an infinite convergent series. By the arithmetic summing of terms of the series, the value of the number can be computed to any required degree of accuracy. For example

SERIES, e AND NATURAL LOGARITHMS   69

π (the ratio of the circumference of a circle to its diameter) is the limiting sum of the series

$$4(1-1/3+1/5-1/7+1/9-1/11+\ldots).$$

Summation of a large number of terms gives $\pi = 3{\cdot}1415927$.

As another example, e (the base of natural logarithms) is the limiting sum of

$$1+1+1/2!+1/3!+1/4!+\ldots.$$

The terms 2!, 3!, etc., are called 'factorial two', 'factorial three', etc. They are the products of all the positive integers between 1 and the number Thus

$$2! = 1.2; \quad 3! = 1.2.3; \quad n! = 1.2.3.4\ldots(n-2)(n-1)n.$$

The value of $n!$ increases very rapidly as the number of terms increase, but tables are available giving values of $n!$, usually as logarithms because of the large numbers involved.

The value of e calculated from this series to seven places of decimals is $2{\cdot}7182818$. An interesting relation between e, π and i, due to Euler, is $e^{\pi i} = -1$, where $i = \sqrt{(-1)}$.

5. Permutations and combinations

In order to evaluate the coefficients of the binomial series (see § 6) it is necessary to consider permutations and combinations.

Permutations. The number of different arrangements of n distinguishable objects taken r at a time is called the *number of permutations*, written ^nP_r. The value of ^nP_r can be deduced as follows. Suppose there are n different pills to go into r pill boxes, each box to receive one pill only. The value of ^nP_r will be the number of different ways in which all the boxes can be filled. Now the first box can receive any one of the n different pills, and there are therefore n different ways in which the box can be filled. For each choice of a pill for the first box, there will be $n-1$ different ways of filling the second box, since the choice of a pill for the first box leaves only $n-1$ pills to choose for the second box. The total number of different ways in which the first two boxes can be filled is therefore $n(n-1)$. For each of these $n(n-1)$ different ways of filling the first two boxes, there will be $n-2$ ways of filling the third one, giving a total of $n(n-1)(n-2)$ different ways of filling the first three boxes. By analogy, there will be $n-r+1$ different ways of filling the rth box for each choice of pills for the preceding boxes. The total number of different ways in which the r boxes can be filled by selection from n different pills is therefore

$$n(n-1)(n-2)\times\ldots\times(n-r+1).$$

This is the product of r factors and is equal to the number of arrangements of n objects taken r at a time, i.e. ^nP_r. Hence

$$^nP_r = n(n-1)(n-2)\times\ldots\times(n-r+1).$$

A more general form for ^nP_r is obtained by multiplying the above expression by
$$\frac{(n-r)!}{(n-r)!}$$
giving
$$^nP_r = \frac{n(n-1)(n-2) \times \ldots \times (n-r+1) \times (n-r)(n-r-1) \times \ldots \times 3.2.1}{(n-r)!}$$
$$= \frac{n!}{(n-r)!} \tag{4.5}$$

As a simple example, suppose there are four different pills a, b, c and d, and two pill boxes, each to be filled with one pill. The first box can be filled with a, b, c or d. For each choice of pill for the first box, the second box can be filled with any one of the remaining pills; thus the following arrangements are possible.

1st box	2nd box	Arrangements
a	b, c or d	ab, ac and ad
b	a, c or d	ba, bc and bd
c	a, b or d	ca, cb and cd
d	a, b or c	da, db and dc

The total number of different arrangements of the four pills in two boxes is therefore $4.3 = 12$. That is,
$$^4P_2 = 4.3 = 12.$$
If $r = n$, i.e. if there are n boxes to receive n different pills, the total number of possible different arrangements is given by
$$^nP_n = n(n-1)(n-2) \times \ldots \times 3.2.1,$$
since when $r = n$, $(n-r+1) = 1$. ^nP_n is therefore the product of all the integers between 1 and n, i.e. it is $n!$.

Combinations. The arrangements of four objects taken two at a time illustrated by the previous example with the pills include a number of groupings such as ab and ba, and ad and da. These groupings contain the same objects but arranged in a different order. The number of selections that can be made by taking n different objects r at a time, each selection containing a *different set of objects* from any other selection, is called the *number of combinations*, written ^nC_r. Each combination consists of r objects, and these can be permuted among themselves to give ^rP_r different arrangements. The total number of arrangements (permutations) of the n objects taken r at a time is therefore $^nC_r.^rP_r$. This is also ^nP_r, so that
$$^nC_r.^rP_r = {}^nP_r,$$
or
$$^nC_r = {}^nP_r/^rP_r.$$

Now, $^rP_r = r!$, and by substituting the value of ^nP_r already deduced, it follows that

$$^nC_r = \frac{n(n-1)(n-2)(n-3) \times \ldots \times (n-r+1)}{r!}. \quad (4.6)$$

In the pill example discussed previously, the number of combinations is given by

$$^4C_2 = 4.3/2! = 6.$$

The combinations are ab, ac, bc, cd, ad and bd.

A simpler expression for ^nC_r is obtained from (4.5):

$$^nC_r = \frac{n!}{r!(n-r)!}. \quad (4.7)$$

EXAMPLE. *Find the number of permutations and combinations of the five letters a, b, c, d, and e taken three at a time.*

$$^5P_3 = 5.4.3 = 60; \quad ^5C_3 = \frac{5.4.3}{1.2.3} = 10.$$

Using ^nC_r in the form (4.7),

$$^nC_r = {}^5C_3 = \frac{5!}{3!2!} = \frac{5.4.3.2.1}{3.2.1.2.1} = 10.$$

The ten combinations are abc, abd, abe, acd, ace, ade, bcd, bce, bde and cde. The sixty permutations are obtained by permuting the three letters of each combination among themselves, and these can be found by writing them out. The six permutations in the first combination, for example, are abc, acb, bac, bca, cab and cba.

It has already been shown (4.7) that

$$^nC_r = \frac{n!}{r!(n-r)!}.$$

But

$$^nC_{n-r} = \frac{n!}{(n-r)!r!} = \frac{n!}{r!(n-r)!}.$$

Hence

$$^nC_r = {}^nC_{n-r}.$$

Now the number of combinations of n objects taken n at a time, i.e. all at once, is 1, the only selection possible. Therefore $^nC_n = 1$; but from (4.7)

$$^nC_n = \frac{n!}{n!(n-n)!} = \frac{n!}{n!0!} = \frac{1}{0!}.$$

Since ^nC_n also equals 1, it follows that $1/0! = 1$. Hence the significance of $0!$, or $(n-r)!$ when $n = r$, is seen to be 1. Thus when $n = r$,

$$(n-r)! = 0! = 1$$

6. The binomial theorem

The binomial theorem gives a series for the expansion of the sum of any two numbers raised to a positive integral power n; thus

$$(a+x)^n = a^n + A_1 a^{n-1} x + A_2 a^{n-2} x^2 + \ldots + A_{n-1} a x^{n-1} + x^n.$$

The series consists of $(n+1)$ terms, and the values of the coefficients A_1, A_2, etc., can be found as follows.

Consider the product of the four factors $a+x$, $b+x$, $c+x$ and $d+x$. Multiplying out and rearranging as a power series in x gives

$$abcd + (abc + abd + bcd + acd)x + (ab + bc + cd + ac + ad + bd)x^2 + \\ + (a+b+c+d)x^3 + x^4.$$

If $a = b = c = d$, the product is $(a+x)^4$, and substituting a for b, c and d gives

$$(a+x)^4 = a^4 + 4a^3 x + 6a^2 x^2 + 4ax^3 + x^4.$$

Therefore the values of the coefficients are

$$A_1 = 4, \quad A_2 = 6, \quad \text{and} \quad A_3 = 4.$$

Values of the coefficients for the general case $(a+x)^n$ can be found in the same way. Consider the product of n factors such as $a+x, b+x, c+x, \ldots$. These factors can be multiplied out and arranged as a power series in x. Each term of this series will be homogeneous, i.e. will contain an equal number n of factors consisting of $a, b, c, \ldots$ or x. For example, the coefficient of x^3 will be made up of the sum of the products consisting of $n-3$ different factors chosen from the n quantities $a, b, c, \ldots$. The number of such products in the coefficient of x^3 will be the number of different ways in which $n-3$ factors can be chosen from the total of n quantities $a, b, c, \ldots$, i.e. $^nC_{n-3}$. Thus

$(a+x)(b+x)(c+x) \ldots$ to n factors
$\quad = [abc \ldots$ to n factors$] +$
$\quad\quad + [$sum of all the different products containing
$\quad\quad\quad n-1$ factors drawn from $a, b, c, \ldots]x +$
$\quad\quad + [$sum of all the different products containing
$\quad\quad\quad n-2$ factors drawn from $a, b, c, \ldots]x^2 + \ldots$
$\quad\quad + x^n.$

If, as previously, all the quantities $b, c, d, \ldots$ are put equal to a, each sum of all the different products containing $n-1$ factors, $a, b, c, \ldots$, will be equal to a^{n-1}, and the number of such products will be equal to the number of ways in which $n-1$ factors can be chosen from a total of n so as to give a different product in each case, i.e. will be equal to $^nC_{n-1}$. The second term of the binomial expansion is therefore $^nC_{n-1} a^{n-1} x$. Similarly, the third term can be shown to be $^nC_{n-2}a^{n-2}x^2$, and, by analogy, a general term, the $(r+1)$th, is $^nC_{n-r}a^{n-r}x^r$. Thus the expansion is

$$(a+x)^n = a^n + {^nC_{n-1}}a^{n-1}x + {^nC_{n-2}}a^{n-2}x^2 + \ldots + {^nC_1}ax^{n-1} + x^n.$$

If these coefficients are expanded by (4.6), then

$$(a+x)^n = a^n + na^{n-1}x + \frac{n}{2!}(n-1)a^{n-2}x^2 + \ldots + nax^{n-1} + x^n. \quad (4.8)$$

The coefficient of the second term A_1 is $^nC_{n-1}$, which equals n, while that of the third term A_2 is $^nC_{n-2}$, which equals $n(n-1)/2!$; the coefficient of the $(r+1)$th term A_r is $^nC_{n-r}$, which equals

$$\frac{n(n-1)(n-2) \times \ldots \times (n-r+2)(n-r+1)}{r!}.$$

There is a total of $n+1$ terms.

The expansion (4.8) is true for all positive integral values of n. If n is fractional or negative, it can be shown (see Appendix XI) that the expansion of $(1+y)^n$ may be represented by the infinite series

$$1 + ny + \frac{n}{2!}(n-1)y^2 + \frac{n}{3!}(n-1)(n-2)y^3 + \ldots.$$

In order that this series shall have a finite sum, it must converge. This convergency is achieved by restricting the value of y so that it lies between $+1$ and -1, written $|y| < 1$. (The symbol '| |' means the magnitude of a simple quantity irrespective of sign.)

The binomial $(a+x)^n$ can be expanded for any value of n by converting it to the form $(1+y)^n$ by dividing by the larger of the two numbers, a or x. Thus, if $a > x$,

$$(a+x)^n = a^n\left(1+\frac{x}{a}\right)^n = a^n\left\{1 + n\frac{x}{a} + \frac{n}{2!}(n-1)\left(\frac{x}{a}\right)^2 + \ldots\right\}. \quad (4.9)$$

An important approximation based on the binomial theorem is that when the magnitude of y is very small, written $|y| \ll 1$:

$$(1+y)^n \simeq 1+ny.$$

The following are some expansions ($|x| < 1$) derived from the binomial theorem

$$(1+x)^{-1} = 1 - x + x^2 - x^3 + x^4 - \ldots.$$
$$(1-x)^{-1} = 1 + x + x^2 + x^3 + x^4 + \ldots.$$
$$\sqrt{(1+x)} = (1+x)^{1/2} = 1 + \tfrac{1}{2}x - \tfrac{1}{8}x^2 + \tfrac{1}{16}x^3 - \ldots.$$
$$(1-x)^{-2} = 1 + 2x + 3x^2 + 4x^3 + \ldots.$$

EXAMPLE. *Find the value of $\sqrt{1\cdot 04}$ to four decimal places by means of the binomial theorem.*

$$\sqrt{1\cdot 04} = (1+0\cdot 04)^{1/2} = 1^{1/2}\left\{1+\tfrac{1}{2}(0\cdot 04)+\frac{\tfrac{1}{2}(\tfrac{1}{2}-1)}{2!}(0\cdot 04)^2+\ldots\right\}$$
$$= 1+\tfrac{1}{2}(0\cdot 04)+\tfrac{1}{2}\cdot\tfrac{1}{2}(-\tfrac{1}{2})(0\cdot 04)^2+\ldots$$
$$= 1+0\cdot 02-0\cdot 0002+\ldots$$
$$= 1\cdot 0198.$$

7. The significance of e and natural logarithms

The irrational number e and the system of natural logarithms based on it enter into a large number of theoretical equations in chemical kinetics, bacteriology, light absorption and statistics. This is chiefly because if the rate of change of a given quantity y with another quantity x is proportional to the magnitude of y at a given instant (compound interest law), then the relation between y and x is of the form (see Chap. 9)

$$y = ae^{bx},$$

or

$$\ln y = A+Bx,$$

where a, b, A and B are constants.

The exponential theorem and e. The expansion of the binomial $(1+1/n)^n$ can be found from (4.9) if the value of n is greater than 1, i.e. if $|1/n| < 1$. Thus

$$\left(1+\frac{1}{n}\right)^n = 1+n\left(\frac{1}{n}\right)+\frac{n(n-1)}{2!}\left(\frac{1}{n}\right)^2+\frac{n(n-1)(n-2)}{3!}\left(\frac{1}{n}\right)^3+\ldots$$

Dividing each of the factors n, $n-1$, $n-2$, ..., in the numerators by n in the denominators gives

$$\left(1+\frac{1}{n}\right)^n = 1+1+\frac{1(1-1/n)}{2!}+\frac{1(1-1/n)(1-2/n)}{3!}+\ldots$$

As n is made very large the quantities $1/n$, $2/n$, ..., become very small compared with 1, and the series can be written

$$\left(1+\frac{1}{n}\right)^n = 1+1+\frac{1}{2!}+\frac{1}{3!}+\ldots \quad (n\to\infty).$$

This is a convergent series whose limiting sum has already been defined in § 4 as the irrational number e, the base of natural logarithms. The importance of e as a base for logarithms lies in the fact that e raised to any power z can be expressed as the limiting sum of an infinite convergent series. Thus

$$e^z = \{(1+1/n)^n\}^z = (1+1/n)^{nz}.$$

SERIES, e AND NATURAL LOGARITHMS   75

Expanding this binomial by (4.9) gives

$$e^z = 1 + nz\left(\frac{1}{n}\right) + \frac{nz(nz-1)}{2!}\left(\frac{1}{n}\right)^2 + \frac{nz(nz-1)(nz-2)}{3!}\left(\frac{1}{n}\right)^3 + \cdots$$

Dividing each of the factors nz, $nz-1$, $nz-2$, ... by n in the denominators and neglecting the quantities $1/n$, $2/n$, ... gives

$$e^z = 1 + z + z^2/2! + z^3/3! + \cdots \qquad (4.10)$$

This series can be shown to be convergent for all finite values of z. For convenience in printing, e^z is frequently written as $\exp(z)$.

An approximation for $(1-m/n)^n$ when n is very large is used in the theory of the Poisson distribution in statistics and is derived in the same way as is (4.10). Thus

$$\left(1-\frac{m}{n}\right)^n = 1 - n\left(\frac{m}{n}\right) + \frac{n(n-1)}{2!}\left(\frac{m}{n}\right)^2 - \cdots$$

$$= 1 - m + \frac{(1-1/n)m^2}{2!} - \cdots$$

$$= 1 - m + \frac{m^2}{2!} - \cdots, \quad \text{when } n \text{ is very large,}$$

$$= e^{-m}, \quad \text{or} \quad \exp(-m).$$

Therefore, when $n \to \infty$,

$$(1-m/n)^n \to \exp(-m). \qquad (4.11)$$

The value of any number a raised to any power z can be expressed as the sum of an infinite convergent series by means of (4.10) and the definition of a logarithm. From the latter (see Chap. 2, § 3)

$$a = e^{\ln a}.$$

Hence $a^z = e^{z \ln a}$. Substituting $z \ln a$ for z in (4.10) gives

$$a^z = 1 + z \ln a + \frac{(z \ln a)^2}{2!} + \frac{(z \ln a)^3}{3!} + \cdots \qquad (4.12)$$

This equation is a statement of the *exponential theorem*.

8. Logarithmic series

If a in (4.12) is put equal to $1+x$, then

$$(1+x)^z = 1 + z \ln(1+x) + \frac{\{z \ln(1+x)\}^2}{2!} + \cdots$$

Providing $|x| < 1$, $(1+x)^z$ can also be expanded by the binomial series for any value of z. Thus

$$(1+x)^z = 1 + zx + \frac{z(z-1)x^2}{2!} + \frac{z(z-1)(z-2)x^3}{3!} + \cdots$$

These two series express the same result for all values of z; they are therefore identities in z, which means that the coefficient of any power of z in one series is equal to the coefficient of the same power of z in the other series (see Chap. 2, § 4). The coefficient of z in the exponential series expansion is $\ln(1+x)$. In order to find the total coefficient of z in the binomial series expansion, it is necessary to multiply out each term; thus it is

$$x + \frac{(-1)x^2}{2!} + \frac{(-1)(-2)x^3}{3!} + \frac{(-1)(-2)(-3)x^4}{4!} + \ldots$$
$$= x - \tfrac{1}{2}x^2 + \tfrac{1}{3}x^3 - \tfrac{1}{4}x^4 + \ldots.$$

The logarithmic series is obtained by equating these two coefficients: thus

$$\ln(1+x) = x - \tfrac{1}{2}x^2 + \tfrac{1}{3}x^3 - \tfrac{1}{4}x^4 + \ldots \qquad (4.13)$$

This relation only holds for values of x between $+1$ and -1, i.e. $|x| < 1$. If x is very small, i.e. $|x| \ll 1$, then $\ln(1+x) \simeq x$.

This series can be used to compute the values of natural logarithms, and by suitable modification, a complete table of natural logarithms can be drawn up.

Problems

1. Find the sum of the first nine terms of the series
$$5 + 8 + 11 + 14 + \ldots.$$
2. Find the sum of the first six terms of the series
$$7 + 14 + 28 + 56 + \ldots.$$
3. Using the binomial series, evaluate the following expressions to four places of decimals:
$$(1 \cdot 01)^5, \quad (1 \cdot 003)^7, \quad (1 \cdot 02)^{3/4}, \quad 1/1 \cdot 04, \quad \sqrt[4]{1 \cdot 002}.$$
4. A forecast of the results of twelve football matches is to be made, and it is decided that eight of these matches will be home wins. In how many different ways can these eight wins be chosen?
5. On how many nights can a watch of four men be drafted from a crew of twenty-four, so that no two watches are identical?
6. Show that $\exp(x) + \exp(-x) = 2$, if x is small.
7. Using the exponential theorem, evaluate $\sqrt[4]{1 \cdot 649}$ to four decimal places, given that $\ln(1 \cdot 649) = 0 \cdot 5$.
8. Using the logarithmic series, evaluate $\ln(1 \cdot 1)$ to four places of decimals.
9. In how many ways can three draws be forecast in ten football matches?

CHAPTER 5

Differential calculus

1. Rate processes

Studies of the rates of chemical or biological reactions often yield valuable information about the process mechanisms. In order to interpret the results of these rate studies it is necessary to be able to determine the velocity of a process at any given instant. Differential calculus, which is the

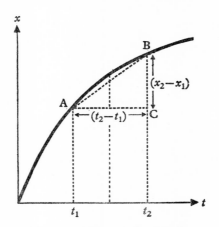

FIG. 5a. Mean reaction velocity

mathematics of changing systems, is the technique which is used to determine this velocity.

Consider a system in which a slow chemical reaction is occurring, e.g. the hydrolysis of ethyl acetate in aqueous alkaline solution

$$CH_3.CO.OC_2H_5 + NaOH \rightleftharpoons CH_3.CO_2Na + C_2H_5OH.$$

The progress of this reaction can be followed by withdrawing samples of reaction mixture at different times, determining the amount of sodium hydroxide remaining titrimetrically, and hence the amount of alkali used up. The amount of sodium hydroxide used is a measure of the amount x of reaction product formed in a given time t. If x is plotted against t, a curve of the type shown in Fig. 5a is obtained.

The mean reaction velocity $\bar{v}$ during any time interval $t_2 - t_1$ is defined

78 CHAPTER 5

as the change in x, i.e. x_2-x_1, during this period divided by the time interval, i.e.

$$\bar{v} = (x_2-x_1)/(t_2-t_1)$$
$$= BC/AC \quad \text{(from Fig. 5a)}.$$

This means that the mean reaction velocity is measured by the slope of the line AB. The time interval t_2-t_1 is called the 'increment in t' and is denoted by the symbol Δt; the corresponding increment in x, i.e. x_2-x_1, is denoted by Δx. Hence

$$\bar{v} = \Delta x/\Delta t.$$

As the reaction proceeds, the mean reaction velocity decreases, ultimately becoming zero when equilibrium or completion of the reaction is reached. If the graph in Fig. 5a is examined, it is obvious that the slope of the line AB becomes smaller and smaller as the time increases. In order to obtain an accurate estimate of the velocity at any instant, the time interval Δt must be made extremely small. If the mean time $\frac{1}{2}(t_2+t_1)$ is kept constant while Δt is decreased, the points A and B in Fig. 5a will approach each other, and the value of $\Delta x/\Delta t$ will change very slightly but remain finite and approach a limiting value as Δt becomes very small. This limiting value of $\Delta x/\Delta t$ as $\Delta t \to 0$ is the true reaction velocity v at a given time t: thus

$$v = \lim_{\Delta t \to 0} \left(\frac{\Delta x}{\Delta t}\right).$$

The abbreviation 'lim' is used to denote a limiting value.

For all values of Δt, $\Delta x/\Delta t$ will be equal to the slope of the line AB. As $\Delta t \to 0$ and A and B ultimately coincide, the slope of AB will then be equal to the slope of the curve at the point where A and B coincide, i.e. time t. The slope of a curve at a point is measured by the slope of the tangent drawn to that point. The true reaction velocity v at any instant can therefore be found by measuring the slope of the tangent to the curve at the point corresponding to this time.

EXAMPLE. *The results of a reaction–velocity experiment (chemical kinetic) are summarised in the following table:*

t (minutes)	10	15	20	25	30	35	40
x (amount of reaction product formed)	26·5	36·5	44·8	52·1	57·1	61·3	64·4

Calculate the mean reaction velocity $\bar{v}$ at a mean time of 25 minutes for values of Δt of 30, 20 and 10 minutes. Find also the true reaction velocity v at 25 minutes.

The required values of $\bar{v}$ can be calculated directly from the values given in the table, chosing values of t_1 and t_2 so that their mean is 25 minutes and their difference is the required value of Δt: thus

t_1	t_2	$\frac{1}{2}(t_1+t_2)$	Δt	x_1	x_2	Δx	$\bar{v}(=\Delta x/\Delta t)$
10	40	25	30	26·5	64·4	37·9	1·26
15	35	25	20	36·5	61·3	24·8	1·24
20	30	25	10	44·8	57·1	12·3	1·23

The required mean reaction velocities are therefore 1·26, 1·24 and 1·23.

The true reaction velocity v at 25 minutes is found by plotting x against t, drawing the tangent to the curve at the point at which $t = 25$, and measuring the slope of the tangent (see Fig. 5b). Hence $v = \text{DE}/\text{EF}$. The reaction velocity is obtained in units of x divided by units of time; if in the graph, the unit of x is represented by a different length from that representing a unit of time, this must be allowed for in the calculation. Thus

$$v = \frac{\text{DE in cm}}{\text{EF in cm}} \cdot \frac{\text{units of } x \text{ per cm}}{\text{units of } t \text{ per cm}}.$$

From Fig. 5b, the value of v is 1·22.

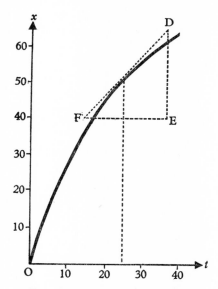

FIG. 5b. Exact reaction velocity

It is seen from the numbers given in the example that as the time increment is shortened, the average mean reaction velocity $\bar{v}$ approaches more closely to the true velocity v.

The graphical method of finding true velocities by measuring the slopes of tangents to a curve is subject to considerable error owing to the random experimental errors in x, and an algebraic method of determining v is preferable. Such a method can be developed by means of calculus if the relation between x and t is known.

In order to determine v it is necessary to measure the change in x occurring during an infinitely small (infinitesimal) period of time. The symbol dx is used to represent the infinitesimal increment in x corresponding to the infinitesimal increment dt in t, and the true reaction velocity is then dx/dt. The ratio dx/dt is called the *differential coefficient* or *derivative* of x with respect to t.

2. Calculus

Differential calculus is a technique which was developed independently by Newton and Leibnitz towards the end of the seventeenth century. The value of calculus lies in the fact that if an algebraic relation is known between two variables it is possible to determine the true rate of change of one variable with another.

In the example in § 1, there are two variables, x and t, x being a function of t. If the algebraic relation $x = \mathrm{f}(t)$ is known, then $\Delta x/\Delta t$ can be evaluated by the increment method. Thus, if t is changed by an increment Δt and if Δx is the corresponding change or increment in x, then the two new values of the variables, $t+\Delta t$ and $x+\Delta x$, will still satisfy the original x, t equation, i.e.

$$x+\Delta x = \mathrm{f}(t+\Delta t).$$

Subtracting

$$x = \mathrm{f}(t),$$

gives

$$\Delta x = \mathrm{f}(t+\Delta t) - \mathrm{f}(t),$$

which, on dividing by Δt, becomes

$$\frac{\Delta x}{\Delta t} \quad (\text{i.e. } \bar{v}) = \frac{\mathrm{f}(t+\Delta t) - \mathrm{f}(t)}{\Delta t}.$$

If Δt is made very small, $\bar{v}$ will approach the true rate of change of x with t, i.e. v. Hence

$$v = \frac{\mathrm{d}x}{\mathrm{d}t} = \lim_{\Delta t \to 0} \left(\frac{\Delta x}{\Delta t}\right) = \lim_{\Delta t \to 0} \left\{\frac{\mathrm{f}(t+\Delta t) - \mathrm{f}(t)}{\Delta t}\right\},$$

and this expression for $\mathrm{d}x/\mathrm{d}t$ gives the rate of change of x with t. A similar relation can be deduced for any pair of variables. Thus, if y is a function of x, i.e. $y = \mathrm{f}(x)$, the value of $\mathrm{d}y/\mathrm{d}x$ for a definite value of x will give the rate of change of y with x. Hence, when $y = \mathrm{f}(x)$,

$$\frac{\mathrm{d}y}{\mathrm{d}x} = \lim_{\Delta x \to 0} \left(\frac{\Delta y}{\Delta x}\right) = \lim_{\Delta x \to 0} \left\{\frac{\mathrm{f}(x+\Delta x) - \mathrm{f}(x)}{\Delta x}\right\}. \quad (5.1)$$

If y increases with x, the differential coefficient will be positive; if y decreases while x increases, the differential coefficient will be negative.

Sometimes the differential coefficient $\mathrm{d}y/\mathrm{d}x$ is known as the *derivative* or *derived function* of y with respect to x; it is simply the rate of change of y with x and is equal to the slope of the tangent to the x, y curve at a point. In general, the differentiation of $\mathrm{f}(x)$ with respect to x may be written as

$$\frac{\mathrm{d}\{\mathrm{f}(x)\}}{\mathrm{d}x} \quad \text{or} \quad \frac{\mathrm{d}}{\mathrm{d}x}\{\mathrm{f}(x)\}.$$

The infinitesimal increments in x and y, i.e. $\mathrm{d}x$ and $\mathrm{d}y$ respectively, are called *differentials*. They are treated as ordinary algebraic quantities, but, of course, the 'd' cannot be separated from the other letter.

DIFFERENTIAL CALCULUS

EXAMPLE 1. *Evaluate dy/dx when $y = 3x^2$.*

Change x by the increment Δx, and if Δy is the corresponding change in y, then the new value of the variables, $x+\Delta x$ and $y+\Delta y$, will still satisfy the equation. Hence
$$y+\Delta y = 3(x+\Delta x)^2$$
$$= 3x^2+6x\Delta x+3(\Delta x)^2.$$
Subtracting
$$y = 3x^2,$$
it follows that
$$\Delta y = 6x\Delta x+3(\Delta x)^2,$$
which, on dividing both sides of the equation by Δx, becomes
$$\Delta y/\Delta x = 6x+3\Delta x.$$
As the increment Δx is made smaller, the value of the first term on the right-hand side of the equation remains unchanged, but the second term progressively decreases, becoming zero in the limit $\Delta x \to 0$. Thus
$$\frac{dy}{dx} = \lim_{\Delta x \to 0} \left(\frac{\Delta y}{\Delta x}\right) = 6x.$$

EXAMPLE 2. *If $y = x^3+1$, find the numerical values of $\Delta y/\Delta x$ at $x = 4$ for increments of x equal to 1, $0\cdot 1$, $0\cdot 01$, $0\cdot 0001$ and for infinitely small values of Δx, i.e. dy/dx.*

If Δy is the increment in y corresponding to the increment Δx in x, then
$$y+\Delta y = (x+\Delta x)^3+1$$
$$= x^3+3x^2\Delta x+3x(\Delta x)^2+(\Delta x)^3+1.$$
Subtracting
$$y = x^3+1,$$
gives
$$\Delta y = 3x^2\Delta x+3x(\Delta x)^2+(\Delta x)^3,$$
which, on dividing throughout by Δx, becomes
$$\Delta y/\Delta x = 3x^2+3x\Delta x+(\Delta x)^2.$$
Now, when $x = 4$,
$$\Delta y/\Delta x = 48+12\Delta x+(\Delta x)^2,$$
and if $\Delta x = 1$, then
$$\Delta y/\Delta x = 48+12+1 = 61;$$
if $\Delta x = 0\cdot 1$, then
$$\Delta y/\Delta x = 48+1\cdot 2+0\cdot 01 = 49\cdot 21;$$
if $\Delta x = 0\cdot 01$, then
$$\Delta y/\Delta x = 48+0\cdot 12+0\cdot 0001 = 48\cdot 1201;$$
if $\Delta x = 0\cdot 0001$, then
$$\Delta y/\Delta x = 48+0\cdot 0012+0\cdot 00000001 = 48\cdot 00120001;$$
and if $\Delta x \to 0$,
$$dy/dx = 48.$$

3. Rules of differentiation

Rules by which any function can be differentiated can be developed by using the increment method, and the derivation of some of these rules is given below. In all cases, Δy and Δx are the corresponding increments in y and x respectively, n, a and b are constants, and u, v and w are functions of x.

Rule 1. Differentiation of a constant, $y = a$. In this case y is independent of the other variable x, so that any change in x will not result in a change in y. The increment in y corresponding to Δx is therefore 0, i.e. $\Delta y = 0$. Hence, for all values of Δx, $\Delta y/\Delta x = 0$, and

$$\frac{dy}{dx} = \lim_{\Delta x \to 0} \left(\frac{\Delta y}{\Delta x}\right) = 0.$$

Thus, if $y = a$, then
$$dy/dx = 0, \tag{5.2}$$

or the derivative of a constant is zero.

Rule 2. Differentiation of $y = ax^n$. The new values of the variables, $y+\Delta y$ and $x+\Delta x$, will still satisfy the equation, so that

$$y+\Delta y = a(x+\Delta x)^n.$$

Subtracting
$$y = ax^n$$
gives
$$\Delta y = a(x+\Delta x)^n - ax^n = ax^n\left\{\frac{(x+\Delta x)^n}{x^n} - 1\right\}$$
$$= ax^n\{(1+\Delta x/x)^n - 1\}.$$

If $\Delta x/x$ is less than 1, $(1+\Delta x/x)^n$ can be expanded by the binomial theorem (see Chap. 4, § 6) for all values of n. Hence

$$\Delta y = ax^n\left\{1 + n\left(\frac{\Delta x}{x}\right) + \frac{n(n-1)}{2!}\left(\frac{\Delta x}{x}\right)^2 + \ldots - 1\right\}.$$

Dividing throughout by Δx gives

$$\frac{\Delta y}{\Delta x} = ax^n\left\{n\frac{1}{x} + \frac{n(n-1)}{2!} \cdot \frac{\Delta x}{x^2} + \ldots\right\}.$$

All the terms within the larger brackets, except the first, contain Δx as a factor, and so in the limit as $\Delta x \to 0$, they will each become zero. Hence

$$\lim_{\Delta x \to 0}\left(\frac{\Delta y}{\Delta x}\right) = ax^n\left(\frac{n}{x}\right) = anx^{n-1}.$$

Thus, if $y = ax^n$, then
$$dy/dx = anx^{n-1}. \tag{5.3}$$

It follows from (5.3) that when $n = 1$, i.e. $y = ax$, then $dy/dx = a$, i.e. the derivative of a term in the first power of x is a constant.

Rule 3. Differentiation of $y = ae^x$ ($= a \exp x$). As previously, the new values of the variables, $y+\Delta y$ and $x+\Delta x$, still satisfy the equation, so that
$$y+\Delta y = a \exp(x+\Delta x).$$

DIFFERENTIAL CALCULUS

Subtracting
$$y = a \exp x$$
gives
$$\Delta y = a \exp(x+\Delta x) - a \exp x.$$
From the theory of indices, it follows that
$$\exp(x+\Delta x) = e^{x+\Delta x} = e^x . e^{\Delta x} = \exp x . \exp \Delta x.$$
Hence
$$\Delta y = a(\exp x . \exp \Delta x - \exp x)$$
$$= a \exp x (\exp \Delta x - 1).$$
Now, $\exp \Delta x$ can be expanded by the series (4.10), giving
$$\Delta y = a \exp x \left(1 + \Delta x + \frac{(\Delta x)^2}{2!} + \ldots - 1\right),$$
which on dividing throughout by Δx gives
$$\Delta y / \Delta x = a \exp x \left(1 + \frac{\Delta x}{2!} + \ldots\right).$$
In the limit $\Delta x \to 0$, the factor within the brackets becomes equal to 1, so that
$$\lim_{\Delta x \to 0} \left(\frac{\Delta y}{\Delta x}\right) = a \exp x.$$
Thus, if $y = ae^x$, then
$$dy/dx = ae^x. \tag{5.4}$$

Rule 4. *Differentiation of* $y = a \ln x$. Again, the new values of the variables, $y+\Delta y$ and $x+\Delta x$, will still satisfy the equation, so that
$$y + \Delta y = a \ln(x+\Delta x).$$
Subtracting
$$y = a \ln x$$
gives
$$\Delta y = a \{\ln(x+\Delta x) - \ln x\}.$$
From the properties of logarithms (see Chap. 2, § 3), it follows that
$$\ln(x+\Delta x) - \ln x = \ln\left(\frac{x+\Delta x}{x}\right) = \ln(1+\Delta x/x).$$
If $\Delta x/x$ is less than 1, the logarithm can be expanded by the series (4.13), giving
$$\Delta y = a\left\{\frac{\Delta x}{x} - \tfrac{1}{2}\left(\frac{\Delta x}{x}\right)^2 + \ldots\right\},$$
which on dividing throughout by Δx gives
$$\frac{\Delta y}{\Delta x} = a\left(\frac{1}{x} - \tfrac{1}{2}\frac{\Delta x}{x^2} + \ldots\right).$$

In the limit $\Delta x \to 0$, the factor within the brackets becomes equal to $1/x$, so that

$$\lim_{\Delta x \to 0} \left(\frac{\Delta y}{\Delta x}\right) = a\left(\frac{1}{x}\right).$$

Thus, if $y = a \ln x$, then

$$dy/dx = a/x. \tag{5.5}$$

Rule 5. *Differentiation of the sum of functions*. If y is the sum of a number of functions u, v, w of the variable x, then

$$y + \Delta y = (u + \Delta u) + (v + \Delta v) + (w + \Delta w).$$

Subtracting

$$y = u + v + w$$

gives

$$\Delta y = \Delta u + \Delta v + \Delta w,$$

which on dividing throughout by Δx gives

$$\Delta y/\Delta x = \Delta u/\Delta x + \Delta v/\Delta x + \Delta w/\Delta x,$$

where Δx is the increment in x which causes the respective increments in u, v and w. The above equation holds for all values of Δx and will therefore hold in the limit $\Delta x \to 0$. Thus, if $y = u + v + w$, then

$$dy/dx = du/dx + dv/dx + dw/dx. \tag{5.6}$$

A similar result is obtained for the sum of any number of functions of x. Also, any of the signs may be negative, e.g. if $y = u - v + w$, then

$$dy/dx = du/dx - dv/dx + dw/dx.$$

Rule 6. *Differentiation of a product of functions*, $y = uv$. If u and v are both functions of x, then

$$y + \Delta y = (u + \Delta u)(v + \Delta v)$$
$$= uv + v \Delta u + u \Delta v + \Delta u \, \Delta v.$$

Subtracting

$$y = uv$$

gives

$$\Delta y = v \Delta u + u \Delta v + \Delta u \, \Delta v,$$

which on dividing throughout by Δx gives

$$\Delta y/\Delta x = v(\Delta u/\Delta x) + u(\Delta v/\Delta x) + \Delta u(\Delta v/\Delta x).$$

In the limit $\Delta x \to 0$, the ratios of the increments will remain finite. The last term in the equation, which is a ratio multiplied by an infinitesimal increment, will in the limit become zero, so that if $y = uv$, then

$$dy/dx = v(du/dx) + u(dv/dx). \tag{5.7}$$

DIFFERENTIAL CALCULUS

Rule 7. *Differentiation of a quotient of functions, $y = u/v$.* If y is the quotient of two functions, u and v, of the variable x, then

$$y + \Delta y = \frac{u + \Delta u}{v + \Delta v}.$$

Subtracting

$$y = \frac{u}{v}$$

gives

$$\Delta y = \frac{u + \Delta u}{v + \Delta v} - \frac{u}{v} = \frac{v(u + \Delta u) - u(v + \Delta v)}{v(v + \Delta v)}$$

$$= \frac{v\,\Delta u - u\,\Delta v}{v(v + \Delta v)},$$

which on dividing throughout by Δx gives

$$\frac{\Delta y}{\Delta x} = \frac{v(\Delta u/\Delta x) - u(\Delta v/\Delta x)}{v(v + \Delta v)}.$$

In the limit $\Delta x \to 0$ and $\Delta v \to 0$, so that $v + \Delta v \to v$. Hence if $y = u/v$, then

$$\frac{dy}{dx} = \frac{v(du/dx) - u(dv/dx)}{v^2}. \qquad (5.8)$$

SUMMARY OF THE RULES OF DIFFERENTIATION

Rule	Function	Derivative dy/dx
1	$y = a$	0
2	$y = ax^n$	anx^{n-1}
3	$y = ae^x$	ae^x
4	$y = a \ln x$	a/x
5	$y = u + v + w$	$du/dx + dv/dx + dw/dx$
6	$y = uv$	$v(du/dx) + u(dv/dx)$
7	$y = u/v$	$\dfrac{v(du/dx) - u(dv/dx)}{v^2}$

The following examples illustrate the way in which the above rules of differentiation can be used to find the derivatives of various functions.

EXAMPLE 1. *Differentiate $y = 4x^3$.*

This can be differentiated directly by Rule 2, putting $a = 4$ and $n = 3$. Thus

$$dy/dx = 4.3x^{3-1} = 12x^2.$$

EXAMPLE 2. *Differentiate $y = x^5 + e^x + 3\ln x + 8$.*

By Rule 5, the differential coefficient is equal to the sum of the derivatives of the separate functions of x. The derivatives of the separate functions can be found by use of Rules 2, 3, 4 and 1, respectively. Hence

$$dy/dx = 5x^4 + e^x + 3/x + 0.$$

EXAMPLE 3. *Differentiate* $y = 3(x^3+4)e^x$.

The right-hand side of the equation is a product of two functions of x. Let $u = 3(x^3+4)$ and $v = e^x$. Then, by Rule 2, $du/dx = 9x^2$; by Rule 3, $dv/dx = e^x$, and hence by Rule 6,

$$dy/dx = e^x \cdot 9x^2 + 3(x^3+4)e^x$$
$$= (3x^3+9x^2+12)e^x.$$

EXAMPLE 4. *Differentiate* $y = (x^3-1)/(x+3)$.

The right-hand side of the equation is a quotient of two functions of x. Let $u = x^3-1$ and $v = x+3$. Then, by Rule 2, $du/dx = 3x^2$, and $dv/dx = 1$. Hence, by Rule 7,

$$\frac{dy}{dx} = \frac{(x+3)3x^2-(x^3-1)1}{(x+3)^2}$$
$$= \frac{3x^3+9x^2-x^3+1}{(x+3)^2} = \frac{2x^3+9x^2+1}{(x+3)^2}.$$

EXAMPLE 5. *Differentiate* $y = 2 \log x + 4$.

Log x means the log of x to the base 10, and before it can be differentiated it must be converted to base e, i.e. from common logarithms to natural logarithms. Thus
$$\ln x = \ln 10 \cdot \log x = 2 \cdot 303 \log x.$$
Hence
$$y = \frac{2}{2 \cdot 303} \ln x + 4.$$

Therefore, by Rules 4 and 1,
$$\frac{dy}{dx} = \frac{2}{2 \cdot 303} \cdot \frac{1}{x} = \frac{0 \cdot 8686}{x}.$$

EXAMPLE 6. *Differentiate* $y = 7/x^2$.

This may be rewritten as $y = 7x^{-2}$, which may be differentiated by Rule 2: thus
$$dy/dx = -14x^{-2-1} = -14x^{-3} = -14/x^3.$$

4. Differentiation of a function of a function (or change of variable)

Functions such as $y = \sqrt{(x^3+7x+3)}$, which cannot be differentiated directly by the rules, can often be converted to a suitable form by changing the variable. The expression $\sqrt{(x^3+7x+3)}$ is a function of x^3+7x+3, which is itself a function of x. If u is put equal to x^3+7x+3, then y is a function of u, and u is a function of x, i.e. $y = f(u)$ and $u = f(x)$. The required derivative dy/dx is found from the identity

$$\frac{dy}{dx} \equiv \frac{dy}{du} \cdot \frac{du}{dx}. \tag{5.9}$$

The differential coefficients dy/du and du/dx can be found by using the rules of differentiation. Since
$$u = x^3+7x+3,$$
then $du/dx = 3x^2+7$. Further, since $y = u^{1/2}$, then $dy/du = \frac{1}{2}u^{-1/2}$.

Using the identity (5.9), it follows that
$$dy/dx = (3x^2+7) \cdot \tfrac{1}{2}u^{-1/2} = (3x^2+7) \cdot \tfrac{1}{2}(x^3+7x+3)^{-1/2}$$
$$= \frac{3x^2+7}{2\sqrt{(x^3+7x+3)}}.$$

The identity (5.9) can be shown to be true for finite increments. For, if Δy and Δu are the respective increments in y and u corresponding to the change Δx in x, then
$$\frac{\Delta y}{\Delta u} \cdot \frac{\Delta u}{\Delta x} \equiv \frac{\Delta y}{\Delta x} \cdot \frac{\Delta u}{\Delta u} \equiv \frac{\Delta y}{\Delta x},$$
and this remains true when the increments are infinitesimal. The identity holds for any number of variables.

5. Relation between the derivatives of inverse functions

If y is a function of x, then x is said to be an *inverse function* of y. The relation between dy/dx and dx/dy is shown by the identity
$$\frac{\Delta y}{\Delta x} \cdot \frac{\Delta x}{\Delta y} \equiv 1,$$
where Δx and Δy are finite increments in x and y respectively and can be cancelled. In the limit, when Δx and Δy both approach zero,
$$\frac{dy}{dx} \cdot \frac{dx}{dy} = 1.$$
Hence
$$\frac{dy}{dx} = \frac{1}{dx/dy}. \tag{5.10}$$

6. Differentiation of implicit functions

An *explicit function* is a relation between two variables in which one is explicitly given as a function of the other, e.g. $y = x^2+3x+4$. An *implicit function* is a relation in which the two variables cannot be easily separated, although its existence implies that one variable is a function of the other, e.g. $x^2+6xy+y^2 = 8$. The derivative of such a function can be found, often in terms of both variables, by differentiating each term of the function with respect to x. This process is illustrated in the following example.

EXAMPLE. *Find the derivative dy/dx when $x^2+6xy^2+y^3 = 8$.*
Differentiating the first term with respect to x gives
$$\frac{d}{dx}(x^2) = 2x.$$
The second term is a product of two functions of x; therefore let $u = 6x$ and $v = y^2$. Then $du/dx = 6$, and
$$\frac{dv}{dx} = \frac{dv}{dy} \cdot \frac{dy}{dx} = 2y\frac{dy}{dx}.$$

Hence, by Rule 6,

$$\frac{d}{dx}(6xy^2) = y^2 \cdot 6 + 6x \cdot 2y\frac{dy}{dx} = 6y^2 + 12xy\frac{dy}{dx}.$$

The third term on differentiation gives

$$\frac{d}{dx}(y^3) = \frac{d(y^3)}{dy} \cdot \frac{dy}{dx} = 3y^2\frac{dy}{dx}.$$

The derivative of the constant term 8 is, by Rule 1, equal to zero. Substituting the derivatives of the separate terms in the original equation gives

$$2x + 6y^2 + 12xy\frac{dy}{dx} + 3y^2\frac{dy}{dx} = 0,$$

which on rearranging gives

$$\frac{dy}{dx} = \frac{-2x - 6y^2}{3y^2 + 12xy} = -\frac{2(x + 3y^2)}{3(y^2 + 4xy)}.$$

7. Applications of differential calculus

The following list summarises some of the more important applications of differential calculus. It is used

  (i) to find the true rates of change at a given instant in the study of changing systems;
 (ii) to determine the slope of a curve at a given point when the equation of the curve is known;
(iii) to establish the rules of integration so that if dy/dx is known to be a function of x, the direct relation between y and x can be found (see Chap. 7 and 9);
 (iv) to locate maximum and minimum values of a function (see Chap. 6);
  (v) to deduce values of the variance of a function of several variates (see Appendix VII); and
 (vi) to express the velocity (dx/dt) and acceleration (d^2x/dt^2) of a moving body in terms of its distance x from a fixed point and time t (see Chap. 6 and 9).

There are many other important applications of differential calculus in theoretical work related to experimental measurements.

8. Newton method for solution of equations

Differential calculus may be used to improve an approximate solution to an equation such as might be obtained graphically or by inspection. If the equation is $f(x) = 0$, then if $y = f(x)$ is plotted against x as in Fig. 5c, the exact solution is the value of x at point S where the curve cuts the x-axis.

Suppose the approximate solution is represented by the value of $x = x_1$ at point Q, then the point on the curve, P, vertically above Q represents the value of y at $x = x_1$, which is $y_1 = f(x_1)$. If the solution were exact, y_1 would be zero.

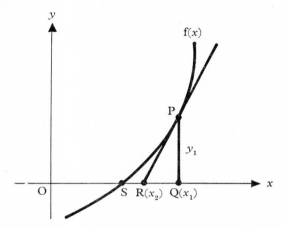

FIG. 5c. Newton method for equations

PR is a tangent drawn to the curve at P and the slope of the tangent will be the slope of the curve at P, that is, dy/dx, at $x = x_1$ or $f'(x_1)$. If R is the point at which the tangent cuts the x-axis then the value of x at R is always a better approximation to the true solution S than is the value of x at Q.

The value of x at R is OR = OQ−RQ = x_1−RQ.
In the $\triangle$PRQ, the slope of PR = $f'(x_1)$ = PQ/RQ,
therefore RQ = PQ/$f'(x_1)$, but PQ is the value of y at P which is $f(x_1)$, giving RQ = $f(x_1)/f'(x_1)$ and

$$x_2 = OR = x_1 - f(x_1)/f'(x_1).$$

If the function is differentiable in the range between S and Q and does not give a zero derivative in this range, then by calculating $f(x_1)$ and $f'(x_1)$ at the value of the approximate solution x_1, a better second approximation x_2 = OR can be obtained and the process may be repeated to any required degree of accuracy.

Repeated or iterative calculations of this type which are very tedious to carry out by other methods are very suitable for solution by computer. The computer repeats the calculations until a predetermined level of agreement between one result and the next is achieved. A programme to effect this is outlined and discussed below.

If the equation $f(x) = 0$ has a number of solutions, the Newton method

9. COMP 3: Fortran programme for Newton solution of a bi-exponential equation

Curves representing plasma drug concentrations at various times after administration of the drug may often be represented by bi-exponential equations of the type

$$c = 9\cdot 60 \exp(-0\cdot 264t) + 6\cdot 27 \exp(-0\cdot 045t).$$

The first exponential term mainly represents the comparatively rapid process of elimination by excretion while the second represents slower processes such as absorption into tissues and the crossing of lipid membranes. In this equation which is for acetylsalicylic acid in man, c is drug concentration in mg l^{-1} and t is time in minutes.

The problem solved by this programme is to find the exact time at which $c = 2$. This is equivalent to solving the equation

$$9\cdot 60 \exp(-0\cdot 264t) + 6\cdot 27 \exp(-0\cdot 045t) - 2 = 0.$$

If the left-hand side is called $f(t)$ then $f'(t)$ is readily obtained by differentiation

$$f'(t) = -9\cdot 60 \times 0\cdot 264 \exp(-0\cdot 264t) - 6\cdot 27 \times 0\cdot 045 \exp(-0\cdot 045t).$$

In the programme, t is called T, $f(t)$ is called FT, and $f'(t)$ is called DFT. Guessing an approximate solution as T = 15 minutes, FT is calculated at T = 15 and so is DFT. The improved approximation TREV calculated by the Newton method is TREV = T−FT/DFT.

The difference between T and TREV is tested and if it is greater than 10^{-6} the calculation is repeated with TREV in place of T. In this way a result accurate to 10^{-6} is obtained. A limit of 20 on the number of iterations is included to avoid excessive computation if the results do not converge reasonably rapidly.

```
C       COMP 3
C       NEWTON SOLUTION OF EQUATION   FT=0
        ITER = 1
        T = 15.
1       FT = 9.60*EXP(-0.264*T) + 6.27*EXP(-0.045*T) -2.0
        DFT = -0.264*9.60*EXP(-0.246*T) - 6.27*0.045*EXP
       1 (-0.045*T)
        TREV = T - FT/DFT
        WRITE(6,2) ITER,TREV
2       FORMAT(1H0,I5,15X,F10.6)
        IF (ABS(T-TREV) .LT. 1.E-6 .OR. ITER .GE. 20) STOP
        ITER = ITER + 1
        T = TREV
        GO TO 1
        END
```

DIFFERENTIAL CALCULUS

In this programme the integer variable ITER is used as a counter and is set to 1 initially; T, the approximate solution, could be read in from a data card but in this case it is put in the programme and no data cards are used. FT and DFT are calculated in the next two statements for T = 15; the statement for FT is numbered as it is referred to later. The improved root TREV is next computed and is written out, together with the value of ITER.

The IF statement tests the difference between T and TREV; the brackets are essential for this type of IF and within them the logical statement is enclosed. ABS( ) means the absolute value or modulus of the difference, that is, the magnitude regardless of sign. The relational operator ·LT· simply means less than; there are five other such operators, ·LE· is less than or equal to, ·EQ· is equal to, ·NE· is not equal to, ·GT· is greater than and ·GE· is greater than or equal to. The full stops before and after these operators are essential and serve to distinguish them to the compiler from other variable names and operators.

1·E−6 is 10^{-6} in E notation, ·OR· means exactly what it says and gives two alternative conditions to be satisfied. ·AND· is another operator to couple conditions and when it is used both conditions have to be satisfied. The second condition is that the iteration counter should be greater than or equal to 20. If either of these conditions is fulfilled then the statement outside the IF brackets is executed. This statement is an example of a STOP which is not physically at the end of the FORTRAN deck. It simply tells the computer to stop calculating.

If neither of the IF conditions are satisfied the programme is continued in the order of the cards; ITER is increased by one and the revised value, TREV, of T is put into the cell labelled T. The GO TO followed by a statement number returns the programme to the calculation of FT with this new value of T. The cycle of calculations continues until one of the conditions of the IF is satisfied, when it stops.

The last card of the Fortran deck, as always, is the signal to the compiler of the END of the source deck.

The output from this programme is given below; the guessed solution $t = 15$ is a poor approximation but the successive approximations rapidly converge to the correct answer.

1	21.646210
2	25.001624
3	25.501384
4	25.518329
5	25.518652
6	25.518658
7	25.518658

Problems

1. Evaluate dy/dx when
   - (i) $y = 3x^5 + 4x^2 - 6x + 3$
   - (ii) $y = (2x^2 - 1)(3x + 2)$
   - (iii) $y = 1/(x^2 + 1)$
   - (iv) $y = 2x/(x+1)^3$
   - (v) $y = (2x+3)/(4x-7)$
   - (vi) $y = (ax^2 + bx + c)(2x + d)$
   - (vii) $y = x^{2b} - bx^3 + 3b$
   - (viii) $y = (x+1)(2x+1)(3x+2)$
   - (ix) $y = \sqrt[3]{(3x)} + \sqrt{(2x)} + 9/x^3$

2. What are the slopes of the curve $y = x^2 + 3x + 4$ at the points where $x = 1, 2, 3$ and 7?

3. Evaluate dy/dx when
   - (i) $y = (x^3 - 2)^2$
   - (ii) $y = 3(6x^2 - 3x^5)^2$
   - (iii) $y = 2 \exp x$
   - (iv) $y = \ln(9x^7)$
   - (v) $y = \ln(x^2 + 4)$
   - (vi) $y = x^4 \exp x$
   - (vii) $y = 2/(3-x)$
   - (viii) $y = (x^2 - 1)/(x+2)$
   - (ix) $y = x \log x$
   - (x) $y = \exp(2x^3 + 3x + 7)$
   - (xi) $x^2 + y^2 = a^2$
   - (xii) $x^3 + y^3 = 5xy$
   - (xiii) $2x^2 - 5xy + y^2 = 3x$
   - (xiv) $x^{2n} + y^{2n} = a$

4. A kinetic study of a chemical reaction gave the following values of x (concentration of reaction product in mole per litre) at various times t (in minutes).

t	x	t	x	t	x
5	1·09	25	4·38	45	6·45
10	2·06	30	4·99	50	6·84
15	2·92	35	5·53	55	7·18
20	3·69	40	6·02	60	7·49

   Plot x against t, and determine the mean reaction velocity at a mean time of 35 minutes for a time interval of 20 minutes. Find also the true reaction velocities (dx/dt) at 15, 25, 35 and 45 minutes, and show that dx/dt is a linear function of x.

5. Plot the curve $y = x^3$ from $x = 1$ to $x = 5$, and determine $\Delta y/\Delta x$ for a mean value of $x = 3$ for values of Δx equal to 2, 1 and 0·5. Determine dy/dx at $x = 3$ by the graphical method, and compare the value with the result obtained algebraically.

6. An approximate solution to the equation

   $$x^5 - 10x^2 - 4 = 0$$

   is $x = 2$. Write a Fortran programme to obtain the exact solution to six places of decimals, using the Newton method.

CHAPTER 6

Higher derivatives and partial differentiation

THE FIRST derivative of a function, say $y = f(x)$, is the differential coefficient dy/dx. Further differentiation of the first derivative with respect to x gives the second derivative d^2y/dx^2. Alternative ways of indicating the second derivative are

$$\frac{d}{dx}\left(\frac{dy}{dx}\right) \quad \text{and} \quad \frac{d^2}{dx^2}\{f(x)\}.$$

Differentiation of the second derivative with respect to x gives the third derivative d^3y/dx^3, and so on. The numbers in the numerator and denominator do not refer to powers of x or d, but are merely the conventional way of indicating the number of times the function has been differentiated. The following example shows how these higher derivatives are obtained by applying the rules of calculus.

EXAMPLE. *Obtain the higher derivatives of the function* $y = x^4 + 3x^3 + 7$.

$$\begin{aligned}
f(x) &= y &= x^4 + 3x^3 + 7. \\
f^I(x) &= dy/dx &= 4x^3 + 9x^2. \\
f^{II}(x) &= d^2y/dx^2 &= 12x^2 + 18x. \\
f^{III}(x) &= d^3y/dx^3 &= 24x + 18. \\
f^{IV}(x) &= d^4y/dx^4 &= 24. \\
f^V(x) &= d^5y/dx^5 &= 0.
\end{aligned}$$

1. Meaning of the first and second derivatives in terms of motion

If a body is moving in a straight line from a fixed point A (see Fig. 6a), then the distance x between that body and the fixed point will be a function of time t. Therefore

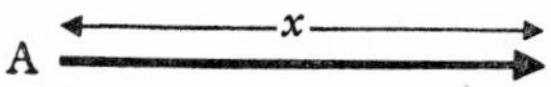

$$x = f(t).$$

FIG. 6a.

The first derivative dx/dt is the rate of change of distance x with time t, i.e. it is the velocity of the body v. Hence

$$v = dx/dt.$$

The second derivative of x with respect to t is the rate of change of velocity with time, i.e. the acceleration of the body, since

$$\frac{d^2x}{dt^2} = \frac{d}{dt}\left(\frac{dx}{dt}\right) = \frac{dv}{dt}.$$

EXAMPLE. *The distance x, in cm, of a body moving in a straight line is given by the equation*
$$x = 2t + 2\cdot 5t^2 - 0\cdot 05t^3.$$
What is the velocity and acceleration when $t = 10$ seconds?

Differentiating the equation relating x and t gives
$$dx/dt = 2 + 5t - 0\cdot 15t^2.$$
Substituting $t = 10$ sec in this equation shows that the velocity at this time is 37 cm s^{-1}. Differentiating the equation again gives
$$d^2x/dt^2 = 5 - 0\cdot 30t.$$
Substituting $t = 10$ sec in this equation shows that the acceleration at this time is 2 cm s^{-2}.

2. Maclaurin's theorem

Equations of the type
$$y = A + Bx + Cx^2 + Dx^3 + Ex^4 + \ldots$$
are often used in physical chemistry. For example the isothermals on pv, p diagrams may be conveniently represented by equations of the form
$$pv = A + Bp + Cp^2 + \ldots,$$
where p is the pressure of the gas and v is its volume; $A, B, C, \ldots$ are constants. One of the uses of Maclaurin's theorem is to evaluate the constants of a power series such as this.

Suppose that y is a function of x, i.e. $y = f(x)$, and let it be assumed that $f(x)$ can be expressed as an ascending power series in x, then
$$y = f(x) = A + Bx + Cx^2 + Dx^3 + Ex^4 + \ldots. \qquad (6.1)$$
If $f(0)$ is the value of $f(x)$ when $x = 0$, then putting $x = 0$ in the above series gives
$$y = f(0) = A.$$
The first derivative of the series is found by differentiating with respect to x: thus
$$\frac{d}{dx}\{f(x)\} = B + 2Cx + 3Dx^2 + 4Ex^3 + \ldots.$$
Putting $f^i(0)$ as the value of the first derivative when $x = 0$ and substituting $x = 0$ in the above equation gives
$$f^i(0) = B.$$
The second derivative is
$$\frac{d^2}{dx^2}\{f(x)\} = 2.1C + 3.2Dx + 4.3Ex^2 + \ldots.$$
If the value of the second derivative is $f^{ii}(0)$ when $x = 0$, then
$$f^{ii}(0) = 2.1C.$$

HIGHER DERIVATIVES AND PARTIAL DIFFERENTIATION

Hence
$$C = \frac{f^{ii}(0)}{2!}.$$

Repeating the differentiation gives
$$\frac{d^3}{dx^3}\{f(x)\} = 3.2.1D + 4.3.2Ex + \ldots$$

Hence
$$f^{iii}(0) = 3.2.1D,$$
and
$$D = \frac{f^{iii}(0)}{3!}.$$

Again repeating the differentiation gives
$$\frac{d^4}{dx^4}\{f(x)\} = 4.3.2E + \ldots$$

Hence
$$f^{iv}(0) = 4.3.2E,$$
and
$$E = \frac{f^{iv}(0)}{4!}.$$

Substituting these values of $A, B, C, D, E, \ldots$ in the original equation (6.1) gives
$$y = f(0) + \frac{f^i(0)}{1!}x + \frac{f^{ii}(0)}{2!}x^2 + \frac{f^{iii}(0)}{3!}x^3 + \frac{f^{iv}(0)}{4!}x^4 + \ldots \quad (6.2)$$

The series on the right-hand side of the expression is a statement of Maclaurin's theorem. Expressed in words, this theorem states that any function $f(x)$ of a variable x can be expressed as a power series in x whose coefficients can be found from the derivatives of the function provided the derivatives of the function are finite when $x = 0$. Maclaurin's theorem is used in Chap. 8 to obtain series expansions for the sine and cosine of an angle. The use of series to summarise experimental measurements is described in Chap. 10, § 2.

EXAMPLE. *Use Maclaurin's theorem to derive the coefficients of the binomial expansion of $(1+x)^n$ for any value of n.*

If $f(x) = (1+x)^n$, then expressing it in the form of a power series in x
$$f(x) = A + Bx + Cx^2 + Dx^3 + \ldots$$

By Maclaurin's theorem, when $x = 0$
$$A = f(0) = (1+x)^n = 1^n = 1.$$

Hence, $A = 1$ for all values of n. Differentiation of $(1+x)^n$ gives
$$f^1(x) = n(1+x)^{n-1}.$$

When $x = 0$, $f^{i}(0) = n$, and therefore $B = n$. Differentiating a second time gives

$$f^{ii}(x) = n(n-1)(1+x)^{n-2}.$$

When $x = 0$, $f^{ii}(0) = n(n-1)$, and therefore $C = n(n-1)/2!$ Similarly,

$$f^{iii}(0) = n(n-1)(n-2),$$

and hence $D = n(n-1)(n-2)/3!$. The expansion is therefore

$$(1+x)^n = 1 + nx + \frac{n(n-1)}{2!}x^2 + \frac{n(n-1)(n-2)}{3!}x^3 + \ldots$$

3. Maximum values

Certain functions show maximum and minimum values, and it is often important to be able to locate points at which these values occur. The

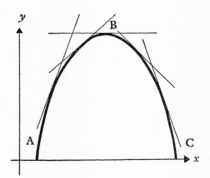

FIG. 6b. Function with a maximum

curve in Fig. 6b represents a function y of x which has a maximum at the point B. The position of this point can be determined by means of calculus by considering the variation of the slope of the curve in the vicinity of B.

The slope of a curve at a point is the slope of the tangent and is also the value of dy/dx at that point. If a series of tangents are drawn to the curve between A and C, it is apparent that their slopes dy/dx are positive from A to B, i.e. y increases as x increases. At the point B, the tangent is parallel to the x-axis and therefore has zero slope, i.e. at B, $dy/dx = 0$; this is the equation used to find the maximum value of a function. From B to C, the slopes of the tangents are negative, i.e. y decreases as x increases.

Since dy/dx is decreasing from positive values and becoming negative as x increases, the rate of change of dy/dx with x in the vicinity of B is negative, i.e.

$$\frac{d}{dx}\left(\frac{dy}{dx}\right) = \frac{d^2y}{dx^2},$$

which is negative in the vicinity of a maximum value.

HIGHER DERIVATIVES AND PARTIAL DIFFERENTIATION

EXAMPLE. *Find the maximum value of y when* $y = 5x - x^2$.

The first derivative $dy/dx = 5 - 2x$. At a maximum value this derivative will be equal to zero. Therefore
$$5 - 2x = 0, \quad \text{or} \quad x = 2 \cdot 5.$$
The corresponding value of y is found by substituting this value of x in the original equation: thus
$$y = 5(2 \cdot 5) - (2 \cdot 5)^2 = 6 \cdot 25.$$
In order to find whether this value of y is a maximum, the expression for dy/dx is differentiated again giving
$$d^2y/dx^2 = -2.$$
This second derivative is negative, and so the value of y at which $dy/dx = 0$, i.e. $6 \cdot 25$, is a maximum value.

A maximum value is shown in the boiling-point–composition curve for a binary mixture of water and hydrogen chloride; the maximum boiling point at 760 mm pressure occurs when approximately 20 per cent of HCl is present.

4. Minimum values

Some functions have a point at which $dy/dx = 0$, but which represents a minimum value of y. Such a function is shown in Fig. 6c. From the

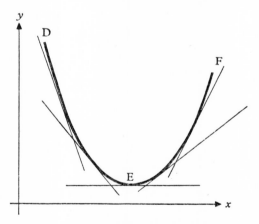

FIG. 6c. Function with a minimum

tangents drawn to the curve, it is seen that dy/dx is negative from D to E, zero at E, and positive from E to F; dy/dx therefore increases as x increases, i.e.
$$\frac{d}{dx}\left(\frac{dy}{dx}\right) = \frac{d^2y}{dx^2},$$
which is positive in the vicinity of a minimum value.

EXAMPLE. *Find the minimum value of y when* $y = x^2 - 6x + 10$.

The first derivative $dy/dx = 2x - 6$, which is equal to zero at a minimum; hence $x = 3$. The corresponding value of y is therefore $9 - 18 + 10$, i.e. 1. Further, $d^2y/dx^2 = 2$, i.e. it is positive, so that the value of y at which $dy/dx = 0$, i.e. 1, is a minimum value.

A minimum value is shown in the boiling-point–composition curve for a binary mixture of water and ethanol; the minimum boiling point at 760 mm pressure occurs when the mixture contains about 90 per cent of C_2H_5OH.

5. Stationary values

Points at which $dy/dx = 0$ are called *stationary values* of a function, since at these points y momentarily ceases to change with x. If at the stationary value, d^2y/dx^2 is negative, then the function has a maximum

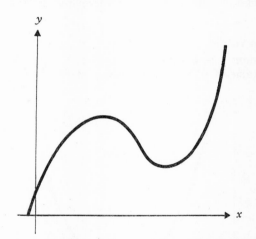

FIG. 6d. Function with a maximum and a minimum

value at this point; if it is positive, then there is a minimum value at this point. Some functions show both maximum and minimum values, and such a function is shown graphically in Fig. 6d. This graph also indicates that maxima and minima are only local and are not necessarily the greatest or least values of the functions.

Another type of stationary value occurs when both dy/dx and d^2y/dx^2 are equal to zero. This is called a *point of horizontal inflection* or *turning point*, and is illustrated in Fig. 6e by the point G. In this particular instance, y is increasing with x until the point G is reached. Here the tangent to the curve becomes parallel to the x-axis. After G, y continues to increase with x. At first, dy/dx is positive, decreasing to zero at the point G and increasing again beyond G; dy/dx therefore has a minimum value at G, at which point

$$\frac{d}{dx}\left(\frac{dy}{dx}\right) = \frac{d^2y}{dx^2} = 0.$$

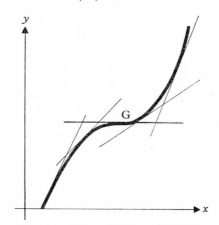

FIG. 6e. Point of horizontal inflection

EXAMPLE 1. *Determine the stationary values of y when*
$$y = 2x^3 - 21x^2 + 60x + 5.$$

The first derivative $dy/dx = 6x^2 - 42x + 60$. At a maximum or minimum this derivative is zero. Therefore
$$6(x^2 - 7x + 10) = 0, \quad \text{i.e.} \quad 6(x-5)(x-2) = 0.$$
Therefore stationary values occur when $x = 5$ or 2.

The second derivative $d^2y/dx^2 = 12x - 42$. When $x = 5$,
$$d^2y/dx^2 = 60 - 42,$$
i.e. a positive value. Hence when $x = 5$, the value of y is a minimum. When $x = 2$, $d^2y/dx^2 = 24 - 42$, i.e. a negative value. Hence when $x = 2$, the value of y is a maximum.

At the minimum,
$$y = 2(5)^3 - 21(5)^2 + 60(5) + 5 = 30.$$
At the maximum,
$$y = 2(2)^3 - 21(2)^2 + 60(2) + 5 = 57.$$

EXAMPLE 2. *Determine the stationary values of y when*
$$y = x^3 - 9x^2 + 27x.$$

Now, $dy/dx = 3x^2 - 18x + 27$. At the stationary values, the first derivative is equal to zero. Hence
$$3x^2 - 18x + 27 = 0, \quad \text{i.e.} \quad 3(x-3)^2 = 0.$$
Therefore a stationary value occurs when $x = 3$. Further, $d^2y/dx^2 = 6x - 18$, so that when $x = 3$, $d^2y/dx^2 = 6(3) - 18 = 0$. Hence the stationary value is a point of inflection. At this point,
$$y = (3)^3 - 9(3)^2 + 27(3) = 27.$$

A point of horizontal inflection is shown in the pressure–volume isotherm of a gas at its critical temperature; the point of inflection is the critical point of the gas.

6. Partial differentiation

The properties of a physical system are usually functions of several variables. For instance, the volume v of a fixed mass of gas is dependent on both the pressure p and the temperature T, i.e. $v = f(p, T)$. The differentiation of a function such as this, in which two or more independent variables occur, is carried out in stages in which one variable is changed while the other variables are kept constant. The rate of variation of v with p at constant temperature is written $(\partial v/\partial p)_T$ and is called the partial derivative or partial differential coefficient of v with respect to p at constant temperature. The derivative $\partial v/\partial p$ has a similar meaning to dv/dp, the difference being that the symbol '∂' indicates that the differentiation has been carried out under restricted conditions. In this instance, the temperature has been held constant, and this is indicated by the letter T outside the brackets; if it is written simply as $\partial v/\partial p$, it indicates the rate of variation of v with p, all other factors liable to affect v being held constant.

Partial derivatives are found by applying the usual rules of differentiation. Variables that are held constant are treated as constants.

EXAMPLE. *If $z = x^3+2x^2y+y^3$, evaluate $\partial z/\partial x$ and $\partial z/\partial y$.*

To evaluate $\partial z/\partial x$, y is treated as a constant, and the differentiation of z with respect to x is carried out in the usual manner. Hence

$$\partial z/\partial x = 3x^2+4xy.$$

It should be noted that since y is kept constant, y^3 becomes zero on differentiation, but this is only true for this particular partial derivative.

Differentiating the original expression with respect to y while x is held constant gives

$$\partial z/\partial y = 2x^2+3y^2.$$

7. Partial and complete differentials

Consider the quantity u, which is a function of three variables x, y and z. If x is changed by a small increment Δx while y and z are kept constant, and Δu_1 is the corresponding increment in u, then, if Δx is small,

$$\Delta u_1/\Delta x \simeq \partial u/\partial x, \quad \text{i.e.} \quad \Delta u_1 \simeq (\partial u/\partial x)\Delta x.$$

Likewise, if y is changed by Δy, causing an increment Δu_2 in u while x and z are kept constant, then, if Δy is small,

$$\Delta u_2/\Delta y \simeq \partial u/\partial y, \quad \text{i.e.} \quad \Delta u_2 \simeq (\partial u/\partial y)\Delta y.$$

Similarly, if z is changed by Δz, causing an increment Δu_3 in u while x and y are kept constant, then, if Δz is small,

$$\Delta u_3/\Delta z \simeq \partial u/\partial z, \quad \text{i.e.} \quad \Delta u_3 \simeq (\partial u/\partial z)\Delta z.$$

HIGHER DERIVATIVES AND PARTIAL DIFFERENTIATION

The total change in u is the sum of these changes; thus
$$\Delta u = \Delta u_1 + \Delta u_2 + \Delta u_3$$
$$\simeq (\partial u/\partial x)\Delta x + (\partial u/\partial y)\Delta y + (\partial u/\partial z)\Delta z.$$
If the increments are infinitesimal, the approximation becomes exact so that
$$\mathrm{d}u = (\partial u/\partial x)\mathrm{d}x + (\partial u/\partial y)\mathrm{d}y + (\partial u/\partial z)\mathrm{d}z. \tag{6.3}$$
$\mathrm{d}u$ is the complete differential of u, being the increment in u caused by infinitesimal changes in any of the variables.

This equation is useful for deriving theoretical relations in, for example, thermodynamics.

EXAMPLE. *The internal energy U of a thermodynamic system can be completely defined by the volume v and temperature T of the system. Derive an expression, in terms of partial derivatives, for the variation of U with temperature at constant pressure.*

Since $U = \mathrm{f}(v, T)$, the total differential of U can be expressed by (6.3) in terms of partial derivatives; thus
$$\mathrm{d}U = (\partial U/\partial v)_T \mathrm{d}v + (\partial U/\partial T)_v \mathrm{d}T.$$
Dividing by the increment in T gives
$$\mathrm{d}U/\mathrm{d}T = (\partial U/\partial v)_T \mathrm{d}v/\mathrm{d}T + (\partial U/\partial T)_v.$$
Here $\mathrm{d}U/\mathrm{d}T$ is the total derivative of U with respect to T and therefore gives the rate of change of U with T under all circumstances. It will hold under conditions of constant pressure provided the same restriction is applied to the other total derivative $\mathrm{d}v/\mathrm{d}T$; so that
$$(\partial U/\partial T)_p = (\partial U/\partial v)_T (\partial v/\partial T)_p + (\partial U/\partial T)_v.$$

8. Higher partial derivatives

If $u = \mathrm{f}(x, y)$, $\partial u/\partial x$ can be evaluated. A new type of higher derivative can then be found by differentiating $\partial u/\partial x$ with respect to y while x is kept constant; this derivative is written
$$\frac{\partial}{\partial y}\left(\frac{\partial u}{\partial x}\right) \quad \text{or} \quad \frac{\partial^2 u}{\partial y \partial x}.$$
If the partial differentiation is carried out in the reverse order, i.e. if $\partial u/\partial y$ is found first and then differentiated with respect to x while y is kept constant, the higher derivative
$$\frac{\partial}{\partial x}\left(\frac{\partial u}{\partial y}\right) \quad \text{or} \quad \frac{\partial^2 u}{\partial x \partial y}$$
is obtained. These two higher derivatives are equal, i.e.
$$\frac{\partial}{\partial y}\left(\frac{\partial u}{\partial x}\right) = \frac{\partial}{\partial x}\left(\frac{\partial u}{\partial y}\right),$$
or
$$\frac{\partial^2 u}{\partial y \partial x} = \frac{\partial^2 u}{\partial x \partial y}. \tag{6.4}$$

Hence the order in which the partial differentiation is carried out does not affect the final result.

This theorem can be proved for the general case, but in order to give some reality to the proof, it will be established here for the particular case of the expansion of a liquid of volume v with change in pressure p and temperature T. Now, by applying (6.3)

$$dv = (\partial v/\partial T)_p dT + (\partial v/\partial p)_T dp. \qquad (6.5)$$

If $\alpha = \partial v/\partial T$, then α is the coefficient of thermal expansion at constant pressure; the value of α will change, however, with change in pressure. If $\beta = \partial v/\partial p$, then β is minus the coefficient of compression of the liquid at constant temperature; the value of β will change, however, with change in temperature. Substituting α and β in (6.5) gives

$$dv = \alpha\, dT + \beta\, dp.$$

If the temperature is changed at constant pressure by ΔT, then the volume change Δv_1 is approximately equal to $\alpha\, \Delta T$, and β becomes $\beta + \Delta\beta$. If the pressure is now changed at constant temperature by Δp, then the volume change Δv_2 will be approximately equal to $(\beta + \Delta\beta)\Delta p$. The total volume change is therefore

$$\alpha\, \Delta T + (\beta + \Delta\beta)\Delta p.$$

Now suppose these changes were carried out in the reverse order. If the pressure is changed at constant temperature by Δp, then the volume change Δv_3 is approximately equal to $\beta\, \Delta p$, and α will become $\alpha + \Delta\alpha$. If the temperature is now changed at constant pressure by ΔT, then the volume change Δv_4 will be approximately equal to $(\alpha + \Delta\alpha)\Delta T$. The total volume change is therefore

$$\beta\, \Delta p + (\alpha + \Delta\alpha)\Delta T.$$

As the final conditions are the same in both cases, the final volumes are the same, and therefore the total volume changes are identical. Hence

$$\alpha\, \Delta T + (\beta + \Delta\beta)\Delta p = \beta\, \Delta p + (\alpha + \Delta\alpha)\Delta T.$$

Therefore
$$\Delta\beta\, \Delta p = \Delta\alpha\, \Delta T,$$
i.e.
$$\Delta\beta/\Delta T = \Delta\alpha/\Delta p.$$

If the changes are infinitesimal, then they can be written in the usual differential form; they are partial differentials because each change is carried out keeping the third variable constant. Hence

$$\partial\beta/\partial T = \partial\alpha/\partial p.$$

Substituting $\partial v/\partial T$ for α and $\partial v/\partial p$ for β gives

$$\frac{\partial}{\partial T}\left(\frac{\partial v}{\partial p}\right) = \frac{\partial}{\partial p}\left(\frac{\partial v}{\partial T}\right),$$

i.e.
$$\frac{\partial^2 v}{\partial T \partial p} = \frac{\partial^2 v}{\partial p \partial T}.$$

This means that the magnitude of the temperature coefficient of compressibility is the same as that of the pressure coefficient of thermal expansion.

9. Optimisation

Partial differentiation is used in the computer process of optimisation in which a key property is expressed as a function of fixed quantities each multiplied by one of a set of coefficients. The optimisation process aims at evaluating the set of coefficients which either maximise or minimise the key property.

This method is widely used in economic planning. In the chemical and other industries it provides a means for exploring and evaluating the most effective use of equipment so as to maximise output or minimise costs.

Optimisation programmes may involve very elaborate computation, depending on the amount of information available. A simple example, in which the results are obtained directly without the need for using inequalities, is provided by the basic energy minimisation method used in molecular orbital theory in quantum chemistry.

In this theory electrons in atomic orbitals are combined together to give electrons in molecular orbitals. If the combination coefficients for the atomic orbital amplitude functions are $a_1, a_2, a_3 \ldots$ the energy of the molecule is

$$E = \frac{a_1^2 H_{11} + a_2^2 H_{22} + a_3^2 H_{33} + \ldots + 2a_1 a_2 H_{12} + 2a_1 a_3 H_{13} + \ldots}{a_1^2 S_{11} + a_2^2 S_{22} + a_3^2 S_{23} + \ldots}$$

where the H and S terms are properties of the electrons in the atomic orbitals and can be regarded as fixed. The set of coefficients which combine them in such a way as to give a minimum value of E are then the values to be expected in the stable, ground state of the molecule.

This set of values of $a_1, a_2 \ldots$ is found by minimising (optimising) E with respect to each coefficient. This is done by differentiating E partially with respect to each coefficient and setting the partial derivative equal to zero for minimum E.

For example, minimising with respect to a_1, the above equation is cross-multiplied to give

$$E(a_1^2 S_{11} + a_2^2 S_{22} + \ldots) = a_1^2 H_{11} + a_2^2 H_{22} + \ldots + 2a_1 a_2 H_{12} + 2a_1 a_3 H_{13} + \ldots$$

This is differentiated partially with respect to a_1; the left-hand side is treated as a product of two variables, E and the contents of the brackets;

all the terms which do not contain a_1 give zero and the result is

$$E(2a_1S_{11}) + \frac{\partial E}{\partial a_1}(a_1^2 S_{11} + a_2^2 S_{22} + \ldots) = 2a_1 H_{11} + 2a_2 H_{12} + 2a_3 H_{13} + \ldots$$

At the minimum, $\partial E/\partial a_1 = 0$ and

$$2a_1 S_{11} E = 2a_1 H_{11} + 2a_2 H_{12} + 2a_3 H_{13} + \ldots$$

rearranging,

$$a_1(H_{11} - ES_{11}) + a_2 H_{12} + a_3 H_{13} + \ldots = 0.$$

If there are N atomic orbitals involved there will be N combination coefficients and minimising E with respect to each of them will give N simultaneous equations. These are however homogeneous in the variables a_1, etc. (Chap. 2, § 11) and so a further relationship is required in order to evaluate the variables. This is provided by the normalisation equation which gives for one electron in a molecular orbital

$$a_1^2 + a_2^2 + a_3^2 + \ldots = 1 \quad \text{or} \quad \sum_{r=1}^{N} a_r^2 = 1.$$

The condition for non-trivial solutions to the set of equations (Chap. 2, § 11) gives a series of values for the energy. These are the stationary energy values corresponding to the quantisation of electron energies in a molecule. For each energy value there is a set of values of the coefficients.

By means of the Scientific Subroutine, EIGEN, these values of energy with their associated sets of coefficients may be determined with, as input data, the quantities relating to the atomic orbitals, H_{11}, H_{12}, S_{11}, etc. From the calculations of E and the coefficients, results related to the physical and chemical properties of the molecule may be computed.

Problems

1. Evaluate d^2y/dx^2 when
   (i) $y = \log x$ (ii) $y = x^4 + 3x^2 + 1$
   (iii) $y = \exp x^2$ (iv) $y = 2/x$.

2. Find the stationary values of the following functions and determine their type (maximum, minimum or point of inflection).
   (i) $y = x^2 + 3x + 4$ (ii) $y = x^4 - 4x + 3$
   (iii) $y = x^3 + 3x^2 - 9x + 1$ (iv) $y = x^3 + 9x^2 + 3x + 2$
   (v) $y = x^3 - 3x^2 + 3x + 2$ (vi) $y = x^3 - 2x^2 + x + 1$

3. The specific gravity s of water at temperature $t°C$ is given by the equation
   $$s = 1 + (5 \cdot 3 \times 10^{-5})t - (6 \cdot 53 \times 10^{-6})t^2 + (1 \cdot 4 \times 10^{-8})t^3.$$
   At what temperature will water have a maximum specific gravity?

4. In dilute aqueous solutions, the product of hydrogen- and hydroxyl-ion concentrations (h and g gramme-equivalents of H^+ and OH^- respectively) is constant and equal to 10^{-14} at $25°$. Find the ratio h/g which gives the minimum total concentration ($h+g$) of both ions. [Let $y = h+g$ and $x = h/g$.]

5. Evaluate $(\partial u/\partial x)_y$ when
   (i) $u = x^5 + 4x^3y^2 + xy^4 + 7$ (ii) $u = x^2 \exp(-y^2)$
   (iii) $u = \ln\{(x^3+3)/(y^2+2)\}$ (iv) $u = 2x^2y^3 + xy^2$
   (v) $u = (x+y)^2$ (vi) $u = 1/(x+y)$

6. The van der Waals equation for a non-ideal gas is
$$(p + a/v^2)(v - b) = RT.$$
The p, v curve at the critical temperature T_c shows a point of horizontal inflection at the critical point ($p = p_c$, $v = v_c$). What are the values of p_c and v_c in terms of R, a and b?

7. Show that if $u = 3x^4 + 7y^2$, $\partial^2 u/\partial x \partial y = \partial^2 u/\partial y \partial x$.

CHAPTER 7

Integration

THE PROCESS of integration may be regarded in two ways: firstly, as the summation of a very large number of infinitesimal elements into which an area, volume or any other quantity can be divided; secondly, as the reverse process of differentiation. From the second aspect, integration is the technique which enables the relation between two variables to be found when the rate of change of one with the other is known.

1. Integration as a summation

This aspect of integration is illustrated in Fig. 7a. In this very simple case the area enclosed between a graph, the x-axis and two ordinates (vertical lines) is to be determined. In the figure, the graph is a straight line

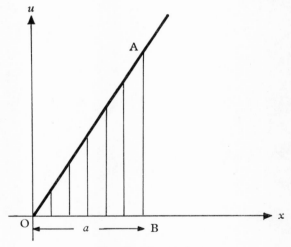

FIG. 7a.   Integration

expressing the relation between two variables, u and x, in which $u = 1·5x$; the two ordinates are at $x = 0$ and $x = a$. Thus the area to be measured is that of the triangle AOB.

INTEGRATION

To determine this area by the summation method, the triangle is divided into a number n of vertical strips of equal width Δx.

$$\text{Area AOB} = \Sigma \text{ strip areas.}$$

The area of the outermost strip is the area of a trapezium of mean height $1 \cdot 5(a - \Delta x/2)$, according to the equation of the graph, and width Δx; this area therefore is $1 \cdot 5(a - \Delta x/2)\Delta x$. Since the ordinate at the centre of the outermost strip but one rises from the point whose abscissa is $a - 3\Delta x/2$, this strip has a mean height of $1 \cdot 5(a - 3\Delta x/2)$; hence the area of this strip is $1 \cdot 5(a - 3\Delta x/2)\Delta x$. The next strip in towards the origin will have an area of $1 \cdot 5(a - 5\Delta x/2)\Delta x$, and the nth strip, i.e. the last one, will have an area of $1 \cdot 5[a - (2n-1)\Delta x/2]\Delta x$. In general, the area of any strip is given by $u\,\Delta x$, where u is the mean height of the strip.

$$\text{Area AOB} = \Sigma \text{ strip areas}$$
$$= 1 \cdot 5\Delta x\{[a - \Delta x/2] + [a - 3\Delta x/2] \ldots + [a - (2n-1)\Delta x/2]\}.$$

The series in the square brackets is an arithmetic progression whose sum can be found from equation (4.1): thus

$$\text{area AOB} = \sum_{x=0}^{x=a} u\,\Delta x = 1 \cdot 5\Delta x \frac{n}{2}[2a - \Delta x - (n-1)\Delta x].$$

From Fig. 7a, $n\,\Delta x = a$, and substituting this value for $n\,\Delta x$ in the above equation gives

$$\sum_{x=0}^{x=a} u\,\Delta x = \frac{1 \cdot 5a}{2}(2a - a) = 0 \cdot 75a^2.$$

If the number of strips is made very large, their width Δx eventually becomes infinitesimal. Hence, as $n \to \infty$, $\Delta x \to \mathrm{d}x$. In the limit, instead of the sign Σ to denote the sum, the integration sign $\int$ is used, i.e. as $n \to \infty$,

$$\sum_{x=0}^{x=a} u\,\Delta x \to \int_0^a u\,\mathrm{d}x.$$

The limits of an integral indicate, unless otherwise stated, values of the variable whose differential occurs in the expression.

Hence when the strips are infinitesimal

$$\text{area AOB} = \int_0^a u\,\mathrm{d}x \qquad (7.1)$$
$$= 0 \cdot 75a^2.$$

The value for the area of AOB can, of course, be deduced directly from the formula for the area of a triangle,

$$\text{area} = \tfrac{1}{2}(\text{base})(\text{perpendicular height}) = \tfrac{1}{2}(\text{OB})(\text{AB})$$
$$= \tfrac{1}{2}.a.1 \cdot 5a = 0 \cdot 75a^2,$$

which is the same answer as is given by the method of summation of strips when $u = 1\cdot 5x$.

Area under a curve. If the u, x plot is a curve, the method of integration can still be used to give an exact value for the area between the curve,

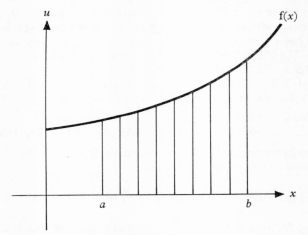

FIG. 7b. Integration with a curve

the x-axis and the two ordinates. A curve expressing some general relation $u = f(x)$ is shown in Fig. 7b.

To find the area enclosed between $x = a$ and $x = b$, the area is divided, as before, into vertical strips of equal width Δx.

The areas of the strips are assumed to be approximately equal to the areas of the trapezia with the same mean heights as those of the central ordinates in the strips. This assumption is an approximation because the top of each strip is not a straight line. The area of a strip ΔI equals $u\,\Delta x$, where u is the mean height of the trapezium. If n is made very large, Δx becomes infinitesimal, and the approximation in the strip area is eliminated, since the top of the strip becomes a straight line. Hence the area of a strip of infinitesimal width dx is

$$dI = u\,dx. \qquad (7.2)$$

The total area becomes the sum of the areas of all such strips between $x = 0$ and $x = a$, and the summation is an integration, since the strips are of infinitesimal width.

The area I between ordinates at $x = a$ and $x = b$, the curve $u = f(x)$ and the x-axis is

$$I = \int_a^b u\,dx. \qquad (7.3)$$

INTEGRATION

This result gives a general method for finding the numerical value of an integral when limits are given (see § 5 on graphical integration). It is, however, an inconvenient process, and in many, though not in all, cases integration can be carried out by regarding it as the reverse process of differentiation.

2. Integration of algebraic functions

The aspect of integration as the reverse of differentiation can be illustrated from the examples used in § 1. From (7.2) and (7.3)

$$dI/dx = u, \quad \text{and} \quad I = \int_a^b u\, dx.$$

The integral I is a function of x, such that when differentiated it gives the function u which is to be integrated. In general terms, putting $I = f(x)$, if $f(x) = \int u\, dx$, then

$$\frac{d}{dx}\{f(x)\} = u.$$

The value of $\int u\, dx$ can therefore be found from a knowledge of differential calculus. It is the function of x which when differentiated gives u. Differentiation of the result of an integration gives the function which is to be integrated. This check should always be applied to the integration of algebraic functions.

As in differentiation, a set of rules can be developed for integrating certain types of functions, but integration is a more difficult process than is differentiation. It is not possible to integrate all functions by the rules, a classical example being $\int \exp(-y^2)\, dy$, which is discussed in Chap. 9, § 3. Refractory integrals such as this are solved either graphically or by computation for given numerical values of the limits.

3. Rules for the integration of algebraic functions

These rules have been built up from differential calculus, and some of the more important will be derived in the following pages. In all these rules, a, n and C are constants.

Rule 1. $\int x^n\, dx$. The integral of $x^n\, dx$ can be solved by using (5.3); this equation shows that the function which when differentiated gives x^n is

$$\frac{x^{n+1}}{n+1}.$$

EXAMPLE. $\int x^3\, dx = \dfrac{x^{3+1}}{3+1} = \tfrac{1}{4}x^4$.

Differentiation of this integral will give the original function, showing that the integration is correct.

Suppose $y = x^3 + 2$. Then differentiation of this function with respect to x will give $dy/dx = 3x^2$. Integration of the latter with respect to x (see Rule 3) will give

$$\int 3x^2 \, dx = \frac{3x^{2+1}}{2+1} = x^3.$$

It is seen that the constant 2 in the original function has now disappeared. The differential coefficient of a constant is zero, and this fact must be allowed for in the integration. Therefore, Rule 1 becomes

$$\int x^n \, dx = \frac{x^{n+1}}{n+1} + C, \tag{7.4}$$

where C is an unknown constant. Hence $\int 3x^2 \, dx$ is $x^3 + C$.

An integral containing an unknown constant is called an *indefinite integral*. The integral becomes *definite* when limits are applied, say a and b. The value of the integral will then be equal to the difference between the values of the integral with the upper (b) and lower (a) limits substituted for x: thus

$$\int_a^b x^n \, dx = \left[\frac{x^{n+1}}{n+1} + C\right]_a^b = \left(\frac{b^{n+1}}{n+1} + C\right) - \left(\frac{a^{n+1}}{n+1} + C\right)$$
$$= \frac{b^{n+1} - a^{n+1}}{n+1}.$$

It should be noted that the unknown constant disappears in the subtraction and therefore the integral is definite. The use of square brackets and the method of expressing the limits should also be noted.

There is an important exception to Rule 1 when $n = -1$. Applying the rule

$$\int x^{-1} \, dx = \frac{x^{-1+1}}{-1+1} = \frac{x^0}{0} = \frac{1}{0} = \infty.$$

This integral, however, is known to be finite when the limits are non-zero and positive. Rule 1 therefore is not applicable in this case.

Rule 2. $\int x^{-1} \, dx$. From (5.5), the function which when differentiated gives x^{-1} is $\ln x$. Therefore the integral of x^{-1} is $\ln x$, i.e.

$$\int x^{-1} \, dx = \ln x + C. \tag{7.5}$$

The unknown constant C has to be introduced as with all indefinite integrals.

Rule 3. $\int ax^n \, dx$. The answer to this integral is obtained from (5.3), since the function which on differentiation gives ax^n is $ax^{n+1}/(n+1)$:

$$\int ax^n \, dx = \frac{ax^{n+1}}{n+1} + C. \tag{7.6}$$

INTEGRATION

It is a general rule of integration that a constant factor may be taken outside the integration sign: thus

$$\int ax^n \, dx = a\int x^n \, dx = a\left(\frac{x^{n+1}}{n+1}\right) + C.$$

EXAMPLES OF RULES 1 AND 3

Evaluation of the following integrals illustrates the application of these rules.

(i) $\int_2^3 x^4 \, dx = \left[\frac{x^5}{5} + C\right]_2^3 = \left(\frac{3^5}{5} + C\right) - \left(\frac{2^5}{5} + C\right) = \frac{3^5 - 2^5}{5} = 42 \cdot 2.$

(ii) $\int_1^3 x^3 \, dx = \left[\frac{x^4}{4} + C\right]_1^3 = \left(\frac{3^4}{4} + C\right) - \left(\frac{1^4}{4} + C\right) = \frac{3^4 - 1^4}{4} = 20.$

It should be noticed in these examples that the constant term C has disappeared in the subtraction of the two terms.

(iii) $\int x^{-2} \, dx = \frac{x^{-2+1}}{-2+1} + C = \frac{x^{-1}}{-1} + C = -\frac{1}{x} + C.$

(iv) $\int 6x^6 \, dx = 6\int x^6 \, dx = \frac{6x^7}{7} + C.$

In this example, $a = 6$ and $n = 6$. It should be noted that the coefficient of x^6, being a constant, has been taken outside the integration sign. The validity of this integral can be checked by differentiation.

(v) $\int_1^3 2x^2 \, dx = 2\int_1^3 x^2 \, dx = 2\left[\frac{x^3}{3} + C\right]_1^3$

$= 2\left(\frac{3^3}{3} + C\right) - 2\left(\frac{1^3}{3} + C\right)$

$= 2\left(\frac{27}{3} + C\right) - 2\left(\frac{1}{3} + C\right)$

$= 17 \cdot 33.$

Rule 4. $\int \ln x \, dx$. The integral of $\ln x \, dx$ is $\{x \ln(x) - x\}$. This may be easily confirmed by differentiating $\{x \ln(x) - x\}$ by application of Rule 6 for the differentiation of a product. If $y = x \ln(x) - x$, or $x\{\ln(x) - 1\}$, then from (5.7) differentiation will give

$$\frac{dy}{dx} = x \cdot \frac{1}{x} + \{\ln(x) - 1\}1$$

$$= 1 + \ln(x) - 1 = \ln x.$$

Therefore

$$\int \ln x \, dx = x \ln(x) - x + C. \tag{7.7}$$

Rule 5. $\int e^x \, dx$ (or $\int \exp x \, dx$). It is seen from (5.4) that the function which when differentiated gives e^x is e^x itself. Therefore

$$\int e^x \, dx = e^x + C,$$

or written in the alternative form

$$\int \exp x \, dx = \exp(x) + C. \tag{7.8}$$

Rule 6. $\int a^x \, dx$. It should be noted that the integral of a^x is not a^x (compare Rule 5), but is, in fact, $a^x/\ln a + C$, i.e.

$$\int a^x \, dx = \frac{a^x}{\ln a} + C. \tag{7.9}$$

In order to prove this result, let $u = a^x$. Then $\ln u = x \ln a$, and hence $x = \ln u/\ln a$; differentiating x with respect to u with $\ln a$ constant gives:

$$\frac{dx}{du} = \frac{1}{u} \cdot \frac{1}{\ln a}$$

or

$$\frac{du}{dx} = u \ln a = a^x \ln a$$

therefore

$$\frac{d}{dx}(a^x) = a^x \ln a$$

and

$$\frac{d}{dx}\left(\frac{a^x}{\ln a}\right) = a^x,$$

proving the integration formula, (7·9).

Rule 7. $\int (u+v+w) \, dx$ (where u, v and w are functions of x). The integral of the sum of a number of functions of x is equal to the sum of the integrals of the separate functions: thus

$$\int (u+v+w) \, dx = \int u \, dx + \int v \, dx + \int w \, dx. \tag{7.10}$$

EXAMPLE.
$$\int (x^2 + 7x + 3) \, dx = \int x^2 \, dx + \int 7x \, dx + \int 3 \, dx$$
$$= \frac{x^3}{3} + \frac{7x^2}{2} + 3x + C.$$

Some further rules for integration are given in Chap. 8 and Appendix III.

4. Reduction of integrals to a standard form

Substitution (changing the variable). The process of algebraic integration can often be carried out by reducing an otherwise intractable integral to a standard form. One of the devices that can be used involves changing the variable. As the new variable must be substituted in both the function to be integrated and the differential, this limits the usefulness of the method.

EXAMPLE 1. *Evaluate* $\int \exp(3x+4) \, dx$.

Let $u = 3x + 4$. Then $du/dx = 3$, i.e. $dx = \frac{1}{3}du$. Substitution for both $(3x+4)$ and dx gives

$\int \exp(3x+4) \, dx = \int \exp u \cdot \frac{1}{3} du = \frac{1}{3} \int \exp u \, du = \frac{1}{3} \exp u + C.$ (Rule 5)

Substituting $(3x+4)$ for u, then

$$\int \exp(3x+4) \, dx = \frac{1}{3} \exp(3x+4) + C.$$

INTEGRATION

EXAMPLE 2. *Evaluate* $\int \exp(-x^2) \, dx$.

Let $u = -x^2$. Then $du/dx = -2x$, i.e. $dx = -\tfrac{1}{2} du/x$. Substitution for both $-x^2$ and dx gives

$$\int \exp(-x^2) \, dx = \int \exp u \cdot -\tfrac{1}{2} du/x = -\tfrac{1}{2}\int \exp u \, du/x.$$

In this example the integral has been complicated rather than simplified, since it now contains two variables, x and u. In general, if u is a linear function of x, then substitution will simplify the integral. If u is not a linear function of x, then substitution may or may not help.

EXAMPLE 3. *Evaluate* $\int x \, dx/(x^2+4)$.

Let $u = x^2+4$. Then $du/dx = 2x$, i.e. $du = 2x \, dx$ and hence $x \, dx = \tfrac{1}{2} du$. Substitution for both (x^2+4) and $x \, dx$ gives

$$\int x \, dx/(x^2+4) = \tfrac{1}{2}\int du/u = \tfrac{1}{2}\int u^{-1} \, du = \tfrac{1}{2} \ln(u) + C.$$

Substituting (x^2+4) for u, then

$$\int x \, dx/(x^2+4) = \tfrac{1}{2} \ln(x^2+4) + C.$$

This is an example in which non-linear substitution succeeds.

Integration by parts. This technique for integrating is expressed by the equation

$$\int u \, dv = uv - \int v \, du, \qquad (7.11)$$

u and v being functions of the same variable x.

This technique is used when $\int v \, du$ is more easily solved than is $\int u \, dv$. The deduction of this formula is based upon Rule 6 for the differentiation of a product; thus from (5.7)

$$\frac{d(uv)}{dx} = u\frac{dv}{dx} + v\frac{du}{dx}.$$

Multiplication of both sides of the equation by dx gives

$$d(uv) = u \, dv + v \, du.$$

Hence

$$u \, dv = d(uv) - v \, du,$$

and

$$\int u \, dv = uv - \int v \, du.$$

EXAMPLE. *Evaluate* $\int x \ln(x) \, dx$.

Let $u = \ln x$. Then $du/dx = 1/x$, i.e. $du = dx/x$. Let $dv = x \, dx$. Then $v = \int x \, dx = \tfrac{1}{2} x^2$, taking the simplest value (introduction of an integration constant here does not affect the final value of the integral, since the constant cancels out). Substitution of these values for v and du in (7.11) gives

$$\begin{aligned}\int x \ln(x) \, dx &= \ln x \cdot \tfrac{1}{2} x^2 - \int \tfrac{1}{2} x^2 \, dx/x \\ &= \tfrac{1}{2} x^2 \ln(x) - \tfrac{1}{2}\int x \, dx \\ &= \tfrac{1}{2} x^2 \ln(x) - \tfrac{1}{2} \cdot \tfrac{1}{2} x^2 + C \\ &= \tfrac{1}{2} x^2 \ln(x) - \tfrac{1}{4} x^2 + C.\end{aligned}$$

Integration by partial fractions. If the function to be integrated is a 'proper' algebraic fraction, i.e. the index of the denominator is higher than that of the numerator, and the denominator can be split into two or more simpler factors, then it will usually be possible to integrate this function using the technique of partial fractions. Thus, suppose $\int dx/(x^2-a^2)$ is to be evaluated. Then

$$\frac{1}{(x^2-a^2)} \equiv \frac{1}{(x-a)(x+a)} \equiv \frac{A}{(x-a)} + \frac{B}{(x+a)}.$$

The terms $A/(x-a)$ and $B/(x+a)$ are called partial fractions, since their sum will equal the original fraction. The values of the coefficients A and B are found by multiplying out the partial fractions: thus

$$\frac{A}{(x-a)} + \frac{B}{(x+a)} \equiv \frac{A(x+a)+B(x-a)}{(x-a)(x+a)} \equiv \frac{1}{(x-a)(x+a)}.$$

Since the denominators of the second and third expressions are the same, the numerators must be identical, i.e. the coefficient of x and the constant term are the same on both sides of the identity, i.e.

$$A(x+a)+B(x-a) \equiv 1.$$

Hence

$$(A+B)x+(A-B)a \equiv 1.$$

Since there is no coefficient of x on the right-hand side of the identity and the constant term is 1, then $A+B = 0$ and $(A-B)a = 1$. Therefore $A-B = 1/a$, or since $A = -B$, then $B = -1/2a$ and $A = 1/2a$. Substitution of these values of A and B gives

$$\frac{1}{(x-a)(x+a)} \equiv \frac{1}{2a(x-a)} - \frac{1}{2a(x+a)}.$$

This identity can be confirmed by multiplying out the right-hand side.

Returning to the original evaluation, it follows that

$$\int dx/(x^2-a^2) = (1/2a)\int dx/(x-a) - (1/2a)\int dx/(x+a)$$
$$= \frac{1}{2a} \ln \frac{(x-a)}{(x+a)} + C.$$

It should be noted that the partial fraction integrals are solved by the substitution method, i.e. by putting, for example, $u = x+a$, so that $du = dx$ and hence $\int du/u = \int u^{-1} du = \ln(u) + C$.

5. Graphical integration

It is not always possible to solve an integral by any of the rules that have been developed, e.g. $\int \exp(-x^2) dx$. However, the numerical value of a definite integral can be found by graphical methods. If the integral concerned is $\int_a^b f(x)\,dx$, then it may be evaluated by plotting f(x) against x and measuring the area enclosed between the curve, the ordinates at $x = a$ and $x = b$, and the x-axis (see Fig. 7c). This area can be found by

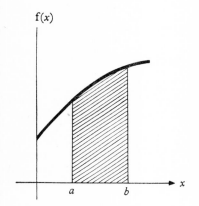

FIG. 7c. Graphical integration

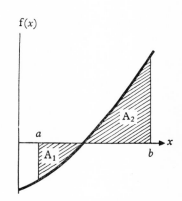

FIG. 7d. Integration with a curve crossing the axis

counting the squares of the graph paper or by using special graph paper of known uniform weight per unit area, cutting out the area representing the integral and weighing it.

The sign convention for graphical integration is important. If the upper limit b is greater than the lower limit a, then the integral is positive if the area lies above the x-axis and negative if below it. If a is greater than b, the signs are reversed. If the curve of f(x) crosses the x-axis between the limits, as in Fig. 7d, the integral has a positive and a negative part, and its net value is the magnitude of the area above the x-axis minus the magnitude of the area below it. Thus, in Fig. 7d, $\int_a^b f(x)\,dx = A_2 - A_1$. If f(x) is many-valued or is discontinuous between the limits, the graphical method cannot be used.

6. Mean value of a function

The mean value of a function, $y = f(x)$ between limits $x = a$ and $x = b$, can be estimated by integration.

Consider the abscissa between the limits divided into strips of equal width as in Fig. 7e so that there are n ordinates, then the mean value of y between the limits is approximately equal to the sum of the ordinates divided by the number of ordinates, n. As n is made larger the approximation improves.

$$\bar{y} = \frac{y_1+y_2+y_3+\ldots+y_n}{n}$$

$$= \sum_{r=1}^{n} (y_r)/n.$$

If numerator and denominator are multiplied by the strip width, Δx

$$\bar{y} = \sum_{r=1}^{n} y_r \Delta x / n\Delta x.$$

If n is now made very large, Δx becomes infinitesimal and is denoted as dx, and the summation becomes an integral between limits. If the limits are a and b, then $n\Delta x$ is the distance between the limits, $(b-a)$. Thus

$$\bar{y} = \int_a^b \frac{y\,dx}{(b-a)} = \frac{1}{(b-a)} \int_a^b y\,dx$$

That is, the mean value of the function between the limits is the integral between the limits divided by the difference between the limits.

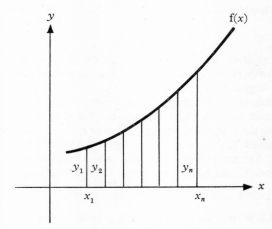

Fig. 7e. Integration by trapezium rule

7. Integration by computation

Any integral with limits, which is finite within the limits, may be evaluated by numerical methods. The simplest method is based on finding the area under the curve by dividing it into strips of equal width—see Fig. 7e. The curve represents the function to be integrated, $f(x)$; $I = \int_{x_1}^{x_n} f(x)\,dx$,

INTEGRATION

is approximately equal to the sum of the areas of the trapezia between the limits. If the width of each strip is h, then its area equals width multiplied by mean height and the integral, I, is

$$I = h[(y_1+y_2)/2+(y_2+y_3)/2+(y_3+y_4)/2+ \ldots$$
$$\ldots +(y_{n-2}+y_{n-1})/2+(y_{n-1}+y_n)/2]$$
$$= h[(y_1+y_n)/2+y_2+y_3+\ldots+y_{n-1}]$$

where y_1, y_2, etc. are the heights of the co-ordinates dividing the trapezia. All the values of y are counted twice except the first and the last, y_1 and y_n.

This trapezium rule is approximate because the top sides of the strips are not straight lines but are sections of the curve. By reducing the width h, the error can be reduced.

For a given value of h a more accurate result is obtained by Simpson's rule, which makes a first order correction for the curvature of the tops of the strips. Simpson's rule requires division of the area into an even number of strips giving an odd number of ordinates, y. The summed area is then

$$I = \frac{h}{3}(y_1+y_n+4\Sigma y_e+2\Sigma y_o).$$

Σy_e is the sum of all the even ordinates and Σy_o is the sum of the odd ordinates, leaving out y_1 and y_n.

Simpson's rule is normally used in integration by computing. If the function to be integrated is known as an algebraic expression, the values of the ordinates can be calculated successively in a DO loop. If the problem is to determine the area under an experimental curve, the ordinates are measured from the curve and entered on a data card to run with the programme.

COMP 4: Programme to integrate a function by Simpson's rule. The integral which is evaluated in this programme is the probability integral, between limits ± 1.

$$\text{PROB} = \frac{1}{\sqrt{(2\pi)}} \int_{-1}^{1} \exp(-u^2/2) \, du.$$

This function cannot be integrated to give an algebraic expression other than an infinite series. The programme starts with $u = -1$ and uses a strip width $H = 0.1$ to give twenty-one values of the ordinate, Y. The sum of the even values of Y, YE, and the sum of the odd values except the first and last, YO, are computed and put into the Simpson's rule formula to give the integral.

Greater accuracy would result from reducing the increment H; it is quite straightforward to devise a programme which would carry out the calculations with successively decreasing values of H until two successive estimates of the integral agreed to within some predetermined limit, for example, 10^{-6}.

```
C         COMP 4
C         PROBABILITY INTEGRAL BY SIMPSONS RULE
          DIMENSION Y(21)
          U = -1.
          H = 0.1
          Y(1) = EXP(-0.5)
          DO 1 I = 2,21
          U = U + H
          USQ = -(U**2)/2.
1         Y(I) = EXP(USQ)
          YE = 0.
          YO = 0.
          DO 2 I = 2,20,2
2         YE = YE + Y(I)
          DO 3 I = 3,19,2
3         YO = YO + Y(I)
          TG = (H/3.)*(Y(1) + 4.*YE + 2.*YO + Y(21))
          PROB = TG/SQRT(2.*3.14159)
          WRITE(6,4) PROB
4         FORMAT(14H1PROBABILITY =,F10.5)
          STOP
          END
```

First of all Y is dimensioned and then the values of the lower limit of U and of the increment H are given. Y(1) is calculated as $\exp(-(-1)^2/2)$ and then the other 20 values of Y are calculated by the first DO; on each cycle, U is increased by the increment H.

In the second DO the even ordinates are summed. In summation the cell for the sum should first be cleared to zero; a statement YE = will clear the register for YE but with YE = YE+Y(I) any rubbish which might be in YE will be added into the sum and to avoid this possibility YE is set to zero. This DO adds the even values of Y by making I the subscript to Y and starting with I = 2, altering I by 2 for each cycle up to I = 20.

The third DO calculates the sum of the odd ordinates by a similar method. They are then used in the Simpson's rule formula to give TG. The quantity required, PROB, is TG divided by $\sqrt{(2\pi)}$. In this programme the numerical value of π has been put in; it could have been obtained from the supplied function ATAN( ) which gives the angle in radians whose tangent is put in the brackets. The angle whose tangent is 1 is $\pi/4$ (Chap. 8) and so $\pi = 4\cdot\text{*ATAN}(1\cdot)$.

The output is

        PROBABILITY =   .68269

COMP 5: Programme to determine the area under a curve by Simpson's rule.

```
C       COMP 5
C       AREA UNDER CURVE BY SIMPSONS RULE
        DIMENSION Y(15)
        READ(5,1) H,Y
1       FORMAT(16F5.0)
        YE = 0.
        YO = 0.
        DO 2 I = 2,14,2
2       YE = YE + Y(I)
        DO 3 I = 3,13,2
3       YO = YO + Y(I)
        TG = (H/3.0)*(Y(1) + Y(15) + 4.*YE + 2.*YO)
        WRITE(6,4) TG
4       FORMAT(12H1INTEGRAL = ,F10.4)
        STOP
        END
```

The area is divided into fourteen strips each of width H and the heights of the fifteen ordinates Y are measured. If the total area under a curve such as that shown in Fig. 7f is required, the first and last ordinates should be zero in order to ensure that the whole area is included.

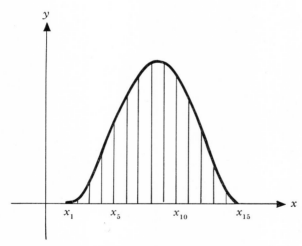

FIG. 7f. Area under a chromatographic peak by Simpson's rule with 15 ordinates, $y_1 = y_{15} = 0$

H and Y in appropriate units are read in from the data card with 5 columns on the card allotted to each value, which must have a decimal point. The summations are carried out as in COMP 4 and the integral evaluated. The WRITE format gives a new page with the title INTEGRAL = the value computed.

The output from COMP 5 is given in the answer to Problem 6 of Chapter 7.

Problems

1. Plot the curve representing $y = 1/x$ between $x = 0.5$ and $x = 12$. By graphical integration, find the value of ln (10) given that $\ln(10) = \int_1^{10} dx/x$.

2. Evaluate
    (i) $\int (x^3 + 3x^2 + 8) \, dx$ (ii) $\int 2x \, dx/(9+x^2)$
    (iii) $\int dx/(x^2 - 9)$ (iv) $\int \exp(-\tfrac{1}{2}x) \, dx$
    (v) $\int x \exp(x) \, dx$ (vi) $\int x \exp(x^2) \, dx$
    (vii) $\int_0^2 (x+1)(x^3-3) \, dx$ (viii) $\int_0^1 \exp(x) \, dx$
    (ix) $\int_1^{10} dx/x$ (x) $\int_1^3 x^2 \ln(x) \, dx$
    (xi) $\int_0^1 dx/(x+1)(x+2)$ (xii) $\int_0^5 (6+8x-3x^2) \, dx$.

3. Calculate the mean values of the following functions between $x = 0$ and $x = 3$.
    (i) $\exp(-x)$ (ii) $\ln(2x+3)$ (iii) $x \exp(-x^2)$.

4. Write a programme to evaluate
$$\int_0^{20} x^2 \ln(x+3) \, dx,$$
using Simpson's rule with 20 strips.

5. Run COMP 4 to evaluate the probability integral between limits ± 0.5, with 20 strips.

6. The fifteen ordinates for the chromatographic peak shown in Fig. 7f with $h = 0.05$ units are
   0·0   2·04   6·24   14·81   25·74   35·13   43·41   47·75   50·31   46·67   35·85   20·83   9·14   3·21   0·0

   Use COMP 5 to evaluate the area under the peak in units of y multiplied by units of x.

CHAPTER 8

Trigonometry

TRIGONOMETRY is the branch of mathematics which deals with the relations between the sides and angles of triangles. It plays an important part in modern chemical theory, since trigonometric functions provide solutions to a number of integrals, particularly those connected with oscillations and wave motion.

1. Trigonometric functions

The most familiar of these functions are the sine, cosine and tangent of an angle. These are defined in terms of the ratios of the sides of a right-angled triangle containing the angle.

In the right-angled triangle ABC (see Fig. 8a), $\widehat{C}$ is a right angle (the symbol $\widehat{C}$ means the angle at C within the triangle, i.e. it is the angle between the lines AC and BC) and $\widehat{B}$ is θ radians. The ratio of the length of

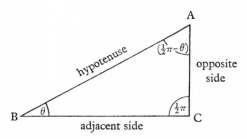

FIG. 8a. Right-angled triangle

the side AC opposite to θ to the length of the hypotenuse AB is independent of the size of the triangle, provided that the angles remain the same, and this ratio is a function only of the magnitude of the angle θ. This ratio is called the *sine* of θ, and is written sin θ; thus

$$\frac{AC}{AB} = \frac{\text{opposite side}}{\text{hypotenuse}} = \sin \theta.$$

Similarly, the ratio of the length of the side BC adjacent to θ to the length of the hypotenuse AB is also a function only of θ; this ratio is called the *cosine* of θ and is written cos θ; thus

$$\frac{BC}{AB} = \frac{\text{adjacent side}}{\text{hypotenuse}} = \cos\theta.$$

The ratio of the length of the opposite side AC to the length of the adjacent side BC is also a function only of θ; this ratio is called the *tangent* of θ and is written tan θ; thus

$$\frac{AC}{BC} = \frac{\text{opposite side}}{\text{adjacent side}} = \tan\theta = \frac{AC/AB}{BC/AB} = \frac{\sin\theta}{\cos\theta}.$$

Tables giving the values of these functions for different values of θ were compiled many years ago for use in navigation and surveying.

2. Relations between trigonometric functions

Application of Pythagoras' theorem (see Appendix II) to the triangle ABC allows the relation between the sine and cosine of an angle to be deduced. Thus

$$(BC)^2 + (AC)^2 = (AB)^2.$$

Dividing throughout by $(AB)^2$ gives

$$\left(\frac{BC}{AB}\right)^2 + \left(\frac{AC}{AB}\right)^2 = 1.$$

Hence

$$(\cos\theta)^2 + (\sin\theta)^2 = 1.$$

By convention, the squares of trigonometric functions, e.g. $(\sin\theta)^2$, are written as $\sin^2\theta$. Therefore

$$\cos^2\theta + \sin^2\theta = 1. \tag{8.1}$$

Alternatively, dividing the original equation throughout by $(BC)^2$ gives

$$1 + \left(\frac{AC}{BC}\right)^2 = \left(\frac{AB}{BC}\right)^2.$$

Hence

$$1 + \tan^2\theta = \frac{1}{\cos^2\theta}. \tag{8.2}$$

The reciprocals of sine, cosine and tangent occur frequently, and these are named *cosecant*, *secant* and *cotangent* respectively, which are abbreviated respectively to cosec, sec and cot. Thus

$$\frac{1}{\sin\theta} = \operatorname{cosec}\theta; \quad \frac{1}{\cos\theta} = \sec\theta; \quad \frac{1}{\tan\theta} = \cot\theta.$$

Therefore (8.2) can be written as

$$1 + \tan^2\theta = \sec^2\theta. \tag{8.3}$$

TRIGONOMETRY

Although $1/\sin\theta$ is $(\sin\theta)^{-1}$, it is not written as $\sin^{-1}\theta$, since this has been given an entirely different meaning, as is explained in § 10. This restriction also applies to $1/\cos\theta$ and $1/\tan\theta$.

3. Angular measure

If a line turns through a complete revolution (see Fig. 8b), it is said, according to an old Babylonian definition of angles, to have turned through 360°. A quarter of a turn is therefore 90°. Each degree is divided into 60 minutes, and each minute is further divided into 60 seconds. For mathematical purposes, it is more convenient to use circular measure for

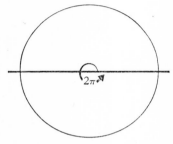

FIG. 8b. A complete revolution

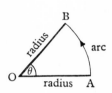

FIG. 8c. Angular measure

angles. The unit of circular measure is the *radian*, which is defined as the angle which is subtended at the centre of a circle by an arc equal in length to the radius of the circle. The size of any angle θ (see Fig. 8c) is measured by the ratio of the curved length of the arc AB to the radius of the circle OA. This ratio is independent of the size of the circle. Thus

$$\theta = \frac{\text{arc AB}}{\text{radius OA}} \text{ radians}.$$

If the line OA rotates, with its centre at O, in an anticlockwise direction through a complete revolution, the angle generated, in radians, is equal to the circumference of the circle divided by the radius. Hence

$$1 \text{ complete revolution } (360°) = \frac{\text{circumference}}{\text{radius}} \text{ radians}$$

$$= \frac{2\pi r}{r} \text{ radians, where } r \text{ is the radius.}$$

Therefore

$$360° = 2\pi \text{ radians}; \quad 180° = \pi \text{ radians}; \quad 90° = \tfrac{1}{2}\pi \text{ radians};$$
$$60° = \tfrac{1}{3}\pi \text{ radians}; \quad 30° = \tfrac{1}{6}\pi \text{ radians}.$$

Since $\pi = 3.142$, then

$$1 \text{ radian} = \left(\frac{360}{2\pi}\right)^\circ = \left(\frac{360}{6.284}\right)^\circ = 57.3^\circ.$$

For the remainder of this chapter it is to be assumed that angles are measured in radians.

4. Variation of sin θ, cos θ, and tan θ with θ

If the line OA (see Fig. 8d) rotates with O as centre in an anticlockwise direction to a position OB, a positive angle θ is generated (by convention, positive angles are generated by anticlockwise rotation). This angle θ is

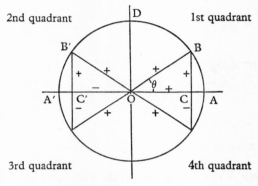

FIG. 8d. Variation of sine and cosine with angle

subtended by the arc AB. The values of the trigonometric functions of θ, e.g. sin θ, can be found by dropping a perpendicular from B to OA, cutting OA at C. Then

$$\sin \theta = \frac{BC}{OB} = \frac{BC}{r}; \quad \cos \theta = \frac{OC}{OB} = \frac{OC}{r}; \quad \tan \theta = \frac{BC}{OC},$$

where r is the radius. As OA rotates to a position OD, the angle θ is increased from 0 to $\frac{1}{2}\pi$ radians, i.e. to a right angle, BC increases from 0 to r, and sin θ therefore increases from 0 to 1. At the same time OC decreases from r to 0, and cos θ therefore decreases from 1 to 0. When $\theta = 0$, tan θ = BC/OC = 0/1 = 0; when $\theta = \frac{1}{2}\pi$, tan $\theta = 1/0 = \infty$.

When the line OA has rotated through an angle greater than $\frac{1}{2}\pi$ it enters the second quadrant, and in completing the circle it passes successively through the 3rd and 4th quadrants. In order to calculate the trigonometric functions of θ in each quadrant, it is necessary to develop a sign convention: thus

(i) r is positive in all quadrants,

(ii) the perpendicular BC is positive if measured above the line AOA' and negative if measured below this line, and
(iii) OC is positive towards A and negative towards A'.

This sign convention is illustrated in Fig. 8d, and the variations in sin θ, cos θ and tan θ with changes in θ are given in Table 1.

TABLE 1
Variations in trigonometric functions

	θ	sin θ	cos θ	tan θ
1st quadrant	$0 \to \tfrac{1}{2}\pi$	$0 \to 1$	$1 \to 0$	$0 \to \infty$
2nd quadrant	$\tfrac{1}{2}\pi \to \pi$	$1 \to 0$	$0 \to -1$	$-\infty \to 0$
3rd quadrant	$\pi \to \tfrac{3}{2}\pi$	$0 \to -1$	$-1 \to 0$	$0 \to \infty$
4th quadrant	$\tfrac{3}{2}\pi \to 2\pi$	$-1 \to 0$	$0 \to 1$	$-\infty \to 0$

After one complete revolution from 0 to 2π, the values repeat themselves. In consequence, trigonometric functions are said to be periodic functions of the variable θ.

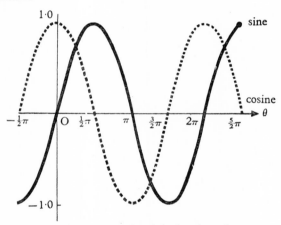

FIG. 8e. Graphs of sin θ and cos θ

A plot of the values of sin θ and cos θ against θ is shown in Fig. 8e; the values of the ratios are shown on the vertical axis and those of θ, in radians, are on the horizontal axis. These graphs should be compared with the values shown in Table 1. The graph of cos θ lags $\tfrac{1}{2}\pi$ radians behind that of sin θ. Hence

$$\sin \theta = \cos (\theta - \tfrac{1}{2}\pi) = \cos (\tfrac{1}{2}\pi - \theta).$$

This result can also be deduced from Fig. 8a, from which it is seen that

$$\cos (\tfrac{1}{2}\pi - \theta) = \frac{AC}{AB} = \sin \theta.$$

Negative angles. If the line OA in Fig. 8d rotates in a clockwise direction, the angle generated θ is negative, and the line OA enters the 4th quadrant first. From this, it can be deduced that

$$\sin(-\theta) = -\sin\theta, \qquad \cos(-\theta) = \cos\theta.$$

Small angles. When θ is small, the perpendicular BC (see Fig. 8d) approximates to the length of the arc AB, so that as $\theta \to 0$, then:

(i) $\qquad \sin\theta = \dfrac{BC}{r} \to \dfrac{\text{arc AB}}{r} = \theta \quad$ (measured in radians),

i.e. as $\theta \to 0$

$$\sin\theta \to \theta. \tag{8.4}$$

(ii) $\qquad \cos\theta = \dfrac{OC}{r} \to \dfrac{r}{r} = 1,$

i.e. as $\theta \to 0$

$$\cos\theta \to 1. \tag{8.5}$$

(iii) $\qquad \tan\theta = \dfrac{\sin\theta}{\cos\theta} = \dfrac{\theta}{1} = \theta,$

i.e. as $\theta \to 0$

$$\tan\theta \to \theta.$$

5. Sine and cosine of the sum and difference of two angles

The sine and cosine of the sum of two angles can be deduced by means of the geometric construction shown in Fig. 8f. In this figure

$$\widehat{DBE} = \widehat{FAE} = \theta,$$

and $\widehat{BAE} = \phi$. Hence $\widehat{BAC} = \theta + \phi$. Right angles are indicated by '$\sqcap$'.

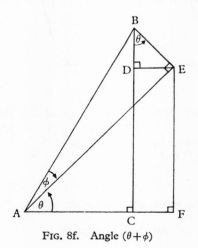

FIG. 8f. Angle $(\theta + \phi)$

***Sin* $(\theta+\phi)$.** In Fig. 8f, $CD = EF = AE \sin\theta$; but $AE = AB\cos\phi$ and hence
$$CD = AB\cos\phi.\sin\theta.$$
Also, $BD = BE\cos\theta$, and $BE = AB\sin\phi$. Hence
$$BD = AB\sin\phi.\cos\theta.$$
Now, in the triangle ABC,
$$\sin(\theta+\phi) = \frac{BC}{AB} = \frac{BD+CD}{AB} = \frac{BD}{AB} + \frac{CD}{AB}$$
$$= \frac{AB\sin\phi.\cos\theta}{AB} + \frac{AB\cos\phi.\sin\theta}{AB}.$$
Therefore
$$\sin(\theta+\phi) = \sin\theta.\cos\phi + \cos\theta.\sin\phi. \tag{8.6}$$

***Cos* $(\theta+\phi)$.** In Fig. 8f, $AF = AE\cos\theta$, and $AE = AB\cos\phi$. Hence
$$AF = AB\cos\phi.\cos\theta.$$
Also, $CF = DE = BE\sin\theta$; but $BE = AB\sin\phi$ and hence
$$CF = AB\sin\phi.\sin\theta.$$
Now, in the triangle ABC,
$$\cos(\theta+\phi) = \frac{AC}{AB} = \frac{AF-CF}{AB} = \frac{AF}{AB} - \frac{CF}{AB}$$
$$= \frac{AB\cos\phi.\cos\theta}{AB} - \frac{AB\sin\phi.\sin\theta}{AB}.$$
Therefore
$$\cos(\theta+\phi) = \cos\theta.\cos\phi - \sin\theta.\sin\phi. \tag{8.7}$$

***Sin* $(\theta-\phi)$ *and cos* $(\theta-\phi)$.** The sine and cosine of the difference of two angles can be deduced by means of the geometric construction shown in Fig. 8g. In this figure $D\widehat{A}F = B\widehat{D}E = \theta$, and $D\widehat{A}B = \phi$. Hence $B\widehat{A}C = \theta-\phi$. By a similar reasoning to that used to obtain the sine and cosine of the sum of two angles it can be shown that
$$\sin(\theta-\phi) = \sin\theta.\cos\phi - \cos\theta.\sin\phi, \tag{8.8}$$
and
$$\cos(\theta-\phi) = \cos\theta.\cos\phi + \sin\theta.\sin\phi. \tag{8.9}$$
Further, if $\theta = \phi$, substitution in (8.6) and (8.7) gives
$$\sin 2\theta = 2\sin\theta.\cos\theta. \tag{8.10}$$
and
$$\cos 2\theta = \cos^2\theta - \sin^2\theta. \tag{8.11}$$

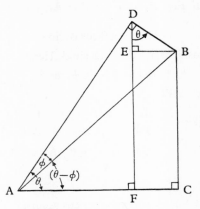

FIG. 8g. Angle $(\theta-\phi)$

6. Sum and difference of sines and cosines

In order to differentiate the trigonometric functions, it is necessary to develop equations for the difference between two sines or cosines. These can be deduced, together with equations for the sums of sines and cosines, by suitable substitutions in (8.6–8.9).

Let $(\theta+\phi) = \alpha$ and $(\theta-\phi) = \beta$. Then, by addition,

$$2\theta = (\alpha+\beta), \quad \text{i.e.} \quad \theta = \tfrac{1}{2}(\alpha+\beta).$$

Similarly, by subtraction,

$$2\phi = (\alpha-\beta), \quad \text{i.e.} \quad \phi = \tfrac{1}{2}(\alpha-\beta).$$

If these values are substituted in (8.6) and (8.8), addition of the resulting equations gives

$$\sin \alpha + \sin \beta = 2 \sin\left(\frac{\alpha+\beta}{2}\right) \cos\left(\frac{\alpha-\beta}{2}\right), \qquad (8.12)$$

while if (8.8) is subtracted from (8.6), the result is

$$\sin \alpha - \sin \beta = 2 \cos\left(\frac{\alpha+\beta}{2}\right) \sin\left(\frac{\alpha-\beta}{2}\right). \qquad (8.13)$$

Further, from (8.7) and (8.9) by similar processes of addition and substitution, it may be shown that

$$\cos \alpha + \cos \beta = 2 \cos\left(\frac{\alpha+\beta}{2}\right) \cos\left(\frac{\alpha-\beta}{2}\right), \qquad (8.14)$$

while substitution and subtraction gives

$$\cos \alpha - \cos \beta = -2 \sin\left(\frac{\alpha+\beta}{2}\right) \sin\left(\frac{\alpha-\beta}{2}\right). \qquad (8.15)$$

TRIGONOMETRY

7. Derivatives of the trigonometric functions

The derivatives of the trigonometric functions are found by the increment method described in Chap. 5, § 2.

$y = \sin \theta$. If θ is increased by $\Delta\theta$ and Δy is the corresponding increment in y, then

$$y + \Delta y = \sin(\theta + \Delta\theta).$$

Subtracting $y = \sin\theta$, it follows that

$$\Delta y = \sin(\theta + \Delta\theta) - \sin\theta.$$

Substituting $(\theta + \Delta\theta)$ for α and θ for β in (8.13) gives

$$\sin(\theta + \Delta\theta) - \sin\theta = 2\cos\left(\frac{\theta + \Delta\theta + \theta}{2}\right)\sin\left(\frac{\theta + \Delta\theta - \theta}{2}\right)$$
$$= 2\cos(\theta + \tfrac{1}{2}\Delta\theta)\sin\tfrac{1}{2}\Delta\theta.$$

As $\Delta\theta$ is made infinitesimal, $(\theta + \tfrac{1}{2}\Delta\theta) \to \theta$. Also, from (8.4), it follows that $\sin\tfrac{1}{2}\Delta\theta \to \tfrac{1}{2}\Delta\theta$. Hence, as $\Delta\theta \to 0$,

$$\Delta y \to 2\cos\theta \cdot \tfrac{1}{2}\Delta\theta,$$

and therefore

$$\Delta y/\Delta\theta \to \cos\theta.$$

Thus, if $y = \sin\theta$,

$$dy/d\theta = \cos\theta. \tag{8.16}$$

$y = \cos \theta$. If θ is increased by $\Delta\theta$ and Δy is the corresponding increment in y, then it can be shown by a similar reasoning to that used for the sine function that

$$\Delta y = \cos(\theta + \Delta\theta) - \cos\theta.$$

Substitution of $(\theta + \Delta\theta)$ for α and θ for β in (8.15) gives

$$\Delta y = -2\sin(\theta + \tfrac{1}{2}\Delta\theta)\sin\tfrac{1}{2}\Delta\theta.$$

As $\Delta\theta \to 0$, $(\theta + \tfrac{1}{2}\Delta\theta) \to \theta$ and $\sin\tfrac{1}{2}\Delta\theta \to \tfrac{1}{2}\Delta\theta$. Hence

$$\Delta y \to -2\sin\theta \cdot \tfrac{1}{2}\Delta\theta,$$

and therefore

$$\Delta y/\Delta\theta \to -\sin\theta.$$

Thus, if $y = \cos\theta$,

$$dy/d\theta = -\sin\theta. \tag{8.17}$$

$y = \tan \theta$. This can be differentiated as a quotient of two functions of θ by application of Rule 7, since

$$\tan\theta = \sin\theta/\cos\theta.$$

Let $\sin\theta = u$ and $\cos\theta = v$. Then from (8.16, 8.17)

$$du/d\theta = \cos\theta, \quad dv/d\theta = -\sin\theta,$$

and, from (5.8),
$$\frac{dy}{d\theta} = \frac{v\,du/d\theta - u\,dv/d\theta}{v^2} = \frac{\cos\theta \cdot \cos\theta - \sin\theta(-\sin\theta)}{\cos^2\theta}$$
$$= \frac{\cos^2\theta + \sin^2\theta}{\cos^2\theta} = \frac{1}{\cos^2\theta}, \quad \text{from (8.1)}$$
$$= \sec^2\theta.$$

Thus, if $y = \tan\theta$,
$$dy/d\theta = \sec^2\theta. \tag{8.18}$$

Higher derivatives. By repeated differentiation of $y = \sin\theta$ and $y = \cos\theta$ using (8.16, 8.17), higher derivatives of the trigonometric functions are obtained.

When $y = \sin\theta$,
$$dy/d\theta = \cos\theta; \quad d^2y/d\theta^2 = -\sin\theta; \quad d^3y/d\theta^3 = -\cos\theta;$$
$$d^4y/d\theta^4 = \sin\theta.$$

When $y = \cos\theta$,
$$dy/d\theta = -\sin\theta; \quad d^2y/d\theta^2 = -\cos\theta; \quad d^3y/d\theta^3 = \sin\theta;$$
$$d^4y/d\theta^4 = \cos\theta.$$

The second derivatives are of particular importance. In both cases it is seen that
$$d^2y/d\theta^2 = -y. \tag{8.19}$$

Hence both $y = \sin\theta$ and $y = \cos\theta$ are functions of θ which satisfy this important differential equation. An equation of the form (8.19) describes the motion of an oscillating body, and its solution, in terms of sines and cosines, introduces these functions into the theory of oscillators (see Chap. 9, § 2).

8. Power series for sine and cosine

The values of the higher derivatives of sine and cosine can be substituted into the equation of Maclaurin's theorem (6.2) to give power series for these functions.

Thus, if $y = f(\theta) = \sin\theta$, then
$$f^i(\theta) = \cos\theta; \quad f^{ii}(\theta) = -\sin\theta; \quad f^{iii}(\theta) = -\cos\theta; \quad f^{iv}(\theta) = \sin\theta.$$

When $\theta = 0$, $f(\theta) = f(0) = 0$ and
$$f^i(0) = 1; \quad f^{ii}(0) = 0; \quad f^{iii}(0) = -1; \quad f^{iv}(0) = 0.$$

Substitution in Maclaurin's theorem gives
$$\sin\theta = \theta - \theta^3/3! + \theta^5/5! - \theta^7/7! + \dots . \tag{8.20}$$

Similarly, it can be shown that
$$\cos\theta = 1 - \theta^2/2! + \theta^4/4! - \theta^6/6! + \dots . \tag{8.21}$$

Exponential forms of sin θ and cos θ. By using the equation for the exponential series (4.10) we get,

$$\exp(i\theta) = 1 + i\theta + i^2\theta^2/2! + i^3\theta^3/3! + i^4\theta^4/4! + i^5\theta^5/5! + \ldots$$

$i = \sqrt{(-1)}$, by definition, therefore $i^2 = -1$, $i^3 = i^2 \times i = -i$, $i^4 = i^2 \times i^2 = 1$, $i^5 = i^4 \times i = i$, the powers of i, i^n, go through a cycle i, -1, $-i$, 1 as n increases by 4. Consequently, substituting for i^n and rearranging the above series in odd and even powers of θ

$$\exp(i\theta) = (1 - \theta^2/2! + \theta^4/4! - \ldots) + i(\theta - \theta^3/3! + \theta^5/5! - \ldots).$$

Comparison with (8.20) and (8.21) gives

$$\exp(i\theta) = \cos\theta + i\sin\theta.$$

Since $\cos\theta$ and $\sin\theta$ are periodic functions of θ exactly repeating their values when θ increases by 2π, $\exp(i\theta)$ is also a periodic function with this same period.

From the above equation it is seen that $\cos\theta$ is the real part of $\exp(i\theta)$ and $\sin\theta$ is the coefficient of the complex part.

It is often advantageous in physical problems involving waves and other periodic effects to use $\exp(i\theta)$ in place of trigonometric functions since it is readily differentiated and integrated.

$$\frac{d}{d\theta}[\exp(i\theta)] = i\exp(i\theta) \qquad \int \exp(i\theta)\,d\theta = \exp(i\theta)/i + C.$$

After integration or differentiation real and complex parts may be separated out by expressing terms containing $\exp(i\theta)$ in terms of sines and cosines.

All the formulae for trigonometric functions and rules for their differentiation and integration may be established by using complex exponents. For example, to prove the equations for $\sin(\theta+\phi)$ and $\cos(\theta+\phi)$, equations (8.6) and (8.7),

$$\begin{aligned}
\cos(\theta+\phi) + i\sin(\theta+\phi) &= \exp[i(\theta+\phi)] = [\exp(i\theta)][\exp(i\phi)] \\
&= (\cos\theta + i\sin\theta)(\cos\phi + i\sin\phi) \\
&= \cos\theta\cos\phi - \sin\theta\sin\phi \\
&\quad + i(\sin\theta\cos\phi + \cos\theta\sin\phi), \text{ since } i^2 = -1
\end{aligned}$$

equating the real and complex parts of the two expressions

$$\cos(\theta+\phi) = \cos\theta\cos\phi - \sin\theta\sin\phi$$
$$\sin(\theta+\phi) = \sin\theta\cos\phi + \cos\theta\sin\phi.$$

9. Integration of the trigonometric functions

The evaluation of the integrals $\sin \theta \, d\theta$ and $\cos \theta \, d\theta$ can be performed directly using a knowledge of the derivatives of sine and cosine. Thus

$$\int \sin \theta \, d\theta = -\cos \theta + C, \tag{8.22}$$

and

$$\int \cos \theta \, d\theta = \sin \theta + C. \tag{8.23}$$

These integrals are proved in the usual way by differentiating the right-hand side of the equation and showing that the result is equal to the function to be integrated.

The integral of $\tan \theta \, d\theta$ may be found by the following method. Since $\tan \theta = \sin \theta / \cos \theta$, then

$$\int \tan \theta \, d\theta = \int \frac{\sin \theta}{\cos \theta} \, d\theta.$$

If $u = \cos \theta$, then $du/d\theta = -\sin \theta$ and hence $\sin \theta \, d\theta = -du$. Therefore

$$\int \tan \theta \, d\theta = \int -du/u = -\ln(u) + C,$$

i.e.

$$\int \tan \theta \, d\theta = -\ln(\cos \theta) + C. \tag{8.24}$$

10. Inverse trigonometric functions

These quantities are written $\sin^{-1} x$, $\cos^{-1} x$, $\tan^{-1} x$, etc., and they are the angles whose sine, cosine, tangent, etc., have the value x. For example, $\sin^{-1}(1)$ is the angle whose sine is 1, i.e. $\frac{1}{2}\pi$ radians. Many other angles, however, also have sines equal to 1, e.g. $\frac{5}{2}\pi$, $\frac{9}{2}\pi$, $\frac{13}{2}\pi$. The inverse functions are therefore many-valued functions of the variable x; the simplest value of an inverse function is called its *principal value*. The importance of these quantities lies in the fact that they appear in the solutions of some standard integrals.

$\int \frac{dx}{\sqrt{(a^2-x^2)}}$. This integral is solved by putting $x = a \sin \theta$, and hence $\theta = \sin^{-1}(x/a)$, i.e. θ is the angle whose sine is x/a. Now,

$$\sqrt{(a^2-x^2)} = \sqrt{(a^2-a^2 \sin^2 \theta)} = a\sqrt{(1-\sin^2 \theta)}$$
$$= a \cos \theta, \quad \text{from (8.1)}.$$

From (8.16), $dx/d\theta = a \cos \theta$, and hence $dx = a \cos \theta \, d\theta$. Therefore

$$\int \frac{dx}{\sqrt{(a^2-x^2)}} = \int \frac{a \cos \theta \, d\theta}{a \cos \theta} = \int d\theta = \theta + C.$$

Therefore
$$\int \frac{dx}{\sqrt{(a^2-x^2)}} = \sin^{-1}(x/a) + C. \qquad (8.25)$$

$\int \frac{dx}{a^2+x^2}$. In this case the integral may be solved by putting $x = a \tan \theta$, and hence $\theta = \tan^{-1}(x/a)$. Now

$$a^2 + x^2 = a^2 + a^2 \tan^2 \theta = a^2 + a^2\left(\frac{1}{\cos^2 \theta} - 1\right), \quad \text{from (8.2)},$$
$$= a^2/\cos^2 \theta.$$

From (8.18), $dx/d\theta = a/\cos^2 \theta$, and hence $dx = a\, d\theta/\cos^2 \theta$. Therefore

$$\int \frac{dx}{a^2+x^2} = \int \frac{a\, d\theta/\cos^2 \theta}{a^2/\cos^2 \theta} = \frac{1}{a}\int d\theta = \frac{\theta}{a} + C.$$

Therefore
$$\int \frac{dx}{a^2+x^2} = \frac{1}{a}\tan^{-1}\left(\frac{x}{a}\right) + C. \qquad (8.26)$$

EXAMPLE 1. *Evaluate* $\int_0^2 \frac{dx}{\sqrt{(4-x^2)}}$.

From (8.25)
$$\int_0^2 \frac{dx}{\sqrt{(4-x^2)}} = \left[\sin^{-1}(\tfrac{1}{2}x)\right]_0^2 = \sin^{-1}(1) - \sin^{-1}(0)$$
$$= \tfrac{1}{2}\pi - 0$$
$$= \tfrac{1}{2}\pi, \quad \text{taking principal values only.}$$

The general solution is $(\tfrac{1}{2}\pi + k\pi)$, where k is an integer.

EXAMPLE 2. *Evaluate* $\int_0^3 \frac{dx}{9+x^2}$.

From (8.26)
$$\int_0^3 \frac{dx}{9+x^2} = \left[\tfrac{1}{3}\tan^{-1}(\tfrac{1}{3}x)\right]_0^3 = \tfrac{1}{3}\{\tan^{-1}(1) - \tan^{-1}(0)\}.$$

The angles whose tangents are 1 and zero are $\tfrac{1}{4}\pi$ and 0 respectively (principal values). Therefore

$$\int_0^3 \frac{dx}{9+x^2} = \tfrac{1}{3} \cdot \tfrac{1}{4}\pi = \tfrac{1}{12}\pi.$$

11. Definite integrals of odd and even functions

A function of a variable such as the sine, which can be expressed as a power series in a variable containing only odd powers, is called an odd function. If the power series contains only even powers of the variable, as in the cosine series, the function is called an even one.

Two important properties of the integrals of odd and even functions between equal and opposite limits are

$$\int_{-a}^{+a} D(x)\, dx = 0, \qquad (8.27)$$

and

$$\int_{-a}^{+a} E(x)\, dx = 2\int_{0}^{a} E(x)\, dx, \qquad (8.28)$$

where $D(x)$ is an odd function of x, and $E(x)$ is an even function.

Proof of these relations follows from the fact that when an odd function is integrated, term by term, the result is an even function, so that when limits are applied the values of each term for $x = +a$ and $x = -a$ are identical and the integral vanishes. For example,

$$\int_{-a}^{+a} \sin\theta\, d\theta = \int_{-a}^{+a} \left(\theta - \frac{\theta^3}{3!} + \frac{\theta^5}{5!} - \ldots \right) d\theta$$

$$= \left[\frac{\theta^2}{2!} - \frac{\theta^4}{4!} + \frac{\theta^6}{6!} - \ldots \right]_{-a}^{+a}$$

$$= \left\{\frac{a^2}{2!} - \frac{(-a)^2}{2!}\right\} - \left\{\frac{a^4}{4!} - \frac{(-a)^4}{4!}\right\} + \cdot$$

$$= 0.$$

When an even function is integrated, the result is an odd function, and the values of each term when the limits are substituted are equal and opposite. Thus

$$\int_{-a}^{+a} \cos\theta\, d\theta = \int_{-a}^{+a} \left(1 - \frac{\theta^2}{2!} + \frac{\theta^4}{4!} - \ldots \right) d\theta$$

$$= \left[\theta - \frac{\theta^3}{3!} + \frac{\theta^5}{5!} - \ldots \right]_{-a}^{+a}$$

$$= \{a - (-a)\} - \left\{\frac{a^3}{3!} - \frac{(-a)^3}{3!}\right\} + \ldots$$

$$= 2a - \frac{2a^3}{3!} + \frac{2a^5}{5!} - \ldots$$

$$= 2\int_{0}^{a} \cos\theta\, d\theta.$$

The equations involving sines and cosines that occur in the theory of wave optics are simplified by application of (8.27, 8.28).

TRIGONOMETRY

12. Trigonometric functions in Fortran

Three trigonometric functions are normally supplied by Fortran compilers.

SIN(θ) and COS(θ) give the sine and cosine of the angle θ. The argument θ must be expressed *in radians* and written as a *real number*.

ATAN( ) is arctan or $\tan^{-1}$ of the argument and it gives the angle in radians whose tangent is shown as a real number within the brackets. The advantage of using $\tan^{-1}$ rather than $\sin^{-1}$ as a supplied function is that the former can take any value between $\pm\infty$, that is, any real value, whereas the latter is limited to values between ± 1.

If the value of π is required to a high degree of accuracy in a calculation, it may be called as $4\cdot\text{*ATAN}(1\cdot)$, since the angle whose tangent is 1, that is $\tan^{-1}(1)$ is $\pi/4$ radians.

Problems

1. The sides of a 30° ($\frac{1}{6}\pi$ radians) right-angled triangle have the proportions shown in the diagram. Write down the values of
    $\sin(\frac{1}{6}\pi)$ $\cos(\frac{1}{6}\pi)$ $\tan(\frac{1}{6}\pi)$ $\cos(\frac{1}{3}\pi)$.
    Calculate the values of
    $\cos(\frac{2}{3}\pi)$ $\sin(\frac{11}{6}\pi)$ $\tan(\frac{5}{6}\pi)$.

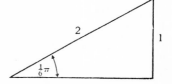

2. If $\theta = \sin^{-1}(\frac{1}{2})$, what are the first four values of the angle θ?
3. Given that $\sin 20° = 0\cdot 342$ and $\cos 20° = 0\cdot 940$, calculate the value of $\sin 50°$.
4. Evaluate:
    (i) $\int_0^2 \frac{dx}{\sqrt{(4-x^2)}}$ (ii) $\int_0^{\frac{1}{4}\pi} \sin 2\theta \, d\theta$ (iii) $\int_0^1 \frac{dx}{9+4x^2}$
    (iv) $\int_0^{\frac{1}{4}\pi} \tan x \, dx$, given that $\cos \frac{1}{4}\pi = \frac{1}{\sqrt{2}}$
    (v) $\int_0^{\frac{1}{6}\pi} \sin(x)\cos(x) \, dx$.
5. Calculate the mean values of $\sin \theta$, $\cos \theta$, $\sin^2 \theta$ and $\cos^2 \theta$ between
   (a) $\theta = 0$ and $\theta = \pi$ (b) $\theta = 0$ and $\theta = 2\pi$.

CHAPTER 9

Differential equations

A DIFFERENTIAL equation contains the differentials of variables as well as the variables themselves; the solution is an equation in which the differential terms have been eliminated. Solutions can usually be found if the variables can be separated, i.e. if the terms including the differentials of one variable can be taken to one side of the equation and the terms including the differentials of the other variable taken to the other side. The two sides are then integrated to give the solution.

When the variables cannot be separated, other methods of solution, such as the power series method, have to be used. Differential equations are of great importance in physical chemistry and occur in the theory of all rate processes. The physical chemist is interested in obtaining particular solutions of a differential equation in which unknown constants introduced by the integration are replaced by known quantities whose values are determined by giving definite values to the variables in the integrated equation. This is illustrated in the following sections, where the usual initial condition for a reaction kinetic equation is that $x = 0$ when $t = 0$. The integrated equation containing unknown constants is called the *general solution* of the equation.

1. Differential equations with separable variables

First-order chemical reactions. In a first-order reaction, e.g. the decomposition or molecular rearrangement of a single substance, the rate is proportional to the concentration of the single reactant. If in such a reaction, a mole per litre is the initial molar concentration of the reactant and x mole per litre of the reactant has decomposed after time t, then the concentration of reactant at time t is $(a-x)$ mole per litre. The rate of reaction is the rate of change of x with t, i.e. dx/dt. Hence

$$dx/dt = k(a-x),$$

where k is a constant for a given temperature and is called the velocity constant of the reaction. This is a differential equation in x and t, which can be solved by separating the variables; thus

$$dx/(a-x) = k\,dt.$$

Therefore

$$\int dx/(a-x) = k\int dt.$$

Now, $dx/(a-x)$ can be integrated by the substitution method (see Chap. 7,

DIFFERENTIAL EQUATIONS

§ 4); thus let $u = a-x$, then $du = -dx$. Substituting these values of $a-x$ and dx gives

$$\int dx/(a-x) = -\int du/u = -\ln(u) + C = -\ln(a-x) + C.$$

Hence
$$-\ln(a-x) + C = kt.$$

The value of the integration constant C can be found if $x = 0$ when $t = 0$. Substituting these values in the integrated equation gives

$$-\ln(a) + C = 0, \quad \text{i.e.} \quad C = \ln a.$$

Substituting this value of C in the integrated equation gives

$$-\ln(a-x) + \ln a = kt.$$

Hence
$$k = \frac{1}{t} \ln\left(\frac{a}{a-x}\right). \tag{9.1}$$

Radioactive decay. The decay of a pure radioactive element is a first-order reaction, since the rate of decay is proportional to the activity at any time. If y is the radioactivity at time t, then the rate of decay is $-dy/dt$. It should be noted that dy/dt is the rate of increase of y with t, so that $-dy/dt$ is the rate of decrease, or decay, of y with t.

The equation relating $-dy/dt$ to y is

$$-dy/dt = \lambda y,$$

where λ is a constant. Separating the variables gives

$$-dy/y = \lambda \, dt.$$

Hence
$$-\int dy/y = \lambda \int dt,$$

and therefore
$$-\ln y = \lambda t + C.$$

If $y = y_0$ when $t = 0$, i.e. when observations begin, then substituting these values in the integrated equation gives

$$-\ln y_0 = C.$$

Substituting this value of C in the integrated equation gives

$$-\ln y = \lambda t - \ln y_0.$$

Hence
$$\lambda = \frac{1}{t} \ln\left(\frac{y_0}{y}\right). \tag{9.2}$$

A radioactive material is characterised by its half-life period τ, i.e. the time taken to decay to half its initial level of radioactivity. This half-life period is related to the constant λ and is independent of the initial level of radioactivity y_0. Substituting $y = \frac{1}{2}y_0$ and $t = \tau$ in (9.2) gives

$$\lambda \tau = \ln(y_0/\tfrac{1}{2}y_0) = \ln 2 = 2 \cdot 303 \log 2 = 0 \cdot 6932.$$

Hence
$$\tau = 0 \cdot 6932/\lambda.$$

Bacterial growth. The maximum rate of growth of a bacterial population follows a law of the same type as the first-order reaction law, the rate of growth dn/dt being proportional to the number of bacteria n at any time t. Hence

$$dn/dt = kn,$$

where k is a constant. Separating the variables and integrating as before gives

$$\ln n = kt + C.$$

If $n = n_0$ when $t = 0$, then substituting in the integrated equation gives

$$\ln n_0 = C.$$

Hence

$$\ln n = kt + \ln n_0.$$

Therefore

$$\ln \left(\frac{n}{n_0}\right) = kt, \quad \text{or} \quad k = \frac{1}{t} \ln \left(\frac{n}{n_0}\right).$$

Second-order chemical reactions. The most common type of chemical reaction is one in which the rate is proportional to the product of the concentrations of the two reactants. Many reactions occur in a series of second-order stages, e.g. reduction of potassium permanganate with oxalic acid.

If a and b mole per litre are the respective initial concentrations of the two reactants, and x mole per litre is the concentration of the reaction product at time t, then the respective concentrations of the reactants at time t will be $a-x$ and $b-x$ mole per litre. The rate of reaction is the rate of formation of the reaction product, i.e. dx/dt. From the theory of second-order reactions, the differential equation

$$dx/dt = k(a-x)(b-x)$$

is obtained, where k is the velocity constant of the reaction. The variables can be separated and the equation integrated; thus

$$\int \frac{dx}{(a-x)(b-x)} = k \int dt.$$

The integral on the left-hand side of the equation can be solved by the method of partial fractions (see Chap. 7, § 4) using the identity

$$\frac{1}{(a-x)(b-x)} \equiv \frac{A}{a-x} + \frac{B}{b-x}.$$

The values of A and B are found by multiplying out the right-hand side of the identity and equating coefficients of like powers of x in the numerators of the two fractions. Hence

$$1 \equiv A(b-x) + B(a-x) \equiv -x(A+B) + Ab + Ba.$$

Therefore, since there is no term in x on the left-hand side of the identity

$$A+B = 0, \quad \text{and} \quad Ab + Ba = 1.$$

Hence, since $A = -B$ and therefore $-Bb+Ba = 1$, it follows that $B(a-b) = 1$. Hence

$$B = 1/(a-b) \quad \text{and} \quad A = -1/(a-b).$$

Substituting these values of A and B in the identity gives

$$\frac{1}{(a-x)(b-x)} \equiv -\frac{1}{a-b}\cdot\frac{1}{a-x}+\frac{1}{a-b}\cdot\frac{1}{b-x},$$

and hence

$$\int\frac{dx}{(a-x)(b-x)} = -\frac{1}{a-b}\int\frac{dx}{a-x}+\frac{1}{a-b}\int\frac{dx}{b-x}.$$

The partial fractions can be integrated as was the integral in the first-order equation; thus

$$\int\frac{dx}{(a-x)(b-x)} = \frac{1}{a-b}\ln(a-x)-\frac{1}{a-b}\ln(b-x)$$

$$= \frac{1}{a-b}\ln\left(\frac{a-x}{b-x}\right).$$

Hence

$$\frac{1}{a-b}\ln\left(\frac{a-x}{b-x}\right) = kt+C. \tag{9.3}$$

If $x = 0$ when $t = 0$, the value of C is found by substituting these values of x and t in (9.3); thus

$$\frac{1}{a-b}\ln\left(\frac{a}{b}\right) = C.$$

Substituting this value of C in (9.3) gives

$$kt = \frac{1}{a-b}\ln\left(\frac{a-x}{b-x}\right)-\frac{1}{a-b}\ln\left(\frac{a}{b}\right).$$

Hence

$$k = \frac{1}{t(a-b)}\ln\left\{\frac{b(a-x)}{a(b-x)}\right\}. \tag{9.4}$$

It is important to note that (9.4) is only true if $x = 0$ when $t = 0$; (9.3) is a general equation for any initial condition.

SECOND-ORDER REACTION IN WHICH $a = b$. If the initial concentrations of the two reactants are equal or if a single reactant is decomposing according to the second-order law, e.g. the thermal decomposition of hydriodic acid, then $a-b = 0$, and (9.3) and (9.4) are no longer valid. In this case the differential rate equation is in the form

$$dx/dt = k(a-x)(a-x) = k(a-x)^2.$$

Hence, separating the variables gives
$$dx/(a-x)^2 = k\,dt,$$
which on integration gives
$$\int dx/(a-x)^2 = kt+C.$$
The integral on the left-hand side of the integrated equation is readily evaluated by substituting u for $a-x$; then, as has been shown previously (see Chap. 7, § 4), $du = -dx$. Therefore
$$\int dx/(a-x)^2 = -\int du/u^2 = -\int u^{-2}\,du = -u^{-1}/(-1) = 1/u = 1/(a-x).$$
Substituting this value in the integrated equation gives
$$1/(a-x) = kt+C.$$
If $x = 0$ when $t = 0$, then $1/a = C$. Hence
$$kt = 1/(a-x) - 1/a,$$
and therefore
$$k = \frac{x}{at(a-x)}. \tag{9.5}$$

Clausius–Clapeyron equation. A differential equation which expresses the rate of change of the vapour pressure of a pure liquid with temperature was deduced from thermodynamic theory by Clausius and Clapeyron; it is
$$dp/dT = Lp/RT^2,$$
where p is the vapour pressure at absolute temperature T, L is the latent heat of evaporation of the liquid and R is the gas constant. This equation assumes that the vapour obeys the ideal gas laws, and this is approximately true for low values of p.

If it is assumed that the latent heat of evaporation L is independent of temperature, the variables can be separated and the equation integrated; thus
$$\int \frac{dp}{p} = \frac{L}{R}\int \frac{dT}{T^2}.$$
Hence
$$\ln p = -\frac{L}{R}\cdot\frac{1}{T}+C. \tag{9.6}$$

If p_1 and p_2 are the vapour pressures at two temperatures T_1 and T_2, then
$$\ln p_1 = -L/RT_1+C, \quad \text{and} \quad \ln p_2 = -L/RT_2+C.$$
Hence
$$\ln\left(\frac{p_2}{p_1}\right) = -\frac{L}{R}\left(\frac{1}{T_2}-\frac{1}{T_1}\right).$$

By measuring values of p at different temperatures, (9.6) can be confirmed experimentally. The plot of $\log p$ against $1/T$ gives a straight line over a limited temperature range.

The assumption that L is independent of T is only reasonably valid over a small temperature range. For a wider temperature range, L can be expressed as a power series in T; thus
$$L = A_1 + B_1 T + C_1 T^2.$$
The differential equation then becomes
$$\frac{dp}{dT} = \frac{p}{RT^2}(A_1 + B_1 T + C_1 T^2) = \frac{p}{R}\left(\frac{A_1}{T^2} + \frac{B_1}{T} + C_1\right),$$
which on separating the variables and integrating gives
$$\ln p = \frac{1}{R}\left(-\frac{A_1}{T} + B_1 \ln T + C_1 T\right) + C,$$
i.e. p is expressed as a series in T.

Equations similar to the Clausius–Clapeyron equation can be deduced for the variation of the equilibrium constant of a reversible reaction (van't Hoff equation) and for variation of the velocity constant of a second-order reaction with temperature (Arrhenius equation).

2. Power series method

A differential equation in which the variables cannot be separated or which contains higher derivatives than the first can sometimes be solved by the power series method. In this, the equation is expressed as a power series, the series is differentiated and the result is substituted back in the original equation. This gives an equation containing powers of only one variable which can be arranged in the form

$$f(\text{one variable}) = 0.$$

In order that this relation shall be true for all values of the variable, the coefficients of each power of the variable must be 0. By equating them all to 0, the coefficients of the original power series can be evaluated.

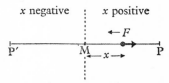

FIG. 9a. Simple harmonic motion

Simple harmonic motion. Simple harmonic motion is the oscillation of a body about a mean position when the force tending to restore it to this position is proportional to the displacement from the mean.

If at any instant the body is a distance x from the mean position M (see Fig. 9a), then the force tending to restore it to M is $-kx$, the minus sign being introduced because the force is acting in the opposite direction to the direction of x. According to Newton's law of motion, the acceleration of a moving body multiplied by its mass m is equal to the force F acting on it; thus

$$F = m\frac{d^2x}{dt^2}.$$

In this case, $F = -kx$, and therefore

$$m(d^2x/dt^2) = -kx,$$

i.e.

$$m(d^2x/dt^2) + kx = 0. \tag{9.7}$$

This is a second-order differential equation in x and t. It can be solved by writing the solution as a power series in t; thus

$$x = A + A_1 t + A_2 t^2 + A_3 t^3 + A_4 t^4 + \ldots.$$

Then

$$dx/dt = A_1 + 2A_2 t + 3A_3 t^2 + 4A_4 t^3 + \ldots,$$

and

$$d^2x/dt^2 = 2A_2 + 3.2A_3 t + 4.3A_4 t^2 + \ldots.$$

Substituting these results in (9.7) gives

$$m(2A_2 + 3.2A_3 t + 4.3A_4 t^2 + \ldots) + k(A + A_1 t + A_2 t^2 + \ldots) = 0.$$

By separating this equation into a single power series in t, it follows that

$$(2A_2 m + kA) + (3.2A_3 m + kA_1)t + (4.3A_4 m + kA_2)t^2 + \ldots = 0.$$

Setting the coefficients of each power of t equal to 0, which is a necessary condition for the above equation to hold for all values of t, a series of equations are obtained by means of which all the even coefficients can be expressed in terms of A and all the odd ones in terms of A_1; thus

$$A_2 = -\frac{1}{2!}\left(\frac{k}{m}\right) A, \quad A_4 = \frac{1}{4!}\left(\frac{k}{m}\right)^2 A, \ldots$$

$$A_3 = -\frac{1}{3!}\left(\frac{k}{m}\right) A_1, \quad A_5 = \frac{1}{5!}\left(\frac{k}{m}\right)^2 A_1, \ldots$$

Substituting these results back in the series for x gives

$$x = A\left\{1 - \frac{1}{2!}\left(\frac{k}{m}\right)t^2 + \frac{1}{4!}\left(\frac{k}{m}\right)^2 t^4 - \ldots\right\} +$$
$$+ A_1\left\{t - \frac{1}{3!}\left(\frac{k}{m}\right)t^3 + \frac{1}{5!}\left(\frac{k}{m}\right)^2 t^5 - \ldots\right\} + \ldots$$

These series are of the form of the cosine (8.21) and sine series (8.20). The first one is, in fact, the expansion of $\cos\{t\sqrt{(k/m)}\}$. If the second series is multiplied by $\sqrt{(k/m)}$, it becomes the expansion of $\sin\{t\sqrt{(k/m)}\}$. The general solution to the original differential equation (9.7) for simple harmonic motion is therefore

$$x = A \cos\{t\sqrt{(k/m)}\} + A_1\sqrt{(m/k)} \sin\{t\sqrt{(k/m)}\}. \tag{9.8}$$

As is to be expected, the solution of a second-order differential equation introduces two undetermined constants, A and A_1. If ν is the frequency of vibration, i.e. the number of vibrations in unit time, then the time taken for one complete vibration is $1/\nu$. After this time period the value of x exactly repeats itself. From the known variations of $\cos\theta$ and $\sin\theta$ with the size of the angle θ, the values of (9.8) will exactly repeat themselves when $t\sqrt{(k/m)}$ increases by 2π, and the value of x exactly repeats itself when t increases by $1/\nu$; these conditions are combined by writing

$$\frac{1}{\nu}\sqrt{\frac{k}{m}} = 2\pi, \quad \text{i.e.} \quad \sqrt{\frac{k}{m}} = 2\pi\nu.$$

Substituting this value of $\sqrt{(k/m)}$ in (9.8) gives

$$x = A\cos(2\pi\nu t) + (A_1/2\pi\nu)\sin(2\pi\nu t).$$

Now, A and A_1 can be eliminated by applying initial conditions to the motion; thus if $x = a$, the extremity of the vibration (point P in Fig. 9a), when $t = 0$, then

$$a = A\cos 0 + (A_1/2\pi\nu)\sin 0$$
$$= A.$$

After a quarter of a vibration, $t = \tfrac{1}{4}(1/\nu)$, and the vibrator is in its mean position, i.e. $x = 0$. Hence

$$0 = a\cos\tfrac{1}{2}\pi + (A_1/2\pi\nu)\sin\tfrac{1}{2}\pi = 0 + A_1/2\pi\nu.$$

Therefore, $A_1 = 0$. For these conditions, therefore, the final complete solution is

$$x = a\cos(2\pi\nu t).$$

If the initial condition, $x = 0$ when $t = 0$, is taken then

$$x = a\sin(2\pi\nu t).$$

This equation describes the motion completely in terms of the amplitude a and the frequency of the oscillator. The formula for the simple harmonic oscillator is used in the theories of wave mechanics. Modern atomic theory treats an electron as a form of three-dimensional oscillator, and by introducing the necessary energy terms into the differential equation describing this oscillator, the equation can be solved in simple cases to give the form of the electron orbital.

3. Partial differential equations

When a differential equation involves more than two variables, it is most usefully expressed in terms of partial derivatives.

Diffusion equation. If a sharp boundary formed between a solvent and solution of concentration c_0 is allowed to diffuse, then the concentration c of solute in any plane at height x above the initial boundary (see

Fig. 9b) is a function of x and also of the time t since the diffusion started. Hence
$$c = f(x, t).$$
In the ideal case, a relation between the two variables can be deduced theoretically as a partial differential equation; thus
$$(\partial c/\partial t)_x = D(\partial^2 c/\partial x^2)_t, \qquad (9.9)$$

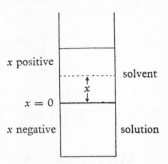

FIG. 9b. Diffusion

where D is a constant at a given temperature and is called the diffusion coefficient. A solution (for proof, see Appendix X) to this equation is
$$c = \tfrac{1}{2}c_0\left\{1 - \frac{2}{\sqrt{\pi}}\int_0^y \exp(-\beta^2)\,d\beta\right\} \qquad (9.10)$$
for the conditions $c = 0$ at $t = 0$ for positive values of x, and $c = c_0$ at $t = 0$ for negative values of x; y is defined by the equation
$$y = \frac{x}{2\sqrt{(Dt)}}.$$
The integral is similar to the probability integral (see Chap. 12). It cannot be expressed in a simpler form algebraically, but its value for any particular value of the upper limit can be found by computing using Simpson's rule or by series or graphical methods. It is written in the form $\int_0^y \exp(-\beta^2)\,d\beta$ to denote that it is the value of this function for any given value of y, just as $\int_0^y \beta^3\,d\beta$ could be written in place of $\tfrac{1}{4}y^4$. When y is infinite, $\int_0^y \exp(-\beta^2)\,d\beta$ has the finite value $\tfrac{1}{2}\sqrt{\pi}$ (see Appendix X).

Equation (9.10) agrees well with concentration changes observed experimentally for a two-component system in a diffusion cell. That it is a solution of (9.9) can be shown by partial differentiation with respect to t and then partial differentiation twice with respect to x; the ratio of the two results obtained, $\partial c/\partial t$ and $\partial^2 c/\partial x^2$, will be found to be equal to D.

4. Exact differentials

An expression of the form $M\,dx + N\,dy$, where M and N are functions of both x and y, is an exact differential if it can be formed by differentiating some function, say u, of x and y, i.e.
$$du = M\,dx + N\,dy.$$
For example, $(2x+y)\,dx + (x+3y)\,dy$ is an exact differential since it is formed by differentiating $x^2 + xy + 1\tfrac{1}{2}y^2$.

A test can be developed to find out whether or not a given expression is an exact differential.

Euler's criterion for exact differentials. Any function u of the variables x and y can be written in the partial derivative form; thus from (6.3)
$$du = (\partial u/\partial x)\,dx + (\partial u/\partial y)\,dy.$$
If $M\,dx + N\,dy$ is an exact differential formed by differentiating u, then
$$du = M\,dx + N\,dy.$$
These two equations express the same result, so therefore
$$M = \partial u/\partial x \quad \text{and} \quad N = \partial u/\partial y.$$
A relation between M and N is then found; thus differentiating $M = \partial u/\partial x$ with respect to y gives
$$\frac{\partial M}{\partial y} = \frac{\partial}{\partial y}\left(\frac{\partial u}{\partial x}\right) = \frac{\partial^2 u}{\partial y\,\partial x},$$
and differentiating $N = \partial u/\partial y$ with respect to x gives
$$\frac{\partial N}{\partial x} = \frac{\partial}{\partial x}\left(\frac{\partial u}{\partial y}\right) = \frac{\partial^2 u}{\partial x\,\partial y}.$$
Since, from (6.4), $\partial^2 u/\partial y\,\partial x = \partial^2 u/\partial x\,\partial y$, it follows that
$$\partial M/\partial y = \partial N/\partial x.$$

This relation provides the test, often called Euler's criterion, for an exact differential. In the case of the example given at the beginning of this section,
$$M = 2x + y, \quad \text{and} \quad N = x + 3y.$$
Hence
$$\partial M/\partial y = 1, \quad \text{and} \quad \partial N/\partial x = 1.$$
Therefore the expression is an exact differential.

EXAMPLE. *Solve the equation* $(3x^2 + 2xy)\,dx + (x^2 + 3y^2)\,dy = 0$.

The first step is to test whether the left-hand side of the equation is an exact differential. Since
$$M = 3x^2 + 2xy, \quad \text{and} \quad N = x^2 + 3y^2,$$
then
$$\partial M/\partial y = 2x, \quad \text{and} \quad \partial N/\partial x = 2x,$$
i.e.
$$\partial M/\partial y = \partial N/\partial x.$$

The expression is therefore proved to be an exact differential, so that the next step is to look for an expression which when differentiated gives the left-hand side of the original equation. This can be found by rearranging the left-hand side to give

$$3x^2\,dx + (2xy\,dx + x^2\,dy) + 3y^2\,dy = 0.$$

The first term is obtained by differentiating x^3, and the third term by differentiating y^3. Examination of the two terms in brackets shows that they result from the differentiation of x^2y. Hence the function which on differentiation gives the left-hand side of the original equation is

$$x^3 + x^2y + y^3 + C,$$

and the solution to the differential equation is

$$x^3 + x^2y + y^3 + C = 0.$$

If Euler's criterion had shown that the left-hand side was not an exact differential, then it would have been pointless to look for a simple expression which gives the left-hand side on differentiation. In some cases, expressions can be converted to exact differentials by multiplying throughout by a function called an *integrating factor*.

Entropy and exact differentials. If a system consisting of 1 mole of an ideal gas obeying the pressure–volume–temperature relation $pv = RT$, where R is a constant, is subjected to infinitesimal changes of volume and temperature, an expression for the heat absorbed q is given by the first law of thermodynamics; thus

$$q = p\,dv + C_v\,dT,$$

where C_v is the molar specific heat of the gas at constant volume and is independent of volume for an ideal gas.

In order to test whether the right-hand side of the above equation is an exact differential, Euler's criterion is applied. In this case, $M = p$ and $N = C_v$. Hence

$$\partial M/\partial T = \partial p/\partial T = R/v,$$

since $pv = RT$, i.e. at constant volume,

$$\partial p/\partial T = R/v.$$

Further, since for an ideal gas the molar specific heat is independent of volume (Joule's law), it follows that

$$\partial N/\partial v = \partial C_v/\partial v = 0.$$

Therefore the right-hand side of the equation for q is not an exact differential, but it can be made so by multiplying throughout by the integrating factor $1/T$. Then

$$q/T = (p/T)dv + (C_v/T)dT.$$

Again applying Euler's criterion to the right-hand side, with $M = p/T$ and $N = C_v/T$, it can be shown that

$$\frac{\partial M}{\partial T} = \frac{\partial}{\partial T}\left(\frac{p}{T}\right) = \frac{\partial}{\partial T}\left(\frac{R}{v}\right) = 0,$$

since the partial differentiation $\partial M/\partial T$ is carried out at constant volume. Further,

$$\frac{\partial N}{\partial v} = \frac{\partial}{\partial v}\left(\frac{C_v}{T}\right) = 0,$$

since C_v is independent of volume and the partial differentiation is carried out at constant temperature. The right-hand side of the equation for q/T is thus an exact differential. It is called the differential of the entropy S of the system, and

$$q/T = (p/T)\mathrm{d}v + (C_v/T)\mathrm{d}T = \mathrm{d}S.$$

Problems

1. Show by partial differentiation that (9.10) is a solution of the differential equation (9.9), and that it agrees with the conditions that when $t = 0$, $c = 0$ if x is positive and $c = c_0$ if x is negative.

2. The relation between the velocity constant k of a second-order reaction and absolute temperature T is given by the Arrhenius equation

$$\mathrm{d}(\ln k)/\mathrm{d}T = E/RT^2,$$

where E is the activation energy of the reaction (which can be assumed to be constant over a limited temperature range) and R is the gas constant (2 cal per °C per mole approximately). Solve this differential equation to give k as a function of temperature, and use the solution to calculate the velocity constant of a reaction at 80°C, given that k is $2 \cdot 8 \times 10^{-3}$ at 40°C and $4 \cdot 62 \times 10^{-2}$ at 65°C.

3. The tip of a tuning fork can be regarded as a body executing simple harmonic motion. If the frequency of vibration is 10^3 vibrations per second and the amplitude of the vibration is 1 mm, derive the equation describing the motion, giving x in terms of t with initial conditions $x = 0$ when $t = 0$.

4. Solve the differential equations
   (i) $(xy^2 + x)\,\mathrm{d}x + (x^2 y - y)\,\mathrm{d}y = 0$
   (ii) $\mathrm{d}y/\mathrm{d}x = -x/y$
   (iii) $x^2(\mathrm{d}y/\mathrm{d}x) + 2xy = \exp(3x)$
   (iv) $\mathrm{d}^2 y/\mathrm{d}x^2 = 3$
   (v) $\mathrm{d}y/\mathrm{d}x = 4yx^2 + 3yx^4$

5. The half-life period of a radioactive element is 25 days. Calculate the value of the constant λ in the differential equation describing its decay, $-\mathrm{d}y/\mathrm{d}t = \lambda y$, where y is the radioactivity at time t.

CHAPTER 10

Equations and series for describing experimental measurements

EXPERIMENTAL measurements can usually be summarised in the form of a table showing numerical values of the measured quantity y for various values of another variable x. There may be a known theoretical relation between y and x and, if this is so, the best method for testing the theory with the measurements is to arrange the relation in the form

$$f_1(y) = b\, f_2(x) + a.$$

Numerical values of $f_1(y)$ and $f_2(x)$ are calculated from the measurements and are plotted against one another. If the theory is correct, the result will be a straight line whose intercept at $f_2(x) = 0$ and slope give the values of the coefficients a and b respectively. For instance, the relation between the vapour pressure p of a pure liquid and its temperature T (over a limited temperature range) given by (9.6) can be checked with experimental measurements by plotting $\log p$ against $1/T$, when the result should be a straight line.

There will usually be some scatter of the experimental results about the straight line, and in accurate work it may be necessary to determine exactly the best-fitting straight line through the points. The method of finding this line, the *regression line*, is described in Chap. 13.

If the theoretical relation between y and x is not known, the first step towards understanding the relation between the two variables is often to develop an empirical mathematical equation relating them. An empirical equation is one derived from experimental measurements without any theoretical background. To find an equation to fit a given set of pairs of values of y and x, the values are plotted as a graph, and, providing that repeated measurements show good agreement with one another, a smooth curve can be drawn through the points. From the shape of the curve, the form of the algebraic relation between y and x can often be deduced.

When repeated measurements show considerable random variation among themselves, the results should be treated by statistical methods.

1. Methods for fitting equations to graphs

In this section some general rules are outlined for fitting algebraic equations to graphs of various shapes.

EQUATIONS FOR DESCRIBING EXPERIMENTAL MEASUREMENTS

Linear graphs. If the plot of pairs of values of y and x is a straight line, then the relation between them is of the form

$$y = bx+a,$$

where a is the intercept on the y-axis at $x = 0$, and b is the slope of the line. Alternatively, if (x_1, y_1) is a particular pair of values of y and x, the equation of the line can be written

$$y-y_1 = b(x-x_1).$$

As before, b is the slope of the line, and this equation satisfies the condition that $y = y_1$ when $x = x_1$.

Slightly curved graphs. If the slope of the y, x plot does not change very greatly, the relation between the variables can usually be described by a quadratic equation of the form

$$y = a+bx+cx^2.$$

The values of a, b and c can be found by substituting pairs of experimental values of y and x in the equation. Alternatively, if a large number of pairs of values are available, the values of dy/dx for a range of values of x can be determined graphically by measuring the slopes of tangents to the y, x curve. If the quadratic equation holds, then the (dy/dx), x plot should be a straight line with slope equal to $2c$ and intercept on the (dy/dx)-axis at $x = 0$ equal to b, since differentiating the original quadratic equation gives

$$dy/dx = b+2cx.$$

Graphs with considerable curvature. If the plot of dy/dx against x does not give a straight line, it may be necessary to add terms in higher powers of x to the empirical quadratic equation, e.g.

$$y = a+bx+cx^2+gx^3.$$

The coefficients can again be determined either by direct substitution of pairs of numerical values of y and x or by repeated graphical differentiation. If the cubic equation fits the results, the plot of d^2y/dx^2 against x should be linear, since differentiation of the cubic equation gives

$$dy/dx = b+2cx+3gx^2,$$

and

$$d^2y/dx^2 = 2c+6gx.$$

Graphs with rapidly changing slopes. In theory, these can always be represented by adding terms in x of powers higher than the third to the cubic equation given above. However, they can often be described more conveniently by logarithmic or exponential equations.

If the graph passes through the origin and its slope increases as x increases, a simple type of 'parabolic' equation may represent the results, e.g.

$$y = ax^n,$$

where a and n (> 1) are constants. If the curve does not pass through the origin, but passes through the point (x_1, y_1), then the equation takes the form

$$y-y_1 = a(x-x_1)^n.$$

This equation satisfies the condition that $y = y_1$ when $x = x_1$. If the slope of the y, x plot decreases as x increases, the required equation may be either of the parabolic equations already given, but with n positive and less than 1. The curve may, however, be better expressed by means of a logarithmic relation, e.g.

$$y = a+b \log x.$$

LOGARITHMIC GRAPH PAPER. In order to find the coefficients of logarithmic or exponential equations, logarithmic and 'log–log' graph papers are used. Logarithmic graph paper has one normal arithmetic scale and the other scale has numbers spaced according to the values of their logarithms, just as on a slide-rule scale.

In order to find whether an equation of the form

$$y = a+b \log x$$

fits a set of values of y and x, a plot is made on the logarithmic graph paper with the arithmetic scale as the y-axis and the log scale as the x-axis. If the equation is satisfactory, a straight line is obtained and from the intercept and slope the values of a and b respectively can be found.

Some care is necessary in determining the intercept and slope. The logarithmic scale never shows zero, since $\log 0 = -\infty$; the scale starts at 1 and continues up to 10. The distance between 1 and 10 is one cycle, and is equal to one logarithmic unit, since $\log 1 = 0$ and $\log 10 = 1$. The upper limit of the cycle is often designated by 1 again instead of by 10, and a further cycle is continued beyond it. The numbers in this second cycle are ten times those in the first cycle.

Since $\log 1 = 0$, the value of a in the logarithmic equation is the value of y when $\log x = 0$, i.e. when $x = 1$, and a is therefore the intercept of the line on the y-axis at $x = 1$. The slope of the line is b, but it is in units of y divided by units of $\log x$. It can be measured by constructing a right-angled triangle in the usual way and measuring the sides, in centimetres (or inches), disregarding the figures shown on the scales; the length of the vertical side is then divided by the length corresponding to each unit of y, and the horizontal side is divided by the length in centimetres corresponding to one logarithmic unit, i.e. the length of one cycle on the logarithmic scale. Thus, from Fig. 10a,

$$a = \text{DE}, \quad \text{and} \quad b = \frac{\text{BC/cm per unit of } y}{\text{AC/length of cycle in cm}}.$$

The same results would be obtained if y were plotted against $\log x$ on ordinary graph paper. The use of logarithmic graph paper avoids the necessity for looking up logarithms in tables.

EQUATIONS FOR DESCRIBING EXPERIMENTAL MEASUREMENTS 151

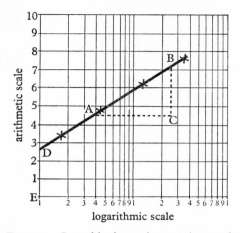

FIG. 10a. Logarithmic graph paper (two cycles)

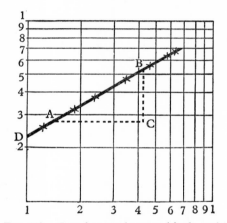

FIG. 10b. Log–log graph paper (single cycle)

LOG–LOG GRAPH PAPER. In order to test whether an equation of the form $y = ax^n$ fits a given set of results, y is plotted against x on log–log graph paper; this paper has both scales spaced logarithmically. If the equation holds, the plot will be a straight line, since

$$\log y = \log a + n \log x.$$

When $x = 1$, $\log x = 0$, and $\log y = \log a$, i.e. $y = a$; a is therefore the value of y when $x = 1$, i.e. the reading on the logarithmic scale at the intercept of the line with the y-axis at $x = 1$. The slope of the line is n, and this is found by constructing a right-angled triangle ABC (see Fig. 10b), measuring its two shorter sides in centimetres and converting the lengths to log units as described previously. In Fig. 10b, a is the reading on the y-axis at the point D, and

$$n = \frac{BC \text{ in cm/length of vertical log cycle}}{AC \text{ in cm/length of horizontal log scale}}.$$

If the y, x curve on ordinary graph paper does not pass through the origin, the relation is modified to

$$y - y_1 = a(x - x_1)^n.$$

A pair of values, x_1 and y_1, of x and y which lie on the smooth y, x curve are taken. Values of $y - y_1$ and $x - x_1$ are calculated from the set of pairs of values of y and x and then plotted against one another on log–log graph paper. Again n is the slope of the plot and a is the value of $y - y_1$ when $x - x_1 = 1$.

Asymptotic graphs. An asymptotic curve is one which becomes linear at its extremities; the linear sections are called the *asymptotes of the*

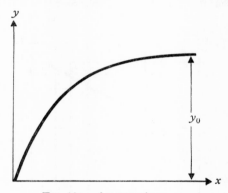

FIG. 10c. Asymptotic curve

curve (see Chap. 3, § 5). If the asymptotes are the x- and y-axis, the relation between the variables will be of the simple 'hyperbolic' form, i.e.

$$yx^n = a, \quad \text{or} \quad y = ax^{-n}.$$

The values of a and n can be found by a plot on log–log paper, since

$$\log y = \log a - n \log x.$$

Hence a is the value of y when $x = 1$, and n is minus the slope of the line. It should be noted that in the case of a hyperbolic function, the slope is negative since $\log y$ decreases as $\log x$ increases.

A familiar example of this type of relation is the pv isotherm of an ideal gas. In this case, $pv = \text{constant}$, i.e.

$$p = av^{-1}.$$

If the y, x plot has the form shown in Fig. 10c, in which the curve

EQUATIONS FOR DESCRIBING EXPERIMENTAL MEASUREMENTS 153

passes through the origin and y approaches a limiting value as x increases, the relation between y and x can often be expressed by a hyperbolic equation, e.g.
$$(x+b)(y_0-y) = a,$$
y_0 being the limiting value of y. The values of a and b can be found by multiplying out the factors to give
$$xy_0+by_0-xy-by = a.$$
Since the line passes through the origin, then $y = 0$ when $x = 0$. Hence $by_0 = a$, and therefore
$$xy_0-xy-by = 0,$$
which on dividing throughout by y gives
$$y_0(x/y)-x-b = 0.$$
Hence
$$x/y = (1/y_0)x+b/y_0.$$
If this type of equation fits the values of y and x, then the plot of x/y against x should be a straight line with a slope of $1/y_0$ and the intercept at $x = 0$ on the (x/y)-axis equal to b/y_0.

Periodic graphs. If the y, x curve has a periodic form, the relation between variables is best expressed by means of the trigonometric functions.

2. Use of series to summarise measurements

If a summary of a set of accurate experimental measurements y at various values of another variable x is required, a series is used.

Simple power series. If no theoretical relation between y and x is known and the y, x plot is a smooth curve, an accurate description of a set of pairs of values of y and x can be given by calculating the coefficients of a power series (polynomial) with as many terms as is required to give the necessary accuracy. Thus
$$y = a+bx+cx^2+gx^3+hx^4+jx^5+\ldots.$$
The coefficients are evaluated by substituting pairs of values of y and x in the equation or by repeated graphical differentiation.

The power series method is used to describe the variation of thermal properties such as specific heat, latent heat, etc., with temperature. For instance
$$Q = Q_0+At+Bt^2+Ct^3+\ldots,$$
where t is the temperature and Q_0 is the value of the property Q at $t = 0$.

More complex series. When a relation between two variables is known for the ideal case, a series can sometimes be developed to express the results for non-ideal conditions. For example, the Clausius–Clapeyron equation (9.6) expresses the variation of vapour pressure p of a pure liquid with absolute temperature T for the ideal case where the latent heat of evaporation Q is independent of temperature, i.e.

$$\frac{dp}{p} = \frac{Q\,dT}{RT^2},$$

or

$$\ln p = -Q/RT + \text{constant}.$$

Over a wide temperature range, Q cannot be regarded as independent of temperature, but it can be expressed as a power series in T; thus

$$Q = Q_0 + AT + BT^2 + CT^3.$$

Substituting this value of Q in the differential form of the Clausius-Clapeyron equation gives

$$\frac{dp}{p} = \frac{(Q_0 + AT + BT^2 + CT^3)\,dT}{RT^2}.$$

Therefore

$$\int \frac{dp}{p} = \frac{Q_0}{R}\int \frac{dT}{T^2} + \frac{A}{R}\int \frac{dT}{T} + \frac{B}{R}\int dT + \frac{C}{R}\int T\,dT.$$

Hence

$$\ln p = -\frac{Q_0}{R}\cdot\frac{1}{T} + \frac{A}{R}\ln T + \frac{B}{R}T + \frac{C}{2R}T^2 + D,$$

where D is the integration constant. This means that values of p over a wide range of values of T can be accurately summarised by a power series of the form obtained by converting the equation just derived to common logarithms; thus

$$\log p = a - b/T + c\log T + gT + hT^2,$$

where $a = D/2\cdot303$, $b = Q_0/2\cdot303R$, $c = A/R$, $g = B/2\cdot303R$ and $h = C/4\cdot606R$.

Another example is the virial series used to express pressure (p)–molar-volume (v) relations for an imperfect gas at constant temperature. The ideal gas equation is

$$pv = RT, \quad \text{i.e.} \quad p = RT(1/v).$$

If the gas is not ideal, the pressure is expressed as a power series in $1/v$; thus

$$p = RT\{1/v + A(1/v)^2 + B(1/v)^3 + \ldots\}.$$

Hence

$$pv = RT(1 + A/v + B/v^2 + \ldots).$$

The coefficients $A, B, \ldots$ are called the second, third, $\ldots$ virial coefficients of the gas.

Fourier series. Fourier's theorem states that any finite single-valued relation between two variables can be expressed as the sum of a series of sines and cosines; thus

$$y = a_0 + a_1\sin x + b_1\cos x + a_2\sin 2x + b_2\cos 2x + \ldots$$

The values of the coefficients $a_0, a_1, b_1, a_2, b_2, \ldots$ can be determined by integration. The value of y is exactly repeated after an increment of 2π in x, so that the series is particularly useful for describing periodic functions.

A well-known application of the Fourier series is in the mathematical analysis of the patterns produced when X-rays are diffracted by crystals. The intensity at any position in the diffraction pattern is a periodic function of position, and by evaluating the coefficients of the Fourier series representing this effect, important information about the molecular structure of crystalline compounds can be obtained.

Tchebychev series. The best polynomial approximation to a given function is provided by the series originally developed in the last century by Tchebychev.

$$f(x) = a_0 + a_1 T_1(x) + a_2 T_2(x) + a_3 T_3(x) + \ldots \quad |x| < 1$$

where $T_r(x) = \cos(r \cos^{-1} x)$.

The first three Tchebychev polynomials are

$T_1(x) = \cos(\cos^{-1} x) = x$, since if $\theta = \cos^{-1} x$, $\cos \theta = x$
$T_2(x) = \cos(2 \cos^{-1} x) = \cos 2\theta = 2\cos^2 \theta - 1 = 2x^2 - 1$
$T_3(x) = \cos(3 \cos^{-1} x) = 4x^3 - 3x$.

They are alternately odd and even functions of x.

The coefficients of the Tchebychev series for a given function may be obtained by integration, like those of the Fourier series. Like the Fourier series the terms are orthogonal, that is, independent, and, unlike the ordinary polynomial, adding a further term to give a better approximation does not alter the coefficients of the previous terms.

The importance of the Tchebychev series in computing is that it enables a function to be stored with a minimum number of coefficients for a given degree of accuracy.

3. Curve fitting by computer

An example of the fitting of a best straight line to a set of experimental data using a library computer programme is given in COMP 9, Chap. 13, § 7 (page 207). This programme may also be used to fit polynomials to experimental curves up to a polynomial of degree 10. After each additional term

$$f(x) = a_0 + a_1 x + a_2 x^2 + a_3 x^3 + \ldots$$

an assessment of the deviation between theoretical values from the polynomial and the observed results is made. Plots showing both theoretical and observed values are also included in the programme which is included in the Biomedical package (BMD).

In COMP 6 below, a useful programme is outlined for locating the minimum point of a quantity when only a limited amount of numerical data is available. An example of this from conformational studies is the location of the value of a parameter such as a bond angle or length at which minimum molecular energy occurs. This will then be the expected ground state value of the parameter. The energy calculations involve a lot of computer time and storage and it is not, in general, reasonable to vary the parameter in very small increments. A broader variation is made and three points which span the minimum are taken, that is, the central energy value E2 is less than the two outside values, E1 and E3. The three points are then fitted by a parabola and the minimum point of the parabola is found.

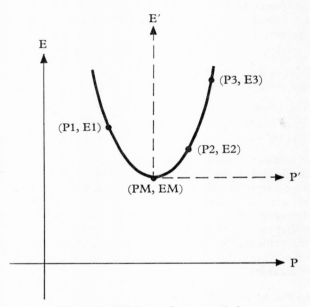

FIG. 10d. Minimum from a parabola

A parabola $y = ax^2$ has a minimum at the origin—see Fig. 3f. If a parabola is fitted to the given three values of E, the energy, and P, the parameter, as shown in Fig. 10d, then the minimum point (PM, EM) may be found by transferring the origin of the co-ordinates to this point. The values on this co-ordinate system are $P' = P - PM$ and $E' = E - EM$. The equation of the parabola is $E' = A(P')^2$ where A is a constant, and the minimum is at $P' = 0$, $E' = 0$.

Translating back to the original co-ordinates, the equation of the parabola is

$$E - EM = A(P - PM)^2$$

and the minimum is at $E = EM$, $P = PM$.

The three pairs of values of P and E (P1, E1; P2, E2; P3, E3) lie on the parabola and so

$$E1 - EM = A(P1 - PM)^2$$
$$E2 - EM = A(P2 - PM)^2$$
$$E3 - EM = A(P3 - PM)^2$$

Substituting the numerical values of P and E, the three equations are solved simultaneously to give values of EM, the energy minimum, and PM, the value of the parameter at which this minimum occurs. The value of A may also be found but it is not usually of much interest.

EM is eliminated from the three equations by subtracting the second from the first and the third from the second; A is then eliminated by dividing the first equation so obtained by the second. On rearranging, an expression for PM is obtained.

$$PM = \frac{E1(P3^2 - P2^2) + E2(P1^2 - P3^2) + E3(P2^2 - P1^2)}{E1(P3 - P2) + E2(P1 - P3) + E3(P2 - P1)}$$

The minimum energy may be derived from PM by dividing the first equation of the original set of three by the second, eliminating A and giving on re-arrangement

$$EM = \frac{E1 - R \cdot E2}{1 - R}$$

where

$$R = \left(\frac{P1 - PM}{P2 - PM}\right)^2$$

With only three sets of data it is not worth dimensioning P and E and so variable names, P1, E1, etc., are given to the values; if P and E had been dimensioned, subscripted variables, the individual values would have been written P(1), E(1), etc.

The statement for PM has continuation cards indicated by numbers in column 6 of the Fortran card. In the print-out, a new page is taken and EM is printed with a title; the slash, /, means a new line for the value of PM.

COMP 6

```
C       COMP 6
C       MINIMUM BY PARABOLA
        READ(5,1) P1,P2,P3,E1,E2,E3
1       FORMAT(6F10.0)
        PM= (E1*(P3**2-P2**2)+E2*(P1**2-P3**2)+E3*
       1 (P2**2-P1**2)) / (2*(E1*(P3-P2)+E2*(P1-P3)+E3*
       2 (P2-P1)))
        R = ((P1-PM)/(P2-PM))**2
        EM = (E1-R*E2)/(1.-R)
        WRITE(6,2) EM,PM
2       FORMAT(17H1ENERGY MINIMUM =,F10.4/23H0PARAMETER AT
       1 MINIMUM =,F10.4)
        STOP
        END
```

Output from this programme is given in the answer to Problem 4 of Chapter 10.

Problems

1. Fit an empirical equation of the form
$$C_p = aT^2 + bT + c$$
to the following experimental measurements of the molar heat capacity of a gas at constant pressure:

T	300	400	500	600	700	800	900	1000
C_p	7·49	7·76	8·05	8·36	8·69	9·04	9·41	9·80

2. Plot the following pressure–molar volume data for a gas at 0°C, and find an equation of the form $pV = A - B/V$ to express the results:

p	1	2	3	4	5	10	20	50
V	22·4	11·2	7·47	5·60	4·46	2·21	1·07	0·40

3. The following data represents the weight in grammes (x/m) of nitrogen dioxide absorbed by 100 g of silica gel at different pressures (p), in mm, of the gas at 15°C:

p	1	2	3	4	5	6	8	10
x/m	0·33	0·82	1·47	2·15	2·84	3·67	5·12	7·22

Using log–log paper, find an equation to represent the results.

4. Use COMP 6 to find the minimum energy (ground state) value of the bond length between aromatic and carboxyl carbon atoms in benzoic acid from the following data. Find also the ground state molecular energy.

E, total energy in eV	−90·9912	−90·9950	−90·9815
P, bond length in picometres	139	154	158

CHAPTER 11

Probability

1. Simple probability

If an event can happen in a ways and fail to happen in b ways, and if all these $a+b$ ways are equally likely to occur, the probability $\mathfrak{p}$ that the event will happen in a single trial is given by

$$\mathfrak{p} = a/(a+b).$$

The probability $\mathfrak{q}$ that it will fail to happen in a single trial is given by

$$\mathfrak{q} = b/(a+b) = 1-\mathfrak{p}.$$

Gothic symbols are used here for simple accurately known probabilities. A probability of 1 means a certainty, and since an event is certain either to happen or to fail to happen in a single trial

$$\mathfrak{p}+\mathfrak{q} = a/(a+b)+b/(a+b) = 1.$$

For many purposes, probabilities are multiplied by 100 and expressed as percentages; thus

$$100\mathfrak{p} = 100a/(a+b) \text{ per cent.} \qquad (11.1)$$

This quantitative definition of probability can be illustrated by the throwing of a poker die. The die is a cube whose six faces are marked as ace, king, queen, jack, ten and nine. If the die is unloaded, i.e. unbiased, the probability of any particular face being uppermost can be calculated from the above equations. Thus, the event of throwing, for instance, an ace can happen in one way but the event of failing to throw an ace can happen in five ways. Hence

$$100\mathfrak{p} = 100 \times 1/(1+5) = 16 \cdot 7 \text{ per cent.}$$

2. Compound probability

If two independent events have probabilities of occurrence in a single trial of p_1 and p_2 respectively, then the compound probability p that both events will happen together is the product of the separate probabilities, i.e.

$$p = p_1 \cdot p_2.$$

For example, the probability of obtaining two aces together in a single throw of two dice is given by

$$p = (1/6)(1/6) = (1/6)^2 = 1/36 = 0 \cdot 0278, \text{ or } 2 \cdot 78 \text{ per cent.}$$

In other words, obtaining an ace with one die does not affect the probability of obtaining another ace with the second die.

The probability of three or more independent events occurring together is the product of their separate probabilities of occurrence; hence, the probability of obtaining, for example, three aces in a single throw of three dice is given by

$$p = (1/6)(1/6)(1/6) = (1/6)^3 = 1/216 = 0.0046, \text{ or } 0.46 \text{ per cent.}$$

The probability of obtaining no aces with a single throw of three dice is given by

$$p = (5/6)(5/6)(5/6) = (5/6)^3 = 125/216 = 0.579, \text{ or } 57.9 \text{ per cent.}$$

If two events are mutually exclusive, i.e. if the occurrence of one event excludes the possibility of the other occurring, the compound probability of one or other of them happening is the sum of their separate probabilities. For example, the probability of obtaining either an ace or a king in a single throw of a die is given by

$$p = 1/6 + 1/6 = 1/3 = 0.333, \text{ or } 33.3 \text{ per cent.}$$

For, if the ace is obtained, this precludes the possibility of obtaining a king in the same throw, and vice versa.

3. Binomial probability

If p is the probability of occurrence of an event in a single trial (derived from a very large number of single trials) and if n trials are performed, then the number of successes expected is np. However, owing to chance, the actual number of successes observed in practice in n trials will vary between 0 and n. The probability of obtaining either 0, 1, 2, 3, ... n successes in n trials can be calculated, and is shown later to be a term of the binomial expansion of $(q+p)^n$, where q (equal to $1-p$) is the probability of failure in a single trial.

When throwing four poker dice, the expected number of aces per throw will be np, i.e. 4/6. However, in a series of such throws, any number of aces between 0 and 4 may be obtained with each throw. The probability p_4 of obtaining four aces simultaneously is given by

$$p_4 = (1/6)^4,$$

while the probability p_0 of obtaining no aces is given by

$$p_0 = (5/6)^4.$$

The probability of obtaining one ace with a particular die, the remaining three dice having a face other than the ace uppermost, is given by

$$p = (1/6)(5/6)(5/6)(5/6) = (1/6)(5/6)^3.$$

This event can occur in four equally likely ways; thus the first die may show an ace on the uppermost face, or the second die, or the third die, or the fourth die, i.e. the total probability p_1 of any die showing an ace on the

uppermost face while the other three fail to show an ace on their uppermost faces is given by
$$p_1 = 4(1/6)(5/6)^3,$$
the coefficient 4 being the number of ways in which a single ace can occur among four dice, i.e. $^4C_1$ ways (see Chap. 4, § 5).

The probability of obtaining two aces and two 'not aces' is similarly given by
$$p = (1/6)(1/6)(5/6)(5/6) = (1/6)^2(5/6)^2.$$
This event can occur in six different, equally likely, ways; thus

1st die	2nd die	3rd die	4th die
ace	ace	not ace	not ace
ace	not ace	ace	not ace
ace	not ace	not ace	ace
not ace	ace	ace	not ace
not ace	ace	not ace	ace
not ace	not ace	ace	ace

Hence the total probability p_2 of obtaining two aces and two 'not aces' in a single throw of the four dice is given by
$$p_2 = 6(1/6)^2(5/6)^2,$$
the coefficient 6 being the number of ways in which two aces can occur among four dice, i.e. $^4C_2$ ways.

In the same way, it can be shown that the probability p_3 of obtaining three aces and one 'not ace' in a single throw of four dice is given by
$$p_3 = 4(1/6)^3(5/6),$$
the coefficient 4 being the number of ways in which three aces can occur among four dice, i.e. $^4C_3$ ways.

From these considerations it can be seen that the probabilities of obtaining no aces or one, two, three or four aces in a single throw of four dice are the successive terms of the binomial expansion of $(5/6+1/6)^4$; thus
$$(5/6+1/6)^4 = (5/6)^4 + 4(5/6)^3(1/6) + 6(5/6)^2(1/6)^2 + 4(5/6)(1/6)^3 + (1/6)^4$$
$$= p_0 + p_1 + p_2 + p_3 + p_4.$$
The total probability of obtaining either no aces or one, two, three or four aces is 1^4, i.e. 1, since every time four dice are thrown together one of these possibilities must occur.

In general, if $\mathfrak{p}$ is the probability of occurrence of the event (i.e. of success) in a single trial, the probability that an event will occur exactly r times in n trials is the $(r+1)$th term of the binomial expansion of $(\mathfrak{q}+\mathfrak{p})^n$. This result can be briefly deduced as follows. The probability of obtaining exactly r successes in n trials is
$$\mathfrak{p}.\mathfrak{p}.\mathfrak{p} \ldots \text{to } r \text{ factors, i.e. } \mathfrak{p}^r,$$
while the probability that the remaining $n-r$ trials in the group of n will all be failures is $\mathfrak{q}^{(n-r)}$. The compound probability of r successes and

$n-r$ failures in any group of n trials is therefore $\mathfrak{q}^{(n-r)}\mathfrak{p}^r$. This result can be obtained in a number of equally likely ways, according to the order in which successes and failures occur. The number of ways is equal to the number of different selections of r things chosen from a total of n, i.e. to ^nC_r. The total probability p_r of obtaining r successes in n trials is therefore given by

$$p_r = {}^nC_r \mathfrak{q}^{(n-r)} \mathfrak{p}^r, \qquad (11.2)$$

which is the $(r+1)$th term of the binomial expansion of $(\mathfrak{q}+\mathfrak{p})^n$ (see Chap. 4, § 6).

EXAMPLE. *Injection of a certain dose of digitalis per unit of body weight into a large number of frogs causes the death of 40 per cent of them. What is the probability that the number of deaths will be between 3 and 5 when this dose is injected into each of a group of ten frogs?*

In this problem, the trial is the injection of digitalis, and the event is the death of a frog. The number of trials n is 10, and the probability of death in a single trial is taken as the proportion of deaths in a large number of trials, i.e. $\mathfrak{p} = 0 \cdot 4$ and $\mathfrak{q} = 0 \cdot 6$. In ten trials the expected number of deaths $(n\mathfrak{p})$ is 4. The probability p_4 of obtaining exactly four deaths in ten trials is the fifth term of the expansion of $(0 \cdot 6 + 0 \cdot 4)^{10}$; hence

$$p_4 = {}^{10}C_4 (0 \cdot 6)^6 (0 \cdot 4)^4 = 0 \cdot 2508, \text{ or } 25 \cdot 08 \text{ per cent.}$$

Therefore, only once in every four sets of ten trials will the expected result of exactly four deaths be obtained.

Similarly, the probabilities of three (p_3) and five (p_5) deaths are given by

$$p_3 = {}^{10}C_3 (0 \cdot 6)^7 (0 \cdot 4)^3 = 0 \cdot 2150, \text{ or } 21 \cdot 50 \text{ per cent.}$$

$$p_5 = {}^{10}C_5 (0 \cdot 6)^5 (0 \cdot 4)^5 = 0 \cdot 2006, \text{ or } 20 \cdot 06 \text{ per cent.}$$

Thus, in two out of every three sets of ten trials the number of deaths will be between ± 1 of the expected number, 4.

This example illustrates the very considerable variation, due to chance, which is to be expected when tests of this type are used to measure the potency of a preparation of a drug.

4. The binomial distribution

Suppose that in the EXAMPLE given above, in which ten frogs are used in the test, the number of positive results in a set consisting of n trials is X_1. If a second set of n trials is performed, the number of positive results is likely to be different from X_1; let it be X_2. If a large number, K sets, of n trials are made, a series of values of X between 0 and n will be obtained. This group of K values of X is called a *binomial distribution*, and each value of X will occur a number of times in the distribution. The frequency of occurrence of each value of X is denoted by f, and the sum of all frequencies, $\sum f$, will be equal to K.

The most frequently occurring value of X will be the most likely value $n\mathfrak{p}$ (if $n\mathfrak{p}$ is an integral number), where $\mathfrak{p}$ is the exact probability of success in a single trial and may be known either from a large number of previous trials, as in the digitalis example, or known from theoretical considerations, as in the throw of a die.

The frequency of occurrence f_X of any value of X in the distribution will be proportional to the probability of occurrence p_X of this value of X in any one set of n trials. If K is large, then

$$f_X = Kp_X, \quad \text{or} \quad p_X = f_X/K,$$

so that p_X can therefore be regarded as the *fractional frequency of occurrence* of a particular value of X, and

$$\Sigma p_X = \Sigma(f_X/K).$$

But $\Sigma f_X = K$, hence $\Sigma p_X = 1$.

The values of p_X are the coefficients of the expansion of the binomial $(\mathfrak{q}+\mathfrak{p}\theta)^n$, where θ is an undefined constant and is introduced as a label for each term of the expansion, the coefficient of θ^r being the $(r+1)$th term of the expansion $(\mathfrak{q}+\mathfrak{p})^n$. This expression is called the *generating function* of the distribution. A variable whose values are controlled by a function in this way is called a *variate*.

The spinning of a coin can be considered as a simple example of a binomial distribution, the trial being a single spin of the coin. If 'heads' uppermost represents success and 'tails' represents failure, then the exact probability of success in a single trial is 0·5. If sets of four spins are taken ($n = 4$), then the number of successes (X) in any set can vary between 0 and 4, with a most likely value ($n\mathfrak{p}$) of 2.

If a large number of sets of trials, e.g. $K = 2000$, are made, the number of times each value of X occurs, i.e. the most likely frequency of occurrence f of each value of X, can be calculated from the expansion of $(0·5+0·5)^4$, and hence it follows that

when $X = 0$ (no 'heads' in 4 trials), $f_0 = 125$;
when $X = 1$ (1 'head'   ,,   ), $f_1 = 500$;
when $X = 2$ (2 'heads'   ,,   ), $f_2 = 750$;
when $X = 3$ (3 'heads'   ,,   ), $f_3 = 500$;
when $X = 4$ (4 'heads'   ,,   ), $f_4 = 125$.
$$\Sigma f = \overline{2000} = K.$$

The expected probability of occurrence p of any value of X in a single set of trials will be the fractional frequency of each value; thus

when $X = 0$, $\quad p_0 = f_0/K = 0·0625$;
when $X = 1$, $\quad p_1 = f_1/K = 0·2500$;
when $X = 2$, $\quad p_2 = f_2/K = 0·3750$;
when $X = 3$, $\quad p_3 = f_3/K = 0·2500$;
when $X = 4$, $\quad p_4 = f_4/K = 0·0625$.
$$\Sigma p = \overline{1·0000}.$$

CHAPTER 11

Mean and variance. The *arithmetic mean* m of the distribution of values of X is the sum of all the values of X in the distribution divided by the number of values. In order to obtain the sum, each value of X is multiplied by its frequency of occurrence, and these products are added together; thus

$$m = \frac{\Sigma fX}{K} = \sum\left(\frac{f}{K}\right)X. \tag{11.3}$$

Hence
$$m = \Sigma pX,$$

or if $f_1, f_2, \ldots$ are the frequencies of occurrence of the values $X_1, X_2, \ldots$ of X and $p_1, p_2, \ldots$ are the corresponding fractional frequencies, then

$$m = \frac{f_1 X_1 + f_2 X_2 + \ldots}{K} = \frac{f_1}{K}X_1 + \frac{f_2}{K}X_2 + \ldots$$
$$= p_1 X_1 + p_2 X_2 + \ldots$$

In Appendix V, the value of m for any binomial distribution is shown to be equal to the most likely value of X, np. If np is not an integer, the most likely integral value of X will be the nearest integer to np.

To obtain a quantitative measure of the scatter of the values of X about this mean, a quantity called the *variance V* is introduced; this is defined as the sum of the squared deviations of the values of X from the mean divided by the number of values of X. By squaring the deviations their signs are eliminated, and a negative deviation contributes as much to the variance as does a positive deviation of equal magnitude. The sum of squared deviations is found by multiplying each of them by the frequency of occurrence of the value of X, and then adding these products; thus

$$V = \frac{f_1(X_1-m)^2 + f_2(X_2-m)^2 + \ldots}{K}$$
$$= p_1(X_1-m)^2 + p_2(X_2-m)^2 + \ldots$$

Hence
$$V = \frac{\Sigma f(X-m)^2}{K}$$
or
$$V = \Sigma p(X-m)^2. \tag{11.4}$$

In Appendix V the variance of any binomial distribution is shown to have the value npq.

EXAMPLE. *If X is the number of deaths resulting from the injection of the dose of digitalis into ten frogs in the example in § 3, page 162, calculate the arithmetic mean and variance of the values of X resulting from a large number of these tests.*

The mean m and variance V can be found readily from the formulae derived in Appendix V. However, to illustrate the calculation of these quantities, they

PROBABILITY

are evaluated here arithmetically. The values of p_X, i.e. $p_0, p_1, p_2, \ldots$, corresponding to the values $0, 1, 2, \ldots$ of X given in the following table, are derived from the successive terms of the expansion of $(0 \cdot 6 + 0 \cdot 4)^{10}$.

X	p	pX	$(X-m)$	$(X-m)^2$	$p(X-m)^2$
0	0·0061	0	−4	16	0·0976
1	0·0403	0·0403	−3	9	0·3627
2	0·1209	0·2418	−2	4	0·4836
3	0·2150	0·6450	−1	1	0·2150
4	0·2508	1·0032	0	0	0
5	0·2006	1·0030	1	1	0·2006
6	0·1115	0·6690	2	4	0·4460
7	0·0425	0·2975	3	9	0·3825
8	0·0106	0·0848	4	16	0·1696
9	0·0016	0·0144	5	25	0·0400
10	0·0001	0·0010	6	36	0·0036
	$\Sigma p = \overline{1 \cdot 0000}$	$\Sigma pX = \overline{4 \cdot 0000}$			$\Sigma p(X-m)^2 = \overline{2 \cdot 4012}$

Thus
$$m = \Sigma pX = 4 \cdot 0, \text{ and } V = \Sigma p(X-m)^2 = 2 \cdot 40.$$

The values of m and V obtained arithmetically are seen to be in agreement with those derived theoretically; thus

$$m = n\mathrm{p} = 10(0 \cdot 4) = 4; \text{ and } V = n\mathrm{p}\mathrm{q} = 10(0 \cdot 4)(0 \cdot 6) = 2 \cdot 4.$$

5. Frequency-distribution histogram

A frequency-distribution histogram is a diagram showing the frequency of occurrence of values of a variate X plotted against the variate. It consists of a series of rectangles, each centred on the ordinate at the value of the variate; the width of each rectangle is proportional to the variate unit and the height of each rectangle is proportional to the fractional frequency of each value of X.

The p, X results given in the table in the example in § 4 above are plotted as a histogram in Fig. 11a. This diagram shows how the values of X in the distribution are grouped about their mean value of 4. The fractional frequency of occurrence of values of X decreases rapidly as the positive or negative deviation from the mean increases.

A histogram can be used to estimate the fraction of the total number of values of the variate between given limits. For example, the fraction of the total values of X between 3 and 5 in Fig. 11a is

$$(p_3 + p_4 + p_5)/\Sigma p,$$

Σp being the sum of all the values of p from p_0 to p_{10}. Since the rectangles of the histogram are all of equal width, this fraction is equal to the shaded area ABCD in the figure divided by the total area under the histogram. The percentage of values of the variate between 3 and 5 is therefore given by

100 × shaded area ABCD/total area under histogram.

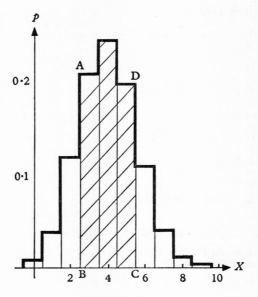

Fig. 11a. Histogram

Normal distribution. If n, the number of trials in each set, is made very large, the number of possible values of the variate X is increased, since X can have values from 0 to n. If the length, in centimetres, of the abscissa of the histogram is kept constant, the width of each of the rectangles will diminish as n increases until the stepped outline of the histogram approximates to a smooth curve which will still show the distribution of values of X about the mean m (see Fig. 11b). This curve represents an algebraic relation between p_X and X. The exact form of this relation when n is large and $\mathfrak{p}$ is moderately large in relation to $1/n$ is deduced from the binomial series in Appendix VI. The relation can be expressed as the normal distribution equation

$$p_X = p_m \exp\{-(X-m)^2/2V\}, \qquad (11.5)$$

where p_X and p_m are the fractional frequencies of occurrence of any value of X and of the mean value m respectively, and V is the variance of the values of X about the mean. One advantage of this relation is that it expresses p_X in terms of the mean and variance of the distribution; the quantities n, $\mathfrak{p}$ and $\mathfrak{q}$ in the binomial generating function from which the distribution arises are eliminated by applying the relations $m = n\mathfrak{p}$, and $V = n\mathfrak{p}\mathfrak{q}$, both of which hold for any binomial distribution.

Equation (11.5) represents a curve similar to that plotted in Fig. 11b, which is symmetrical about the mean, $X = m$. As the deviations from the mean increase either positively or negatively, the values of p_X decrease rapidly to zero.

Owing to the large number of possible values of X in a normal distribution, the fractional frequency of occurrence of values of X between given limits is of greater interest than is the frequency of occurrence of individual values. The former quantities can be found, as in the case of the histogram,

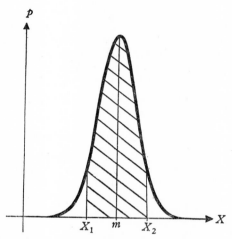

FIG. 11b. The normal distribution

by measuring suitable areas under the curve. If P is the fraction of the total values of X between the limits X_1 and X_2, then in Fig. 11b

$$P = \frac{\text{Shaded area}}{\text{Total area under the curve}}.$$

Since the equation of the curve is known, these areas can be found by integration (see Chap. 7, § 1); thus

$$P = \int_{X_1}^{X_2} p_X \, dX \bigg/ \int_{-\infty}^{\infty} p_X \, dX. \qquad (11.6)$$

The limits $\pm \infty$ are used for the total-area integral in order to include all possible values of X.

If the number of values in the distribution is the very large number K, then PK of these values will be between the limits X_1 and X_2, and $(1-P)K$ values will be outside. If a value of X is chosen completely at random from the whole distribution, the event that it will be between X_1 and X_2 can happen in PK equally likely ways and fail to happen in $(1-P)K$ ways. Therefore, from the definition of probability (see § 1), the probability of obtaining a value of X between X_1 and X_2 when a completely random choice is made from the whole distribution is

$$\frac{PK}{PK+(1-P)K}, \text{ i.e. } P.$$

This result is the basis of the application of the theory of statistics to experimental results. As can be seen from (11.6), the probability of obtaining by random choice a value between given limits of a normally distributed variate X can be found by the purely theoretical process of integrating the normal distribution equation between these limits. This equation cannot be integrated by simple algebra, but the numerical value of the integral for given values of the limits X_1 and X_2 can always be found by computation (see Chap. 7, § 7), and tables giving the values of the integral are available (see page 284); the value between infinite limits in the denominator of (11.6) is finite (see Appendix VI).

6. Experimental measurements and distributions

The 'ten-frog' tests described in the examples in § 3 and § 4 can be regarded as a crude measurement of the potency of a given dose of digitalis; the mean result is four deaths, but when the test is repeated several times, different results scattered about this mean are obtained. This scatter can be considered as due to the unavoidable experimental error of the measurement and is estimated quantitatively by the variance. In the same way, repeated measurements in physics and chemistry give differing results scattered about a mean with a distribution which is often approximately normal. If the quantity measured is a physical property instead of an observation of the number of successes in a given number of trials, the significance of n, p and q in the generating function of the distribution becomes obscure. However, sets of repeated results can be interpreted by means of (11.5) without knowing anything about these three quantities. This equation gives the relation between p_X and X in terms of the mean and variance of the repeated measurements, and both these quantities can be estimated from the set of measurements themselves.

The normal distribution equation is really only applicable to the very large distributions resulting from a very large number of repeated measurements. In practice, an experiment can only be repeated a limited number of times, and in the theory of statistics the limited number of values of the variate X so obtained is regarded as a *sample* of the theoretical *universe* of repeats which would form a proper normal distribution.

The Poisson distribution. The Poisson distribution is a binomial distribution in which n is very large, but the probability of occurrence of an event in a single trial p is very small. This distribution, therefore, describes the frequency of occurrence of an unusual event. The classical application of the Poisson distribution is to the analysis of the number of deaths due to horsekicks in the Prussian army. More recent, and undoubtedly more useful, applications are to the study of counting techniques such as those used in bacteriology, in blood analysis and also in the quantitative estimation of radioactivity (Geiger–Müller counter). In these measurements, only a small proportion of the total number of particles present are counted by the device used and so the rare event is that a particle happens to be counted.

The form of the Poisson distribution is derived from the binomial distribution. If the probability of occurrence p in a single trial is accurately known, the probability of occurrence p_r of an event r times in n trials is given by the binomial term

$$p_r = {}^nC_r q^{(n-r)} p^r$$
$$= \frac{n(n-1)(n-2) \times \ldots \times (n-r+1)}{r!} q^{(n-r)} p^r.$$

If n is very large and r fairly small, the factors $(n-1), (n-2), \ldots (n-r+1)$ can be written as approximately equal to n. There are r of these factors in the above expression, so that their product is approximately n^r. Since $q^{(n-r)} = q^n/q^r$ and $q = 1-p$, then

$$p_r \simeq \frac{n^r}{r!} \cdot \frac{(1-p)^n}{(1-p)^r} p^r = \frac{(np)^r}{r!} \cdot \frac{(1-p)^n}{(1-p)^r}.$$

If p is very small and r is fairly small, then $(1-p)^r \simeq 1$; this approximation cannot be made for $(1-p)^n$ as n is very large. Further, since this distribution is binomial, the mean value m of the variate is given by $m = np$. Using this relation to eliminate p in the expression for p_r gives

$$p_r = \frac{m^r}{r!} \left(1 - \frac{m}{n}\right)^n.$$

If n is very large, then from (4.11) it follows that

$$(1 - m/n)^n \simeq e^{-m}.$$

Hence

$$p_r = m^r e^{-m}/r!. \qquad (11.7)$$

This is the equation giving the probability of occurrence of an unusual event r times (where r is 0, 1, 2, 3, ...) in a large number of trials. If $r = 0$, then $p_0 = m^0 e^{-m}/0!$; since $0! = 1$ (see Chap. 4, § 5), $p_0 = e^{-m}$. Thus

for $r = 0$, $\quad p_0 = e^{-m}$;
for $r = 1$, $\quad p_1 = me^{-m}$;
for $r = 2$, $\quad p_2 = m^2 e^{-m}/2!$;
for $r = 3$, $\quad p_3 = m^3 e^{-m}/3!$, and so on.

The values of p are said to form a *Poisson series*; it is only necessary to know the mean value m of the variate in order to define this series.

CHAPTER 11

Since the Poisson distribution is a binomial, its variance V is npq. But as p is very small, q, which is $1-p$, is approximately equal to 1. Hence $V = np = m$. The variance is therefore equal to the mean. This property is often used to characterise a Poisson distribution.

EXAMPLE. *Show that the results of Rutherford and Geiger, given in the table below, for 2608 repeated counts of the number of alpha particles falling on a counting device from a given source in 7·5 sec form a Poisson distribution.*

(In the table, X is the number of particles registered in a count and is only a small fraction of the large number of particles emitted from the source, f is the frequency with which this number of particles was obtained in the 2608 counts, p is the fractional frequency of occurrence of each value of X, i.e. $f/2608$, and $e^{-m}m^X/X!$ is the value of p calculated from the Poisson series.)

X	f	p	pX	$e^{-m}m^X/X!$
0	57	0·0219	0·0000	0·0208
1	203	0·0778	0·0778	0·0806
2	383	0·1469	0·2938	0·1561
3	525	0·2013	0·6039	0·2015
4	532	0·2040	0·8160	0·1949
5	408	0·1564	0·7820	0·1509
6	273	0·1047	0·6282	0·0997
7	139	0·0533	0·3731	0·0539
8	45	0·0173	0·1384	0·0261
9	27	0·0104	0·0936	0·0112
10	10	0·0038	0·0380	0·0043
11	4	0·0015	0·0165	0·0015
12	1	0·0004	0·0048	0·0005
13	1	0·0004	0·0052	0·0001
	$\Sigma f = 2608$	$\Sigma p = 1·0001$	$\Sigma pX = 3·8713$	$1·0021$

The calculated values of p in the last column are seen to be in reasonable agreement with the observed values in the third column, and the values of X form a Poisson distribution. For evaluating $e^{-m}m^X/X!$, the value of m is given by ΣpX, i.e. 3·8713, while e^{-m} is obtained by making use of logarithms; thus

$$\ln(e^{-m}) = -m.$$

Hence

$$\log(e^{-m}) = -m/2\cdot3026 = -3\cdot8713/2\cdot3026 = -1\cdot6813$$
$$= \bar{2}\cdot3187.$$

Therefore

$$e^{-m} = \text{antilog}\,(\bar{2}\cdot3187) = 0\cdot02083.$$

Problems

1. Calculate the probability of obtaining exactly three aces in eight throws of a poker die.
2. A dose of a drug per unit of body weight which was sufficient to kill 20 per cent of frogs in a large number of trials is injected into a group of ten animals. What is the probability that the number of deaths will be within ± 1 of the expected value (the mean)?

A similar test is carried out with ten frogs using a dose which was sufficient to kill 50 per cent of them in a large number of trials. What is the probability that the number of deaths in the group of ten animals will be within ± 1 of the expected value?

3. Three poker dice are thrown together 216 times; how many times would you expect to obtain exactly two aces in one throw?

4. The fractional frequencies of occurrence (p_X) of values of a variate X are given by the expansion of $(\frac{2}{3}+\frac{1}{3})^6$, i.e. $q = \frac{2}{3}$, and $p = \frac{1}{3}$.

Plot a histogram showing the distribution of the values of X about their mean. Evaluate the mean and variance of the distribution.

5. The number of bacteria seen in equal areas of a microscope field are given in the following table:

No. of bacteria	0	1	2	3	4	5
No. of areas	24	47	36	18	4	2

Show by calculating χ^2 (see page 198) that these results agree with the Poisson distribution.

CHAPTER 12

Statistical analysis of repeated measurements

WHEN AN experimental measurement is repeated under apparently identical conditions and varying results are obtained, the extent of the scatter of these results about their mean value gives a measure of the random error involved in the measurement. In chemical and physical experiments, the scatter is usually small. In quantitative biology, however, it may be very large; it then becomes particularly important to be able to assess the reliability of the mean of a set of measurements. By means of statistics this reliability can be expressed in the form of numerical limits of error for a given probability level. In other words, as a result of statistical analysis of a set of repeated measurements, it can be stated that in 19 cases out of 20 ($P = 0.95$) or in 99 cases out of 100 ($P = 0.99$) the true result is within certain calculated limits (*limits of error*) of the observed mean. The main purpose of this chapter is to show how these limits of error can be computed.

1. Mean and variance

The results of a set of N repeated measurements of a quantity X, denoted by $X_1, X_2, X_3, \ldots$, can best be summarised by calculating their arithmetic mean m and their variance V about this mean. Mean and variance have been defined by (11.3) and (11.4); if each individual value of X occurs only once in the set, the f_X values in these equations are all 1. Hence

$$m = \Sigma(X)/N, \quad \text{and} \quad V = \Sigma\{(X-m)^2\}/N.$$

If the mean is calculated from a small number of results, a better estimate of the variance than is given by (11.4) is obtained by dividing the sum of the squared deviations by $N-1$ rather than by N. This introduces the idea of the number of *degrees of freedom* ϕ into statistical calculations; ϕ can be defined as the number of independent comparisons which can be made between the numerical results. When the mean is calculated from the results, one degree of freedom is used up, leaving only $N-1$ degrees for the calculation of variance. For example, if the mean m of three results X_1, X_2, X_3 is calculated, only two deviations from this mean, say X_1-m and X_2-m, are independent, the third one X_3-m being fixed since

$$m = \tfrac{1}{3}(X_1+X_2+X_3),$$

i.e.

$$3m = X_1+X_2+X_3,$$

therefore
$$2m - X_1 - X_2 = X_3 - m,$$
and
$$X_3 - m = -(X_1 - m) - (X_2 - m).$$

The sum of the squared deviations includes only two independent comparisons of the results, and the denominator of the variance equation should be 2 rather than 3; thus

$$V = \Sigma\{(X-m)^2\}/\phi = \Sigma\{(X-m)^2\}/(N-1),$$

the second expression for V being used when m is calculated from the values of X.

The expression for V can be put in a more convenient form for calculations if the squared deviations are expanded; thus

$$V = \frac{\Sigma\{(X-m)\}^2}{N-1} = \frac{(X_1-m)^2 + (X_2-m)^2 + \ldots}{N-1}$$

$$= \frac{(X_1^2 - 2mX_1 + m^2) + (X_2^2 - 2mX_2 + m^2) + \ldots}{N-1}$$

$$= \frac{(X_1^2 + X_2^2 + \ldots)}{N-1} - \frac{2m(X_1 + X_2 + \ldots)}{N-1} + \frac{(m^2 + m^2 + \ldots)}{N-1}$$

$$= \frac{\Sigma(X^2) - 2m\Sigma(X) + Nm^2}{N-1}.$$

But $m = \Sigma(X)/N$, i.e. $\Sigma(X) = Nm$. Hence

$$V = \frac{\Sigma(X^2) - 2Nm^2 + Nm^2}{N-1}$$

$$= \frac{\Sigma(X^2) - Nm^2}{N-1}.$$

This equation is used in the form

$$V = \frac{\Sigma(X^2)}{N-1} - \frac{\{\Sigma(X)\}^2}{N(N-1)},$$

or

$$V = \frac{\Sigma(X^2) - \Sigma^2(X)/N}{N-1}, \tag{12.1}$$

$\Sigma^2(\ )$ being used to denote $\{\Sigma(\ )\}^2$. The numerator of this expression is the sum of the squared deviations of the values from the mean and is written here and subsequently as $\Sigma\, d^2$

$$V = \frac{\Sigma\, d^2}{N-1}.$$

The sums $\Sigma(X^2)$ and $\Sigma^2(X)$ can be rapidly evaluated from a set of numerical values of X by means of a calculating machine. Errors introduced by the need to round off the value of m in calculating deviations $X-m$ are avoided by using (12.1) for computing variance.

It should be noted that the transformation

$$\Sigma\{(X-m)^2\} = \Sigma(X^2) - Nm^2 = \Sigma(X^2) - \Sigma^2(X)/N = \Sigma\, d^2$$

is frequently used in subsequent sections; when $\Sigma\, d^2$ is to be evaluated from numerical data the third of the above expressions should be used

$$\Sigma\, d^2 = \Sigma(X^2) - \Sigma^2(X)/N.$$

2. Standard deviation

The units of variance are the square of those of the mean. In order to describe experimental errors, it is desirable to have a measure of the scatter of a set of results about their mean which has the same dimensions as has the mean. The square root of the variance, called the *standard deviation s*, is used for this purpose; thus

$$s = \sqrt{V}. \tag{12.2}$$

The ratio of the standard deviation to the mean, usually expressed as a percentage, is called the *coefficient of variation* of single results about the mean.

3. Statistics and experimental errors

Systematic errors in a set of measurements cannot be detected by the application of a simple statistical analysis. A systematic error is one which occurs in the same way in each measurement; a simple example is the error in a series of volumetric determinations caused by the use of a pipette which delivers 20·1 ml instead of 20·0 ml. Statistical examination of repeated results obtained with this pipette will not reveal this error; the systematic error will not be discovered until results are obtained with a standardised pipette. Before results are submitted to statistical analysis, systematic errors should be eliminated as far as possible.

In considering quantitative results a convention has been developed which distinguishes between the *accuracy* and *precision* of a set of repeated measurements. Although these two words are, according to the dictionary, synonymous, the experimentalist's convention considers that the word accuracy refers to the mean of the results, while precision refers to the scatter about the mean. If the results are accurate, their mean is near to the true value of the measurement, although their scatter could be large; if the results are precise, it implies that their scatter about the mean is small, but because of a systematic error their mean could differ appreciably from the true value. *It is better to be roughly accurate than precisely wrong.*

If systematic errors are eliminated, the standard deviation of a set of repeated measurements may be used to give an assessment of the random error of the mean. More definite conclusions about the reliability of a mean can be obtained by considering the scatter of a series of means, each obtained from a number N of repeated measurements, about the true value of the measurement. This true value can be defined as the mean of a very large number, or *universe*, of results, such a mean being denoted by the symbol μ. From a very large number of results the true, or universe, standard deviation σ of single results about μ can be calculated. It should be noted that in statistics, the Roman letters m and s are used to denote a mean and a standard deviation, respectively, of a small set of numbers; the Greek letters μ and σ are used to denote the corresponding quantities for a very large or theoretical set of numbers.

If a universe of single results is divided into groups or *samples*, each containing a set of N values, the means of the samples, $m_1, m_2, m_3, \ldots,$ will differ from the universe mean μ, and they will be scattered about μ with a standard deviation equal to the standard error of the sample mean, $\sigma/\sqrt{N}$ (see Appendix VIII).

In practice, only a limited number of repeats can be made for a measurement, and from these only a sample mean m and a sample standard deviation s can be calculated. The best estimate of the standard deviation of m about the true result μ is then $s/\sqrt{N}$. This quantity, called the *standard error* of a mean, provides an estimate of the error of the mean. It takes into account the number of results used to calculate the mean. From the definition of standard error given above, it is seen that if N is doubled, the standard error is reduced by a factor of $1/\sqrt{2}$, or 0·71.

4. Standard error

Standard error is the quantity used to calculate the *limits of error* of a mean. These limits give the range within which the true value of the result will lie; they cannot be stated with certainty but they can be estimated for given probability levels. For most biological assays, where the scatter of results about their mean is large, a probability level of 0·95 is taken, meaning that in 19 cases out of 20, the true result will lie within the limits. When standard errors are small, as in most physical and chemical experiments, a probability level of 0·99 (99 cases out of 100) is often used. In any application it is important to state the level of probability used in calculating the limits.

The calculation of limits of error from standard errors is based on the theory of the normal distribution, which is outlined in § 5, and discussed in Appendix VI.

It should be noted that standard deviation measures the variation of single results about the true mean (if $N-1$ is used in the denominator of the variance equation), while standard error is an estimate of the variation of a mean about a hypothetical true value or universe mean.

5. The normal distribution

The normal distribution has been described in Chap. 11 as a limiting case of the binomial distribution. The equation of the normal distribution curve (11.5) can be expressed in a form which has as parameters the mean and variance of the numbers in the distribution.

Large sets of repeated experimental measurements are often distributed about their mean in a way which can be expressed by the normal distribution equation. In dealing with sets of repeated measurements, the significance of the quantities $\mathfrak{p}$, $\mathfrak{q}$ and n in the generating function of the distribution becomes obscure, but fortunately their values are not required in the practical applications of normal distribution theory.

The equation of the normal distribution can be given in terms of standard deviation; thus

$$p_X/p_\mu = \exp\{-(X-\mu)^2/2V\} = \exp\{-(X-\mu)^2/2\sigma^2\},$$

where p_X is the fractional frequency of occurrence of any value of X in the distribution, and p_μ is the fractional frequency of occurrence of the mean μ. The distribution is assumed to contain a very large number of values of the variate X, and therefore μ and σ are used for the mean and standard deviation respectively.

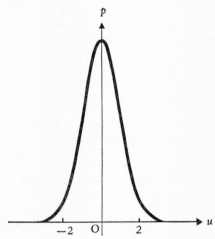

Fig. 12a. The normal distribution

In order to simplify this equation, a linear function of X, u, is introduced into the equation, u being called the *normal deviate* of a value of X and is the ratio of the deviation of this value X from the mean to the standard deviation, i.e.

$$u = (X-\mu)/\sigma. \tag{12.3}$$

STATISTICAL ANALYSIS OF REPEATED MEASUREMENTS 177

Hence
$$p_X/p_\mu = \exp(-\tfrac{1}{2}u^2). \tag{12.4}$$

Graphs representing normal distributions can be plotted from (12.4); they are curves such as that shown in Fig. 12a and are symmetrical about the ordinate at $u = 0$. It was shown in Chap. 11, § 5, that if a value of the variate is picked at random from the normal distribution, then the probability P that it will lie within given limits, X_1 and X_2, of the mean can be calculated by integrating the normal distribution equation

$$P = \int_{X_1}^{X_2} p_X \, dX \bigg/ \int_{-\infty}^{\infty} p_X \, dX.$$

X_1 and X_2 are converted to their corresponding normal deviates u_1 and u_2. Since μ and σ are constants for a given distribution, it follows from (12.3) that
$$du = dX/\sigma, \quad \text{i.e.} \quad dX = \sigma \, du.$$
Substituting this value for dX and the value from (12.4) for p_X gives

$$P = \sigma p_\mu \int_{u_1}^{u_2} \exp(-\tfrac{1}{2}u^2) \, du \bigg/ \sigma p_\mu \int_{-\infty}^{\infty} \exp(-\tfrac{1}{2}u^2) \, du,$$

where σ and p_μ are constants for a given distribution and so can be written outside the integration signs and cancelled in the ratio. The denominator in the ratio is shown in Appendix VI to have the value $\sqrt{(2\pi)}$. Hence

$$P = \frac{1}{\sqrt{(2\pi)}} \int_{u_1}^{u_2} \exp(-\tfrac{1}{2}u^2) \, du. \tag{12.5}$$

Therefore the probability of choosing a value of X within equal and opposite limits $\pm u_1 \sigma$ of the mean is given by

$$P = \frac{1}{\sqrt{(2\pi)}} \int_{-u_1}^{u_1} \exp(-\tfrac{1}{2}u^2) \, du,$$

or
$$P = \sqrt{\left(\frac{2}{\pi}\right)} \int_0^{u_1} \exp(-\tfrac{1}{2}u^2) \, du, \tag{12.6}$$

$\exp(-\tfrac{1}{2}u^2)$ being an even function of u (see Chap. 8, § 11). The integral in the alternative form of (12.6) is often called the *probability integral*; its value for given numerical values of u_1 can be found by expanding $\exp(-\tfrac{1}{2}u^2)$, integrating term by term, and substituting the values of the limits u_1 and 0 in each term. Alternatively integration by Simpson's rule can be used (see Chap. 7, § 7); a programme for the integration of this function has been given in COMP 4, Chap. 7 (page 117). Some numerical values of P for given values of u_1 and vice versa are given in the following table:

u_1	1	2	3		P	0·90	0·95	0·99
P	0·683	0·955	0·997		u_1	1·64	1·96	2·58

These figures mean that if a value of the variate X is picked at random from a normal distribution, there is a 0·683 (or 68·3 per cent) probability that its normal deviate u_1 will be within the limits ± 1, a 0·955 (or 95·5 per cent) probability that it will be within ± 2, and a 0·997 (or 99·7 per cent) probability that it will be within ± 3. Alternatively, if round-figure probabilities are chosen, there is a 0·95 (or 95 per cent) probability that u_1 will be within $\pm 1·96$, and a 0·99 (or 99 per cent) probability that it will be within $\pm 2·58$. Each probability level is associated with a definite normal deviate u_1.

6. Limits of error of a mean

Consider the distribution of sample means, m_1, m_2, m_3, ..., each derived from N individual results, about the true mean μ of a very large distribution. Although the individual values of X may not be normally distributed about μ, the values of sample means will, in nearly all cases, be normally distributed about μ providing the samples are chosen at random from the whole distribution. The standard deviation of the sample-mean distribution about μ is $\sigma/\sqrt{N}$ (see Appendix VIII), where σ is the standard deviation of single values of X about μ. The normal deviate u of any sample mean is therefore given by

$$u = \frac{(m-\mu)}{\sigma/\sqrt{N}}. \qquad (12.7)$$

If a sample mean is picked at random from the values of m_1, m_2, m_3, ..., it can be deduced from the table given in § 5 that there will be a 0·95 (95 per cent) probability that u will be between $\pm 1·96$, i.e. that the difference between sample mean and true mean is less than $1·96\sigma/\sqrt{N}$, since $m-\mu = u\sigma/\sqrt{N}$.

The mean of a small set of repeated experimental measurements can be regarded as a randomly chosen sample mean of a universe of repeats, and the result given above can be used to state that there is a 0·95 probability that it diverges from the true value of the result by an amount less than $1·96\sigma/\sqrt{N}$. In other words, $\pm 1·96\sigma/\sqrt{N}$ are the limits of error for the mean value at the 0·95 probability level. At the 0·99 probability level, the limits of error are $\pm 2·58\sigma/\sqrt{N}$.

In attempting to establish limits of error for an experimental mean from a limited number of repeats, a difficulty arises, however, from the fact that the universe standard deviation σ is not usually known, the best estimate of it being the sample standard deviation s. If the limits of error are calculated from the standard error $s/\sqrt{N}$ instead of from $\sigma/\sqrt{N}$, the numerical factor in the limits of error must be increased to allow for the possible difference between s and σ, and this is discussed in § 7.

7. The factor t

By studying the distribution of sample standard deviations, s_1, s_2, s_3, ..., about the universe value σ, 'Student' was able to compute from

STATISTICAL ANALYSIS OF REPEATED MEASUREMENTS 179

the mathematics of the normal distribution, a set of factors for use with the standard error in the calculation of limits of error. These factors, called t, allow for effects due to the use of s in place of σ. The values of t vary with the number of degrees of freedom ϕ in the calculation of s and with the probability level chosen for the limits. Use of t gives wider limits at a given probability level than are given by the normal deviate u, but when ϕ is large (> 30), the value of t becomes almost equal to that of u. A table of values of t for the 0·95 and 0·99 probability levels for various values of ϕ is given in Appendix XII, Table 2.

The limits of error of a sample mean are calculated from the sample standard deviation s by use of the formula

$$\text{limits of error} = \pm ts/\sqrt{N}, \qquad (12.8)$$

which makes allowance for the deviation of s from σ. These limits can then be taken as a reliable assessment of the limits of experimental error in the mean result of a set of repeated measurements, providing systematic errors have been eliminated.

The following examples illustrate the method for calculating limits of error from numerical data. The first example deals with the results of a biological assay and the second with a set of physico-chemical measurements.

EXAMPLE 1. In the assay of a digitalis preparation, the volume of solution (the dose) per unit of body weight required to stop the heart of a guinea-pig is determined. At the same time, the volume for a standard preparation of the drug is found using another guinea-pig under identical conditions. Similar pairs of experiments are repeated a number of times. The ratio of the volume of the standard to that of the test preparation is calculated for each pair of results, and this ratio is inversely proportional to the potency ratio of the preparations (the higher the volume, the lower is the potency). Ideally, the value for the volume ratio should be the same for each pair of results, but owing to uncontrollable variation between animals, the actual values of the ratio differ appreciably from one another. The mean value of the ratio gives the best estimate of the true potency ratio, and the standard deviation of the other results about this mean can be used to calculate values for the limits of error of the mean ratio.

In the results tabulated below, X is the volume ratio (standard/test) for a series of ten pairs of observations.

X	X^2
0·770	0·592900
0·720	0·518400
0·754	0·568516
0·740	0·547600
0·720	0·518400
0·795	0·632025
0·940	0·883600
0·765	0·585225
0·860	0·739600
0·855	0·731025
$\Sigma(X) = 7·919$	$\Sigma(X^2) = 6·317291$

Therefore
$$m = \Sigma(X)/N = 7\cdot919/10 = 0\cdot7919.$$
$$V = \frac{\Sigma d^2}{N-1} = \frac{6\cdot317291 - (7\cdot919)^2/10}{9}$$
$$= 0\cdot0051372.$$

The variance V is a small difference between two relatively large numbers, so that approximations should not be made in the calculations until the standard deviation s has been calculated. $\Sigma(X)$ and $\Sigma(X^2)$ can be evaluated simultaneously by means of a calculating machine, while s is found from V by using square root tables. In order to avoid errors in the decimal place of s, it is best to use indices; thus
$$V = 0\cdot0051372 = 51\cdot372 \times 10^{-4}.$$
Hence
$$s = \sqrt{51\cdot372 \times 10^{-2}} = 7\cdot167 \times 10^{-2} = 0\cdot07167.$$
The standard error of the mean e_m is equal to $s/\sqrt{N}$, i.e.
$$e_m = 0\cdot07167/\sqrt{10} = 0\cdot02266.$$
The limits of error are found by using the value of t corresponding to $P = 0\cdot95$ and $\phi = 9$; this value of t is $2\cdot26$. Hence
$$\text{limits of error} = \pm 2\cdot26 \times 0\cdot0227 = \pm 0\cdot0513 \quad (P = 0\cdot95).$$
The potency ratio (test/standard) is equal to the mean value of X, and so the final result of the assay is given by
$$\text{potency ratio } (T/S) = 0\cdot792 \pm 0\cdot051 \quad (P = 0\cdot95).$$

When limits of error are given the probability level used should always be stated. The above result means that it is highly probable that the true potency ratio of the preparations lies between the limits $0\cdot843$ and $0\cdot741$, i.e. the experimental error of the mean is $\pm 0\cdot051$, or $\pm 6\cdot4$ per cent. It should be noted that these limits of error refer to the mean; the errors of the individual comparisons will be higher (see page 185).

EXAMPLE 2. In the table below, X is the magneto-optic rotation of water in degrees per ampere of current passing through an electromagnet which surrounds a polarimeter tube containing water. These results were obtained at varying current strengths; half of the results were obtained with the current flowing in a clockwise direction and the other half with an anticlockwise current. Ideally, all the values of X should be the same, but owing to errors in reading both the rotation and the current strength, there is some variation of the results about their mean value.

X	X^2
(rotation/current)	
0·391	0·152881
0·377	0·142129
0·392	0·153664
0·376	0·141376
0·385	0·148225
0·377	0·142129
0·384	0·147456
0·380	0·144400
0·382	0·145924
0·381	0·145161
0·384	0·147456
0·382	0·145924
$\Sigma(X) = \overline{4\cdot591}$	$\Sigma(X^2) = \overline{1\cdot756725}$

Therefore
$$m = 4\cdot 591/12 = 0\cdot 38258.$$
$$V = \frac{1\cdot 756725 - (4\cdot 591)^2/12}{11} = 25\cdot 9 \times 10^{-6}.$$

Hence
$$s = \sqrt{25\cdot 9 \times 10^{-3}} = 5\cdot 09 \times 10^{-3}.$$
$$e_m = (5\cdot 09 \times 10^{-3})/\sqrt{12} = 1\cdot 47 \times 10^{-3}.$$

The limits of error are calculated on a $P = 0\cdot 99$ basis, the appropriate value of t being $3\cdot 11$ ($\phi = 11$, $P = 0\cdot 99$), as is found from Appendix XII, Table 2. Therefore

limits of error = $\pm 3\cdot 11 \times 0\cdot 00147 = \pm 0\cdot 00457.$ ($P = 0.99$)

The final result for the magneto-optic rotation of water determined in this apparatus, expressed in degrees per ampere, is therefore $0\cdot 3826 \pm 0\cdot 0046$. It should again be noted that the limits refer to the mean; the individual results are seen in the table above to lie outside these limits in certain cases. The final result means that there is a 99 per cent probability that the true rotation lies between $0\cdot 3780$ and $0\cdot 3872$ degree per ampere, this true rotation being the theoretical mean of a very large number (a universe) of repeated measurements.

COMP 7: Computer programme for mean, variance, standard deviation and error, and limits of error. In COMP 7 the calculation in example 1 is carried out by a computer programme. The data are read in as elements of the subscripted variable X, together with the number of results, N, and the tabulated value of t for the appropriate number of degrees of freedom and probability level, TTAB.

```
C         COMP 7
C         STATISTICAL ANALYSIS OF REPEATED MEASUREMENTS
          DIMENSION X(10)
          READ(5,1) N,TTAB,X
 1        FORMAT(I5,11F6.0)
          TOTX = 0.
          TOTX2 = 0.
          DO 2 I = 1,N
          TOTX = TOTX + X(I)
 2        TOTX2 = TOTX2 + X(I)**2
          EN = N
          XM = TOTX/EN
          SSQX = TOTX2 - (TOTX**2)/EN
          VX = SSQX/(EN - 1.)
          SDX = SQRT(VX)
          VXM = VX/EN
          SEXM = SQRT(VXM)
          REXM = TTAB*SEXM
          WRITE(6,3) N,XM,VX,SDX,SEXM,REXM
 3        FORMAT(5H1N = ,I5/8H0MEAN = ,F10.6/12H0VARIANCE = 
         1 ,F10.6/22H0STANDARD DEVIATION = ,F10.6/18H0STANDARD
         2 ERROR = ,F10.6/19H0LIMITS OF ERROR = ,F10.6)
          STOP
          END
```

The sum of numbers, TOTX, and the sum of squared numbers, TOTX2 are evaluated by the DO loop, but these variables are first set to zero as in COMP 4, § 7 of Chap. 7 (page 117). The calculations are then performed on these results to give the sum of squared deviations, Σd^2, called SSQX, the mean, XM, the variance of results about the mean, VX, the standard deviation, SDX, the variance of the mean, VXM, the standard error of the mean, SEXM, and the limits of error for the mean, REXM.

The values required are printed out with a FORMAT which requires two continuation cards. The numbers in column 6 of these cards have nothing to do with the arrangement of the print-out, they are simply indications of continuation. Each item printed has been given a title by means of an appropriate H format and a new page has been called for the print-out by the 1 after the first H, subsequently carriage control zeroes have been put after H to call for double spacing.

The data card for this problem is

```
10    2·26  0·770  0·720  0·754  0·740  0·720  0·795  0·940
      0·765  0·860  0·855
```

The number 10 must be placed at the right-hand side of the I field, that is in columns 4 and 5, the other numbers can be placed anywhere within their F6·0 fields but TTAB must come first according to the FORMAT; decimal points are required for all the numbers except N but the zeroes in front of the decimal points in the values of X could be omitted.

The print-out with the above data is

```
N =        10
MEAN =        .791900
VARIANCE =        .005137
STANDARD DEVIATION =     .071674
STANDARD ERROR =     .022665
LIMITS OF ERROR =     .051224
```

8. Large sets of repeated measurements

If a large number of repeated results (> 30) are obtained, the calculation of the mean and variance is simplified, with only a slight loss in accuracy, by classifying the results into groups and finding the frequency of occurrence f of values in each group. The group width, i.e. range of values in the group, should be so chosen to give ten to fifteen groups. The mean and variance are then calculated using (11.3) and (11.4), taking X as the central value of each group. Then

$$m = \Sigma(fX)/N,$$

and

$$V = \frac{\Sigma\{f(X-m)^2\}}{N-1} = \frac{\Sigma(fX^2)}{N-1} - \frac{\Sigma^2(fX)}{N(N-1)}.$$

STATISTICAL ANALYSIS OF REPEATED MEASUREMENTS 183

The values of m and V so found are slightly less accurate than are those found from the individual results, the classification having introduced a definite bias which makes the variance larger than it should be. This can be remedied by using *Sheppard's correction* $h^2/12$ (h being the group width in units of X), which is subtracted from the value of V calculated from the classified results.

EXAMPLE. *Fifty repeated determinations of the phosphorus content of samples of a natural phosphatide fraction gave the following results:*

3·92	4·14	4·41	4·27	3·46	4·15	4·51	3·63
4·03	4·27	4·07	3·82	4·47	4·29	4·07	4·33
3·97	4·03	3·51	3·92	4·11	3·99	4·09	3·63
3·84	4·42	3·79	4·02	3·94	4·11	3·90	3·53
3·83	3·77	4·12	3·61	4·03	3·82	3·65	3·92
4·23	4·02	3·91	3·88	4·21	4·61	3·71	4·32
3·93	3·95						

Calculate the mean result and limits of error for the mean.

The smallest result is 3·46 and the highest is 4·61, so that the *range* of the results is 4·61–3·46, i.e. 1·15; in order to obtain between 10 and 15 groups, a suitable group width is therefore 0·10. A table is made showing the group limits, and the number of results falling in each group gives the group frequency f. The products fX and fX^2 are calculated, taking the central value for the group as X; thus

Group limits	Group frequency f	Central values of group X	fX	fX^2
3·40 to 3·49	1	3·445	3·445	11·8680
3·50 to 3·59	2	3·545	7·090	25·1341
3·60 to 3·69	4	3·645	14·580	53·1441
3·70 to 3·79	3	3·745	11·235	42·0751
3·80 to 3·89	5	3·845	19·225	73·9201
3·90 to 3·99	10	3·945	39·450	155·6303
4·00 to 4·09	8	4·045	32·360	130·8962
4·10 to 4·19	5	4·145	20·725	85·9051
4·20 to 4·29	5	4·245	21·225	90·1001
4·30 to 4·39	2	4·345	8·690	37·7581
4·40 to 4·49	3	4·445	13·335	59·2741
4·50 to 4·59	1	4·545	4·545	20·6570
4·60 to 4·69	1	4·645	4·645	21·5760
	$\Sigma(f) = 50$		$\Sigma(fX) = 200·550$	$\Sigma(fX^2) = 807·938$

Therefore
$$m = \Sigma(fX)/N = 200.550/50 = 4.011,$$
and
$$V = \frac{\Sigma(fX^2) - \Sigma^2(fX)/N}{N-1}$$
$$= \frac{807.938 - (200.550)^2/50}{49}$$
$$= 0.072088.$$

Sheppard's correction $h^2/12$ is $(0.09)^2/12 = 0.0007$. Hence the corrected variance is 0.0714. The value of s using the uncorrected variance is 0.268, and with the correction, s is 0.267, the difference being negligible for most practical purposes. Taking s as 0.268, the standard error e_m is given by

$$e_m = s/\sqrt{N} = 0.0379.$$

The value of t for $P = 0.95$ and $\phi = 49$ is 2.01, so that the limits of error are $\pm 2.01 \times 0.0379 = \pm 0.076$. It should be noted that these limits apply to the mean of 50 results; individual measurements clearly often lie outside the limits. This result means that there is a 95 per cent probability that the true phosphorus content of the material lies between 3.935 and 4.087.

9. Skewed distributions

When a very large number of repeated measurements are made and the results are classified into small groups, the plot of the fractional group frequency p (i.e. f/N) against the central value for each group X sometimes differs from the normal distribution curve. The deviations from normality can be of two kinds: (i) *kurtosis*, in which the relative height of the maximum differs from that for a normal distribution, and (ii) *skewness*, in which the ordinate at the maximum point does not bisect the curve (see Fig. 12b).

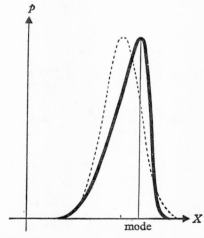

FIG. 12b. Skewed distribution (*broken curve represents a normal distribution*)

STATISTICAL ANALYSIS OF REPEATED MEASUREMENTS   185

In a skewed distribution, the most frequently occurring result or group is called the *mode*, and this is the value of X at the maximum of the p, X curve. In a normal distribution, the mode is equal to the mean, but in a skewed distribution they differ. One measure of the degree of skewness is given by the expression

$$\text{skewness} = (\text{mean} - \text{mode})/\sigma,$$

where σ is the standard deviation of the results about the mean.

If random samples of equal numbers of values of the variate X are drawn from quite highly skewed distributions, it is found that the distribution of sample means, $m_1, m_2, m_3, \ldots$ about the universe mean μ is approximately normal. The existence of some skewness in a distribution, therefore, does not invalidate the use of the properties of the normal distribution in deriving limits of error for a sample mean (the values of t are, of course, derived from the normal distribution).

10. Limits of error for a single measurement

It is sometimes of interest to be able to state the limits of error for a single result, the mean value μ and standard deviation σ of a large number of such results being accurately known from previous work.

If the distribution of single results about μ is normal, the limits are $\pm u\sigma$, where u is the value of the normal deviate for the probability level chosen, e.g. 1·96 for $P = 0.95$ and 2·58 for $P = 0.99$. If an estimate of σ is made from a limited number of results, t should be used in place of u in calculating the limits.

If the distribution of single results is skewed, the factor for the limits of error should be increased. The Camp-Meidell expression gives the probability that a result will lie within $\pm u\sigma$ of the distribution mean for all distributions whose skewness is less than 1 (assuming the distribution curve shows only one maximum value of p, i.e. has only one mode); this expression is

$$P \geqslant 1 - 1/(2 \cdot 25 u^2). \tag{12.9}$$

From this relation, the maximum value of u for $P = 0.95$ is 2·98 and for $P = 0.99$ it is 6·67. For practical purposes, if σ is accurately known, limits of $\pm 3\sigma$ can be taken as the approximate limits of error of a single result ($P = 0.95$). It should be noted that σ is used for limits of error for a single result, whereas $\sigma/\sqrt{N}$ is used for limits of error for a mean.

11. Cumulative frequency

The number of values of a variate X in a distribution which are less than or equal to a given value X_1 is called the *cumulative frequency* of X_1. If this number is divided by the total number of values of X, the *fractional cumulative frequency* $\mathfrak{F}$ is obtained. For example, the values of $\mathfrak{F}$ for the grouped frequency data on page 183 are shown overleaf.

X_1 (upper group limit in cumulative frequency table)	f	Cumulative frequency (equal to the sum of all the values of f up to and including the value in the same row)	$\mathfrak{F}$ (cum. freq./N) ($N = 50$)
3·49	1	1	0·02
3·59	2	3 (i.e. 1+2)	0·06
3·69	4	7 (i.e. 1+2+4)	0·14
3·79	3	10 (i.e. 1+2+4+3)	0·20
3·89	5	15	0·30
3·99	10	25	0·50
4·09	8	33	0·66
4·19	5	38	0·76
4·29	5	43	0·86
4·39	2	45	0·90
4·49	3	48	0·96
4·59	1	49	0·98
4·69	1	50	1·00

If $\mathfrak{F}$ is plotted against X, a sigmoid curve is obtained.

The theoretical value of $\mathfrak{F}$ for a given value of the normal deviate u can be found by integration; thus the fraction of the total number of values of the variate equal to or less than any value X_1 whose normal deviate is u_1, is the fraction of the total number of values of X which lie between the limits $-\infty$ and u_1 ($-\infty$ is taken as the lower limit so as to include all possible values of X less than X_1). This probability can be deduced from (12.5), with the lower limit equal to $-\infty$ and the upper limit equal to u_1; thus

$$\mathfrak{F} = \frac{1}{\sqrt{(2\pi)}} \int_{-\infty}^{u_1} \exp\left(-\tfrac{1}{2}u^2\right) du. \qquad (12.10)$$

This result is shown graphically in Fig. 12c; $\mathfrak{F}$ is the ratio of the shaded area to the left of the vertical line at $X = X_1$ to the total area under the curve. When $X = \mu$, this vertical line bisects the curve and so $\mathfrak{F} = 0.5$.

When values of $\mathfrak{F}$ calculated from the integral (12.10) are plotted against u_1, a sigmoid curve is obtained (see Fig. 12d) which has a point of oblique inflection ($d^2y/dx^2 = 0$, but $dy/dx \neq 0$ at a point of oblique inflection) at the value $u_1 = 0$, i.e. at the point where $X_1 = \mu$ and $\mathfrak{F} = 0.5$.

It is seen from (12.10) and from Fig. 12d, which is derived from it, that $\mathfrak{F}$ is a single-valued algebraic function of u_1; thus, for any given value of u_1 there is a single definite numerical value of $\mathfrak{F}$, and vice versa. A few of these values, derived from the probability integral, are shown in the table below.

u	$\mathfrak{F}$	u	$\mathfrak{F}$
−3	0·0014	0	0·500
−2	0·023	0·5	0·691
−1·5	0·067	1	0·841
−1	0·159	1·5	0·933
−0·5	0·309	2	0·977
		3	0·9986

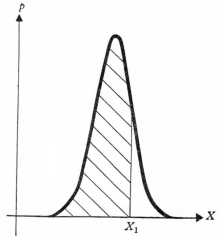

FIG. 12c. Cumulative frequency integral

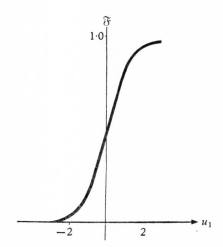

FIG. 12d. Cumulative frequency curve

12. Probability graph paper

Probability graph paper (see Fig. 12e) is used to find approximate values of the mean and standard deviation of fairly large sets of repeated results. The results are classified into groups, and the percentage cumulative frequency ($100\mathfrak{F}$) of each group is estimated. Assuming that the distribution is normal, each value of $100\mathfrak{F}$ corresponds to a normal deviate

u_1, whose theoretical value can be found from tables of $\mathfrak{F}$ and u based on (12.10). If these theoretical values of u_1 are plotted against the upper values of the variate X_1 for each group, a straight line should result, since

$$u_1 = (X_1 - \mu)/\sigma,$$

in which μ and σ are constants for a given distribution, and hence u_1 is a linear function of X_1.

In order to avoid the necessity for consulting tables giving the values of u_1 for each value of $100\mathfrak{F}$, probability graph paper has been developed. This has one arithmetic (or sometimes logarithmic) axis while the other axis shows values of the percentage cumulative frequency $100\mathfrak{F}$ spaced

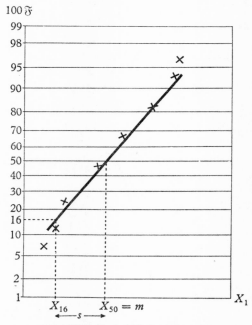

FIG. 12e. Probability graph paper

according to the corresponding values of u_1, just as with logarithmic graph paper, the numbers on the log-axis are spaced according to their logarithms. The use of probability paper only saves the labour of looking up values of u_1 in tables; exactly the same results are obtained by finding the values of u_1 from tables for each value of $100\mathfrak{F}$ and plotting u_1 against X_1 on ordinary graph paper.

If the distribution is normal, the plot on probability paper is linear. If the plot is not quite linear, the best straight line is drawn through the

STATISTICAL ANALYSIS OF REPEATED MEASUREMENTS 189

points in the centre of the plot, e.g. between $100\mathfrak{F} = 20$ and 80 per cent, since these are the most reliable points.

The mean and standard deviation of the values of X_1 are found from the plot as follows. The mean is simply the value of X_1 at the point on the line at which $100\mathfrak{F} = 50$ per cent. For finding the standard deviation, use is made of the facts that the value of $100\mathfrak{F}$ corresponding to $u = -1$ is, from (12.10), approximately 16 per cent and that the value of X_1 corresponding to $100\mathfrak{F} = 50$ per cent is the mean μ. Therefore, if X_{16} is the value obtained from the graph for X_1 at $100\mathfrak{F} = 16$ per cent, then

$$u_{16} = (X_{16} - \mu)/\sigma = -1,$$

and

$$\sigma = \mu - X_{16} = X_{50} - X_{16}.$$

The standard deviation σ is therefore the difference obtained from the graph between the values of X_1 at $100\mathfrak{F} = 50$ per cent and $100\mathfrak{F} = 16$ per cent.

Probits. The quantal response type of biological assay depends on the cumulative frequency, normal deviate, variate ($100\mathfrak{F}, u_1, X_1$) relations described in this section. This type of assay is outlined in Chap. 14.

Problems

1. In a biological assay, the following values of the potency ratio of two preparations of a drug were obtained:

1·05	1·17	1·06	0·98	1·11
1·09	1·02	1·20	1·03	1·07

    Calculate the mean potency ratio, the standard deviation about this mean and the standard error of the mean. Estimate the limits of error of the mean for $P = 0.95$. Do you consider that the preparations differ significantly in potency?

2. Calculate the percentage cumulative frequency $100\mathfrak{F}$ for each group in the table on page 183. Plot $100\mathfrak{F}$ against X_1 on probability paper, and find the approximate values of the mean and standard deviation from this plot. Compare these with the computed values.

3. The weights, in grammes, of 50 rats were

30	47	37	29	38	32	42	32	30	34
34	32	33	37	36	39	33	45	40	35
43	41	35	32	41	36	27	28	35	30
38	28	41	37	34	41	36	32	30	37
31	31	35	28	25	26	49	34	34	33

    Classify these results into groups and calculate the mean weight and the standard deviation both by computation and by means of probability paper. Within what limits would you expect the weight of a further single rat of this type to lie? (A group width of 3 is convenient.)

CHAPTER 13

Comparison of data by statistical methods. Tests of significance

IF TWO MEAN results differ from one another, two hypotheses can be put forward to explain the discrepancy; either it is due to chance experimental error, or it is due to a significant difference. The first hypothesis, that of *no* difference, is known as the *null hypothesis*. The second hypothesis, that there *is* a difference, is known as the *alternative hypothesis*. The null hypothesis is tested at a chosen *level of significance*. As a result of the test the null hypothesis will either be accepted, or it will be rejected and the alternative hypothesis will be tentatively accepted in its place. In this chapter, four different types of tests of significance are outlined:

(i) The normal deviate test, which is used to determine whether the mean of a small set of results differs significantly from the universe mean of a large number of similar results.
(ii) The t test, which is used to decide whether the means of two small sets of repeated measurements differ significantly from one another.
(iii) The variance ratio test, which is used to determine whether the scatter of one set of results about their mean is significantly different from the scatter of another independent set of similar results about their mean.
(iv) The χ^2 test, which is used to decide whether a series of frequencies of occurrence differ significantly from those predicted by a theory.

1. Normal deviate test

This test is used to decide whether a sample mean m of N results differs significantly from a known universe mean μ, the universe standard deviation σ being known. This situation arises in industrial work when the mean chemical analysis of a product or the efficiency of a batch process is known from a large number of observations. If, following some alteration to the process, a series of results are obtained which differ from this mean, the normal deviate test can be used to decide whether this is due only to random variation.

The normal deviate u_1 of the sample mean m from the universe mean μ is calculated using the expression

$$u_1 = \frac{m-\mu}{\sigma/\sqrt{N}}, \qquad (13.1)$$

where $\sigma/\sqrt{N}$ is the standard error of the means of groups of N results. If the difference between m and μ is due only to chance, i.e. the null hypothesis is assumed, this means that m is a sample mean drawn from a normal distribution of such means about μ. The probability that a randomly chosen sample mean will have a normal deviate within $\pm u_1$ is given by (12.6). The probability that a randomly chosen sample mean will be *equal to* or *outside the limits* $\pm u_1$ is P', where P' is equal to $1-P$. If the calculated value of u_1 is large, P' will be small; this can be seen from Fig. 13a, in which the value of P' is proportional to the *sum* of the two shaded areas. Furthermore, if P' is small, the probability that a randomly chosen sample mean will have a value as large as u_1 will be small; thus the probability

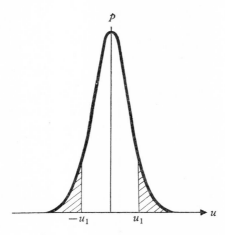

FIG. 13a. Test of significance

that the sample mean has, in fact, been drawn from the normal distribution is small, and the null hypothesis that the difference between the sample mean and the true mean is due only to chance is not likely to be correct. Thus P' can be regarded as a measure of the *level of significance* of the null hypothesis. In expressing *limits of error* the probability P is the probability that a quantity will be *within* certain limits; in *tests of significance*, the probability P' is the probability that a quantity will lie *outside* the limits. For given limits, the sum of the probabilities that the quantity will be either within the limits or outside them is 1, i.e. it is a certainty that one or other of these hypotheses will be correct. Thus

$$P+P' = 1,$$

and hence

$$P' = 1-P.$$

In the statistical tables in Appendix XII, both P and P' are given.

It is possible to find P' for any value of u_1 from the normal deviate tables (see Appendix XII, Table 1). If u_1 is large, P' will be small and the level of significance of the null hypothesis is low; this means that m is unlikely to have come from the normal distribution of sample means, and that the difference between m and μ is significant. If u_1 is small, P' is large, and the level of significance of the null hypothesis is high; m is therefore likely to have come from the normal distribution, and the difference between m and μ is not significant and can be reasonably attributed to chance variation of a sample mean from μ.

The exact level of significance chosen for acceptance or rejection of the null hypothesis varies with the type of problem considered. As a rough general rule, a probability of 0·05 is taken as the dividing line between acceptance and rejection of the hypothesis. If $P' < 0\cdot05$, the null hypothesis is rejected, and the difference between m and μ is considered to be significant. If $P' > 0\cdot05$, the null hypothesis is accepted, and the difference between m and μ is ascribed to chance. The value of u_1 corresponding to $P' = 0\cdot05$ is $\pm 1\cdot96$. If u_1 is outside these limits, $m-\mu$ is significant; if u_1 is within these limits, $m-\mu$ is not significant.

Other tests of significance are based on similar principles. A quantity $[u, t, F, \chi^2$ (see following sections)] which has a theoretical limiting value for a given probability level, derived theoretically, is calculated from two sets of experimental data. As in the normal deviate test, if the value calculated from the observations exceeds the theoretical value at a given probability level, usually $P' = 0\cdot05$, which is found from statistical tables, the null hypothesis is unlikely to be correct, and there is a significant difference between the two sets of results. If the calculated value of the quantity is less than the theoretical value, the difference between the two sets of experimental results can be considered as due to chance.

EXAMPLE. *The purity of a product was checked by analysis of a sample from each batch. The analysis for nitrogen, averaged over a long period, was 4·02 per cent, with a standard deviation of 0·06 per cent. Four successive batches in one week had a mean nitrogen content of 4·20 per cent. Does this indicate a significant change in the composition of the product?*

By inspection, the difference between m and μ, i.e. 4·20–4·02, does appear to be significant. Calculation of u_1 confirms this; since

$$u_1 = \frac{4\cdot20 - 4\cdot02}{0\cdot06/\sqrt{4}} = 6\cdot0.$$

The theoretical value of u_1 for a probability of 0·05 is 1·96. The observed value of 6·0 exceeds this considerably, and therefore the null hypothesis is extremely unlikely to be correct. Hence the difference between m and μ is highly significant, and the nitrogen content of the product has changed appreciably, indicating some fault in the process.

2. The t test

A more common problem is to decide whether the difference between

COMPARISON OF DATA BY STATISTICAL METHODS

two sample means is significant when the universe mean μ and standard deviation σ are unknown. In this case the normal deviate of the difference u_1 cannot be calculated, and the best estimate of it has to be found from the given samples.

If m_A and m_B are means of samples of N_A and N_B values respectively, then in order to decide whether the difference $m_A - m_B$ is significant, the null hypothesis is first assumed, i.e. that the difference is due only to chance, and that the true value of $m_A - m_B$ is zero. The quantity $m_A - m_B$ is then considered to be normally distributed about a universe mean of zero; the universe standard deviation is not known, and the best estimate of it has to be found from the variances of both the samples.

If $X_1, X_2, X_3, \ldots$ are the individual values in the sample of mean value m_A, and $X_1', X_2', X_3', \ldots$ are those in the sample of mean value m_B, then the overall variance V of the two samples is given by

$$V = \frac{(X_1-m_A)^2+(X_2-m_A)^2+\ldots+(X_1'-m_B)^2+(X_2'-m_B)^2+\ldots}{\phi}$$

$$= \frac{\Sigma d_A^2 + \Sigma d_B^2}{N_A+N_B-2};$$

since the two means have been calculated from the total of N_A+N_B results, ϕ is N_A+N_B-2.

The overall standard deviation s is $\sqrt{V}$, and the standard errors of the two means are $s/\sqrt{N_A}$ and $s/\sqrt{N_B}$ respectively. The standard deviation of the difference between the means about its universe value of zero is, from Appendix VII,

$$\sqrt{(s^2/N_A+s^2/N_B)} = s\sqrt{(1/N_A+1/N_B)} = s\sqrt{\frac{N_A+N_B}{N_A N_B}}.$$

The difference between the means divided by this standard deviation is not a normal deviate, since the above quantity is only an estimate of the universe standard deviation of $m_A - m_B$ about a universe mean of zero. It is, in fact, a value of t with the number of degrees of freedom used in the calculation of s, i.e. N_A+N_B-2, and

$$t = \frac{|m_A-m_B|}{s\sqrt{(1/N_A+1/N_B)}} \tag{13.2}$$

$$t = |m_A-m_B|\sqrt{\frac{N_A N_B(N_A+N_B-2)}{(N_A+N_B)(\Sigma d_A^2+\Sigma d_B^2)}}.$$

Theoretical limiting values of t have been tabulated for different probability levels and values of ϕ (see Appendix XII, Table 2). The probability level P' generally used for acceptance or rejection of the null hypothesis is 0·05, as in the normal deviate test. If the magnitude of the observed value of t is greater than the theoretical one for $P' = 0·05$, then the null hypothesis is usually rejected, and the difference between the means is considered to be significant.

EXAMPLE. *Two analysts each determine the iron content of a material and their results on ten samples are shown in the columns headed X_A and X_B respectively. Do their mean results differ significantly?*

Sample	X_A	X_A^2	X_B	X_B^2
1	4·40	19·3600	4·62	21·3444
2	4·62	21·3444	4·57	20·8849
3	4·43	19·6249	4·85	23·5225
4	4·60	21·1600	4·94	24·4036
5	4·55	20·7025	4·67	21·8089
6	4·43	19·6249	4·50	20·2500
7	4·46	19·8916	4·55	20·7025
8	4·39	19·2721	4·35	18·9225
9	4·75	22·5625	4·90	24·0100
10	4·71	22·1841	4·84	23·4256
Total	$\Sigma(X_A) = 45·34$	$\Sigma(X_A^2) = 205·7270$	$\Sigma(X_B) = 46·79$	$\Sigma(X_B^2) = 219·2749$
Means	$m_A = 4·534$		$m_B = 4·679$	
$\Sigma^2(X)/N$		205·5716		218·9304
$\Sigma(X^2) - \Sigma^2(X)/N = \Sigma d^2$		0·1554		0·3445

Both sets of results show scatter and the two mean results differ. The question to be answered is: is there a significant difference between the mean results of the two analysts?

To decide this, the value of t is calculated. The sums are simplified for computation, as in the deduction of equation (12.1): thus

$$s^2 = V = \frac{\Sigma d_A^2 + \Sigma d_B^2}{N_A + N_B - 2}$$

$$= \frac{0·1554 + 0·3445}{18} = 0·02777.$$

Hence

$$s = 0·1666.$$

Now, substituting the calculated values in (13.2) gives

$$t = \frac{4·679 - 4·534}{0·1666\sqrt{(1/10 + 1/10)}} = 1·95.$$

For $\phi = 18$, the value of t (see Appendix XII, Table 2) is 2·10 for $P' = 0·05$. The calculated value of t is less than this, and therefore it is concluded that the difference between the two mean results can be attributed to chance and that no significant difference between the results obtained by the two analysts can be established. If t had been greater than 2·10, it would have indicated a significant difference between the two analysts' results, suggesting that one of them had a faulty technique, and in this case the results of the analyst with the higher variance are possibly the less reliable. However, the difference in means might be due to a systematic error by one analyst which would not be revealed by the variance criterion. In fact, one of the analysts might be precisely wrong in his estimations.

COMP 8: Computer programme for t. In COMP 8 below, the two sets of data together with the numbers of results for carrying out the

COMPARISON OF DATA BY STATISTICAL METHODS

t-test for difference of means, are read in. The numbers of results are required in both integer form for the DO loop and in real form for calculating the sum of squares. They are read in as integers and then in order to avoid mixed mode (integer and real number) arithmetic, which is rejected by some Fortran compilers, they are also given real names. A statement such as ENA = NA is a legitimate one in Fortran, giving an integer an additional real name.

In COMP 8, the sums of numbers and squares are found by DO loops, the programme being put in a general form to cover cases where NA differs from NB. The only Fortran cards which then have to be altered for different problems are the DIMENSION and the FORMAT for reading the data. This latter FORMAT is designed so that all the values for group A of the values are on one card; the slash, /, means that the values for group B are on a separate card.

```
C       COMP 8
C       T TEST
        DIMENSION XA(10),XB(10)
        READ(5,1) NA,XA,NB,XB
1       FORMAT(I5,10F6.0/I5,10F6.0)
        TOTA = 0.
        TOTB = 0.
        TOTA2 = 0.
        TOTB2 = 0.
        DO 2 I = 1,NA
        TOTA = TOTA + XA(I)
2       TOTA2 = TOTA2 + XA(I)**2
        DO 3 J = 1,NB
        TOTB = TOTB + XB(J)
3       TOTB2 = TOTB2 + XB(J)**2
        ENA = NA
        ENB = NB
        AM = TOTA / ENA
        BM = TOTB / ENB
        SSQA = TOTA2 - (TOTA**2)/ENA
        SSQB = TOTB2 - (TOTB**2)/ENB
        T=(BM-AM)*(SQRT((ENA*ENB*(ENA+ENB-2.))/((ENA+ENB)*
       1 (SSQA+SSQB))))
        WRITE(6,4) NA,AM,SSQA,NB,BM,SSQB,T
4       FORMAT(6H1NA = ,I5//6H AM = ,F10.6//8H SSQA = ,F20.6
       1 //6H NB = ,I5//6H BM = ,F10.6//8H SSQB = ,F20.6
       2 //14H VALUE OF T = ,F10.6)
        STOP
        END
```

In this programme the sums of numbers and squares are first set to zero and are then evaluated by means of DO loops. The sums of squared deviations from the means

$$SSQA = \Sigma d_A{}^2 = \Sigma X_A{}^2 - \Sigma^2 X_A/N_A$$

are computed and are put into the formula for t using the second form of equation (13.2). The statement for this calculation involves some elaborate bracketing and in order to check that all the brackets have been closed, the numbers of brackets facing each way should be counted. In this case there are eight of each type.

In the WRITE statement most of the useful information from the calculation is included in the list. The FORMAT gives titles to the numbers and requires continuation indicated by numbers in column 6 of the Fortran card.

The print-out from this programme is given in the answer to Problem 6 of this chapter.

3. Variance ratio test

The variance ratio test is another method by which two samples of values of a quantity can be compared with one another. In the t test the means are compared, whereas in the variance ratio test the two sample variance values are compared by determining their ratio (always the larger variance divided by the smaller variance); this ratio is denoted by F.

If the null hypothesis is assumed, the two samples are both considered to be drawn from the same normal distribution. The distribution of sample variances about the universe variance can be computed from the theory of normal distribution, and so theoretical limiting values of the sample variance ratio F can be determined for given probability levels. If the samples contain different numbers of values, the value of F will be a function of the degrees of freedom of both the variances. A table of limiting values of F is given in Appendix XII, Table 4.

EXAMPLE. The variance ratio test can be applied to the data given for the example in § 2. From the two analysts' results,

$$V_A = \frac{\Sigma d_A^2}{N_A - 1} = \frac{0.1554}{9};$$

$$V_B = \frac{\Sigma d_B^2}{N_B - 1} = \frac{0.3445}{9}.$$

The variance ratio F (larger/smaller) is given by

$$F = \frac{0.3445}{0.1554} = 2.22.$$

The theoretical value of F at $P' = 0.05$ with 9 degrees of freedom in both variances is 3·2 (interpolated from Table 4, Appendix XII), and therefore, since the calculated value of F is considerably less than the theoretical value, the difference between V_A and V_B is not significant, i.e. it does not indicate that the technique of one analyst is significantly better than that of the other.

The value of F can be calculated by adding some statements to COMP 8. First the two variances are calculated by

```
VA = SSQA/(ENA-1·)
VB = SSQB/(ENB-1·)
```

Then, in order to obtain the greater variance divided by the lesser, an arithmetical IF may be used

```
      IF(VA−VB) 5,6,7
5     F = VB/VA
      GO TO 8
6     F = 1·
      GO TO 8
7     F = VA/VB
8     WRITE(6,4) NA,AM,SSQA,SSQB,T,F
4     FORMAT—as for COMP 7 with addition of
      //14H VALUE OF F = ,F10·6
```

The arithmetical IF is discussed under COMP 10 in Chap. 14, § 1 (page 220). Its operation is that if the quantity in the brackets is negative, execution of the programme goes to the first of the three statement numbers given after the brackets; when this quantity is zero, execution jumps to the second statement number; when it is positive, execution goes to the third statement number.

4. Analysis of variance

Variance is an additive property; the total variance V of a set of results consists of the sum of the variances $V_1, V_2, V_3, \ldots$ due to the different independent effects. It is often informative to break down the total variance of a set of results into components associated with different causes. The results are tabulated in a suitable form, the variances within groups and between groups are estimated, and their ratio F is determined. By comparing this value with theoretical values of F for a normal distribution, the decision can be made as to whether the variances differ significantly.

An example of the analysis of variance is given in Chap. 14, § 3.

5. The χ^2 test

This test should not be applied to quantitative measurements; it is only suitable for comparing the frequencies of occurrence of qualitative observations. The symbol χ^2 is a measure of the deviations of a number of observed frequencies from expectation; it is defined by the equation

$$\chi^2 = \sum \frac{(O-E)^2}{E}, \qquad (13.3)$$

where O is the observed frequency of occurrence of an event or an object in a given classification, and E is the expected frequency from some hypothesis. Summation is made over all classes into which the events or objects may fall. The magnitude of χ^2 then gives an indication of the significance of the hypothesis. If χ^2 is very large, this significance is low, and the hypothesis is probably wrong; if χ^2 is small, the hypothesis is probably correct.

If the frequencies of occurrence in each class follow a binomial distribution, χ^2 can be shown to be equal to $\Sigma\{(X-m)^2\}/V$, where X is the observed frequency, m is the mean or expected frequency, and V is the variance of X about m. This expression has the form of a sum of squared normal deviates; hence

$$\Sigma \frac{(X-m)^2}{V} = \Sigma \frac{(X-m)^2}{\sigma^2} = \Sigma u^2,$$

and, in fact, theoretical values of χ^2 are calculated from the normal distribution properties as a sum of normal deviates drawn at random from a normal distribution, i.e.

$$\chi^2 = u_1{}^2 + u_2{}^2 + u_3{}^2 + \ldots,$$

or

$$\chi^2 = \Sigma(u^2) \qquad (13.4)$$

Just as the maximum values of a normal deviate u_1 likely to be obtained by random choice from a normal distribution can be calculated for a given probability level (see Appendix XII, Table 1), so can maximum values of a sum of squared normal deviates (squared to eliminate effects due to sign), each drawn at random from the distribution, be computed for given probability levels. These values are the theoretical values of χ^2, the number of squared normal deviates summed being the degrees of freedom in the calculation.

Whereas the relation between u_1 and P is comparatively simple [see (12.6)], the relation between χ^2 and P is more complex. However, by suitable mathematical treatment of the normal distribution equation, these theoretical maximum values of χ^2 can be computed for different values of P and ϕ. Some of these values of χ^2 are given in Appendix XII, Table 3.

In the χ^2 test, the 0·05 level for P' is often taken as the dividing line between significant and chance differences. The number of degrees of freedom ϕ of the calculated χ^2 is the number of observations minus the number of quantities which have to be calculated from these observations in order to estimate χ^2.

If $\phi = 1$, then $\chi^2 = u^2$, the square of the normal deviate for that probability level. For $P' = 0·05$, $\chi^2 = u_1{}^2 = (1·96)^2 = 3·84$. If the observed value of χ^2 is greater than the theoretical value for the 0·05 level, then the hypothesis on which the values of E have been calculated is not likely to be correct.

When there are only two classes of data for the χ^2 summation, a better comparison between a calculated χ^2 with a low value of ϕ and the theoretical quantity calculated from the normal distribution is given by applying Yates' correction, which involves reducing the magnitude of $O-E$ by 0·5. In order to apply the χ^2 test of significance to a set of data, it is necessary to have at least six observations in each class.

COMPARISON OF DATA BY STATISTICAL METHODS

The use of χ^2 to check the fit between observed frequencies of occurrence of events and the expected frequencies derived from some hypothesis can be illustrated by using the figures given in the EXAMPLE on radio-active counting on page 170. In the table below, X is the number of particles registered in a particular count, O is equal to f (table, page 170), the observed frequency of occurrence of each count, and E is the value expected when a Poisson distribution ($E = 2608\mathrm{e}^{-m}m^X/X!$) is assumed; 2608 is the total frequency and $\mathrm{e}^{-m}m^X/X!$ is the proportion of that frequency expected for each value of X.

X	0	1	2	3	4	5	6	7	8	9	10	11	12	13
O	57	203	383	525	532	408	273	139	45	27	10	4	1	1
E	54	210	407	526	508	394	260	141	68	29	11	4	1	0

χ^2 is calculated from $\Sigma[(O-E)^2/E]$, the last three values being omitted as they contain less than six observations.

The sum gives $\chi^2 = 12\cdot13$. There are 10 degrees of freedom, as the mean has to be calculated in order to estimate the expected frequencies and three results have not been used. The theoretical value of χ^2 for $P = 0\cdot95$ and $\phi = 10$ is 18·3 (obtained from Table 3, Appendix XII), and therefore the differences between O and E are no greater than could be attributed to chance. The observed frequencies can therefore be considered to form a Poisson distribution.

6. Linear regression

Regression gives the best-fitting relation between two quantities, one of which is liable to chance error. In experimental work the values of one quantity Y are usually measured for given values of another quantity X. For example, in the calibration of an absorptiometer for a given analysis, Y is the reading of the instrument, while X is the concentration of the solution giving this value of Y. In cases of this type, there is negligible experimental error in X, but there may be an appreciable error in Y. If Y is plotted against X, the points often show scatter about a straight line or smooth curve due to this error. The problem is to determine the best fitting line or curve through the points; this is done by regression analysis.

The simplest case, and the only one dealt with here, is when the relation between Y and X should theoretically be a straight line, but in practice a set of scattered points are obtained due to experimental error in Y.

Method of least squares. The scatter diagram obtained when values of Y, a quantity liable to random error, are plotted against X, a quantity with negligible error, is shown in Fig. 13b. The best-fitting straight line through these points can be drawn roughly by eye. However, its position is more accurately determined by regression analysis, particularly if the scatter is large.

The regression line Y/X is the straight line which minimises the sum of the squares of the vertical deviations from the straight line (the deviations are squared in order to eliminate their sign), and this is considered to be the best-fitting straight line through the points.

Suppose the equation of the regression, or least squares, line is
$$y = a + bx.$$

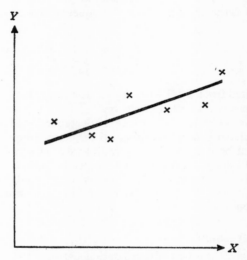

FIG. 13b. Regression line Y/X

Then, the problem is to determine the values of a and b which minimise the sum of the squares of the vertical deviations. If the experimental points are $(X_1, Y_1), (X_2, Y_2), (X_3, Y_3), \ldots$, then if y_1 is the value of Y calculated from the equation of the regression line when $x = X_1$, the vertical deviation Δ_1 of the point (X_1, Y_1) from the line is given by
$$\Delta_1 = Y_1 - y_1 = Y_1 - (a + bX_1).$$
Hence
$$(\Delta_1)^2 = (Y_1 - a - bX_1)^2 = Y_1^2 + a^2 + b^2 X_1^2 - 2aY_1 - 2bX_1Y_1 + 2abX_1.$$
Similar values are obtained for the squares of the vertical deviations of other points. Then if $z = (\Sigma \Delta^2)$, i.e. the sum of all the squared deviations,
$$z = \Sigma(Y^2) + Na^2 + b^2 \Sigma(X^2) - 2a \Sigma(Y) - 2b \Sigma(XY) + 2ab \Sigma(X),$$
where N is the number of points. For a given set of points, all the sums $\Sigma(\ )$ in the above equation will be constants, and the value of z will vary with the values given to a and b. The conditions for z to be a minimum (see Chap. 6, § 4) are
$$(\partial z/\partial a)_b = 0, \quad \text{and} \quad (\partial z/\partial b)_a = 0.$$

From the equation for z, it follows that for the regression line

$$(\partial z/\partial a)_b = 2Na - 2\,\Sigma(Y) + 2b\,\Sigma(X) = 0.$$

Therefore

$$\Sigma(Y) = b\,\Sigma(X) + aN,$$

that is,

$$\Sigma(Y)/N = b\,\Sigma(X)/N + a.$$

Now, $\Sigma(Y)/N$ is the arithmetic mean $\bar{Y}$ of all the values of Y (it should be remembered that these are not repeated values of the same result), while $\Sigma(X)/N$ is the mean $\bar{X}$ of all the values of X, hence

$$\bar{Y} = b\bar{X} + a,$$

i.e. the values $y = \bar{Y}$ and $x = \bar{X}$ fit the equation of the regression line. This means that the line passes through the point $(\bar{X}, \bar{Y})$. In order to fix the line completely, it is now only necessary to determine the slope b, which is found from the second condition for a minimum value of z; that is, for the regression line

$$(\partial z/\partial b)_a = 2b\,\Sigma(X^2) - 2\,\Sigma(XY) + 2a\,\Sigma(X) = 0.$$

Therefore

$$\Sigma(XY) = a\,\Sigma(X) + b\,\Sigma(X^2).$$

But $a = \bar{Y} - b\bar{X} = \Sigma(Y)/N - b\,\Sigma(X)/N$, so that

$$\Sigma(XY) = \Sigma(X)\,\Sigma(Y)/N - b\,\Sigma^2(X)/N + b\,\Sigma X^2.$$

Hence

$$b = \frac{\Sigma(XY) - \Sigma(X)\,\Sigma(Y)/N}{\Sigma(X^2) - \Sigma^2(X)/N}. \tag{13.5}$$

The final result for the regression line equation is therefore

$$y - \bar{Y} = b(x - \bar{X}),$$

a being eliminated by the result that when $y = \bar{Y}$, then $x = \bar{X}$. The slope b is called the *regression coefficient* Y/X, and its numerical value is computed from the set of values of Y and X after determining $\Sigma(X)$, $\Sigma(Y)$, $\Sigma(XY)$ and $\Sigma(X^2)$.

Covariance. The numerator of the regression coefficient is related to a quantity called the *covariance of two variates* σ_{xy}, which is defined by the equation

$$\sigma_{xy} = \frac{\Sigma\{(X-\bar{X})(Y-\bar{Y})\}}{\phi},$$

where X and Y are the corresponding values of the variates, and ϕ is the number of degrees of freedom in the calculation.

The covariance can be expressed in several different ways. For instance, since $\bar{X}$ is fixed for a given set of values of X, it can be taken outside the summation; thus

$$\Sigma\{(X-\bar{X})(Y-\bar{Y})\} = \Sigma\{X(Y-\bar{Y})\} - \bar{X}\,\Sigma(Y-\bar{Y}).$$

By the definition of a mean, $\bar{Y} = \Sigma(Y)/N$, so that

$$\Sigma(Y-\bar{Y}) = (Y_1-\bar{Y})+(Y_2-\bar{Y})+ \ldots$$
$$= \Sigma(Y)-N\bar{Y} = N\bar{Y}-N\bar{Y} = 0.$$

Therefore
$$\Sigma\{(X-\bar{X})(Y-\bar{Y})\} = \Sigma\{X(Y-\bar{Y})\},$$
and
$$\sigma_{xy} = \frac{\Sigma\{X(Y-\bar{Y})\}}{\phi}.$$

In the same way, it can be shown that

$$\sigma_{xy} = \frac{\Sigma\{Y(X-\bar{X})\}}{\phi}.$$

Another alternative form can be deduced, since

$$\Sigma\{(X-\bar{X})(Y-\bar{Y})\} = \Sigma(XY)-\bar{X}\,\Sigma(Y)-\bar{Y}\,\Sigma(X)+N\bar{X}\bar{Y}.$$

Substituting $N\bar{Y}$ for $\Sigma(Y)$ and $N\bar{X}$ for $\Sigma(X)$, the two central terms on the right-hand side of the equation each become $N\bar{X}\bar{Y}$, giving

$$\Sigma\{(X-\bar{X})(Y-\bar{Y})\} = \Sigma(XY)-N\bar{X}\bar{Y},$$

and hence
$$\sigma_{xy} = \frac{\Sigma(XY)-N\bar{X}\bar{Y}}{\phi}.$$

Therefore
$$\left.\begin{aligned}\phi\sigma_{xy} &= \Sigma\{(X-\bar{X})(Y-\bar{Y})\} \\ &= \Sigma\{X(Y-\bar{Y})\} \\ &= \Sigma\{Y(X-\bar{X})\} \\ &= \Sigma(XY)-N\bar{X}\bar{Y}.\end{aligned}\right\} \quad (13.6)$$

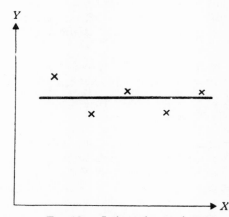

FIG. 13c. Independent variates

COMPARISON OF DATA BY STATISTICAL METHODS

If a scatter diagram is drawn in which the variate Y is independent of X, the slope b of the regression line will not be significant, and the line may be parallel to the X-axis (see Fig. 13c); the variations in Y are only random variations about their mean $\bar{Y}$. If $b = 0$, then $\Sigma(XY) - \Sigma(X)\Sigma(Y)/N = 0$, since this is the numerator of (13.5), and therefore $\sigma_{xy} = 0$. This is a general result stating that if two variates are independent, their covariance is zero.

Correlation coefficient. Before carrying out a regression calculation, a test is made to determine whether there is a *significant linear correlation* between the two sets of figures. This can be done by calculating the *correlation coefficient r*. This quantity, unlike the regression coefficient b, is symmetrical with respect to both X and Y and has a magnitude of 1 when there is a perfect linear relation between the two variables. If X and Y are independent, then $r = 0$. The correlation coefficient is defined by the equation

$$r = \frac{\Sigma\{(X-\bar{X})(Y-\bar{Y})\}}{\sqrt{\{\Sigma(X-\bar{X})^2 \Sigma(Y-\bar{Y})^2\}}}$$

$$= \frac{\Sigma(XY) - \Sigma(X)\Sigma(Y)/N}{\sqrt{[\{\Sigma(X^2) - \Sigma^2(X)/N\}\{\Sigma(Y^2) - \Sigma^2(Y)/N\}]}}. \quad (13.7)$$

Theoretical values for the magnitude of r can be computed for given probability levels and degrees of freedom ($\phi = N-2$). If the magnitude of r calculated from a set of data is greater than the theoretical value for a chosen probability level, then there is a significant correlation between X and Y; if the magnitude of r is less than the theoretical value, there is no significant correlation between X and Y and it is not worth proceeding with a regression analysis.

Some theoretical values of r for $P' = 0.05$ are given in Table 1.

TABLE 1

Theoretical values of the correlation coefficient ($P' = 0.05$)

Degrees of freedom ϕ	2	3	5	7	10	15	20	30	50	100
Correlation coefficient r	0.95	0.88	0.75	0.67	0.58	0.48	0.42	0.35	0.27	0.20

If a calculated value of r is less in magnitude than the value given in Table 1 for the appropriate value of ϕ, then the correlation between X and Y is unlikely to be significant.

Variance about regression. The variance V_Y of a set of points about their regression line is the sum of the squared vertical deviations from the line divided by the number of degrees of freedom in the calculation, usually $N-2$, where N is the number of points, since two degrees of freedom are used up in calculating the means $\bar{X}$ and $\bar{Y}$. Hence

$$V_Y = \frac{\Sigma(Y-y)^2}{\phi}.$$

Substituting $y = \bar{Y}+b(X-\bar{X})$ gives

$$\Sigma(Y-y)^2 = \Sigma\{Y-\bar{Y}-b(X-\bar{X})\}^2$$
$$= \Sigma(Y-\bar{Y})^2 + b^2\, \Sigma(X-\bar{X})^2 - 2b\, \Sigma\{(Y-\bar{Y})(X-\bar{X})\}.$$

From (12.1), it follows that

$$\Sigma(Y-\bar{Y})^2 = \Sigma Y^2 - \Sigma^2(Y)/N = \Sigma \mathrm{d}_Y{}^2 \quad \text{and} \quad \Sigma(X-\bar{X})^2 = \Sigma \mathrm{d}_X{}^2$$

while from (13.5, 13.6), it follows that

$$\Sigma\{(Y-\bar{Y})(X-\bar{X})\} = \Sigma(XY) - \Sigma(X)\,\Sigma(Y)/N$$
$$= b\,\{\Sigma(X^2) - \Sigma^2(X)/N\} = b\, \Sigma \mathrm{d}_X{}^2.$$

Hence
$$\Sigma(Y-y)^2 = \Sigma \mathrm{d}_Y{}^2 - b^2\, \Sigma \mathrm{d}_X{}^2,$$
and therefore
$$V_Y = (\Sigma \mathrm{d}_Y{}^2 - b^2\, \Sigma \mathrm{d}_X{}^2)/\phi. \tag{13.8}$$

Error of prediction of X from a regression line. In physico-chemical analysis, the measuring instrument is usually calibrated with solutions of known concentration. A regression line of *(instrument reading Y)/(concentration X)* is used as a calibration curve for determining the concentration of an unknown solution from the reading it gives with the instrument. It is then of interest to be able to calculate limits of error for the value of the concentration so obtained.

A formula by means of which the variance V_X of a value of X determined in this way from an observed reading Y using the regression line equation is deduced in Appendix IX; the formula is

$$V_X = \frac{V_Y}{b^2}\left\{1 + \frac{1}{N} + \frac{(Y-\bar{Y})^2}{b^2}\cdot\frac{1}{\Sigma(X-\bar{X})^2}\right\}.$$

From this formula, limits of error can be calculated as $\pm t\sqrt{V_X}$. The value of ϕ used to obtain t is $N-2$, where N is the number of points on which the regression line equation is based. Care should be taken in using this equation when the regression coefficient is small or when the errors are relatively large.

EXAMPLE. *Calculate the regression line equation for the following data, which relate the turbidity (Y) of solutions of an alkaloid to which phosphomolybdic acid has been added to the concentration (X). Estimate σ_{xy}, V_Y and V_X, and determine the limits of error of a value of X, calculated by means of the regression equation, from a solution which gave a turbidity reading of 25.*

X	2	4	6	8	10
Y	15	12	20	22	30

COMPARISON OF DATA BY STATISTICAL METHODS

It is seen that Y tends to increase with X, but there is considerable scatter of the results, which gives no great confidence in this process as a method for quantitative assay.

In order to calculate the regression line equation, the appropriate summations are made; thus

X	X^2	Y	Y^2	XY
2	4	15	225	30
4	16	12	144	48
6	36	20	400	120
8	64	22	484	176
10	100	30	900	300
$\Sigma(X) = 30$	$\Sigma(X^2) = 220$	$\Sigma(Y) = 99$	$\Sigma(Y^2) = 2153$	$\Sigma(XY) = 674$

Since $N = 5$, then $\bar{X} = 6$ and $\bar{Y} = 19.8$.

The correlation coefficient r is first calculated from (13.7); thus

$$r = \frac{674 - 30 \times 99/5}{\sqrt{\{(220 - 30^2/5)(2153 - 99^2/5)\}}}$$
$$= 0.911.$$

This value is greater than the theoretical value of r for $P' = 0.05$ and $\phi = 3$, i.e. 0.88; there is therefore a significant correlation between X and Y, and so it is worth proceeding with the regression analysis.

The regression coefficient is given by

$$b = \frac{\Sigma(XY) - \Sigma(X)\Sigma(Y)/N}{\Sigma(X^2) - \Sigma^2(X)/N}$$

$$= \frac{674 - 594}{220 - 180} = 2.00.$$

The regression line equation is

$$(y - \bar{Y}) = b(x - \bar{X}).$$

Hence
$$y - 19.8 = 2.00(x - 6),$$

that is
$$y = 2.00x + 7.8.$$

The variance V_Y of Y about the regression line is given by

$$V_Y = (\Sigma d_Y^2 - b^2 \Sigma d_X^2)/\phi$$
$$= \tfrac{1}{3}\{2153 - 1960 \cdot 2 - (2 \cdot 0)^2(220 - 180)\} = 10.9.$$

The variance V_X of a value of X predicted from the regression line when $Y = 25$ is given by

$$V_X = \frac{V_Y}{b^2}\left[1 + \frac{1}{N} + \frac{(Y - \bar{Y})^2}{b^2} \cdot \frac{1}{\Sigma\{(X - \bar{X})^2\}}\right].$$

Now,
$$\Sigma(X - \bar{X})^2 = \Sigma(X^2) - \Sigma^2(X)/N$$
$$= 220 - 180 = 40.$$

Therefore
$$V_X = \frac{10.9}{(2 \cdot 0)^2}\left\{1 + \frac{1}{5} + \frac{(25 - 19 \cdot 8)^2}{(2 \cdot 0)^2} \cdot \frac{1}{40}\right\} = 3.73.$$

Hence $\sqrt{V_X} = 1.93$. Limits of error are then calculated, using the value of t for $\phi = 3$ and $P = 0.95$; these limits are $\pm 3.18 \times 1.93 = \pm 6.14$.

The value of X calculated from the regression line equation for $Y = 25$ is 8·6. Therefore, the concentration of the unknown solution determined from the turbidity data is given by

$$X = 8.6 \pm 6.1. \qquad (P = 0.95.)$$

The wide limits of error confirm that the method as used to obtain these results is unsuitable for quantitative assay.

7. Computer programme for linear and curved regression

For the calculation of the regression line through a series of scattered points and for the calculation of the best fitting curve representing a polynomial

$$y = a_0 + a_1 x + a_2 x^2 + a_3 x^3 + \ldots + a_k x^k$$

through the points, if curvature appears to be significant, a programme in the Biomedical Computer package (BMD) may be used. The programme BMD05R is for polynomial regression and the value of k, the index of x in the last term may be set at any value up to 10. With $k = 1$, a linear regression is obtained. With higher values of k the data is analysed successively for each degree of polynomial up to the value set, each additional term altering the coefficients of all the previous terms. A plot is also given of the theoretical values from the equation against the experimental values put in as data.

Unlike the Scientific Subroutines, the Biomedical programmes are complete, with detailed instructions for input and a set form for the output. The user specifies the degree of polynomial required, whether any transformation of data is required before the regression analysis is begun, the number of pairs of values to be analysed, and the FORMAT in which they are presented on data cards. This FORMAT is flexible except that the data must be entered in successive pairs of X, Y (abscissa, ordinate) values.

No Fortran cards are required with the Biomedical programmes but additional job control cards are required to obtain the programme from the tape or disc on which it is stored. There is a special card on which details of the calculation required are given and then there is a variable FORMAT card which shows how the data are to be read from the data cards.

In COMP 9, the complete card deck to carry out a linear regression analysis of the data in the last example is given; the data given in the output are also listed.

COMP 9: Linear regression by BMD05R. Complete card deck for CDC 6600 computer, University of London.

Job Control Cards

```
JOB(      )                                    Your name
ATTACH(BINBIOM,BINBIOM)
COPYN(,DUMMY,BINBIOM)
RETURN(BINBIOM)
DUMMY
```

End of record card

```
BMD05R,*,BINBIOM
```

End of record card

Data Cards

```
PROBLM 9 5 10100                                             1
(10F5.0)
2.   15.   4.   12.   6.   20.   8.   22.   10.   30.
FINISH
```

End of file card

               No programme cards are required

The details of the Biomedical programmes are published in two volumes by the University of California Press. They are mostly statistical and cover a wide variety of data handling methods.

In this polynomial regression programme the output consists of the means of the two sets of results and their correlation coefficient. If the degree of polynomial required is greater than one, the computation is carried out for each successive degree and there is a print-out of the intercept at $X = 0$ and the regression coefficients together with their standard errors and an analysis of variance. When the final degree specified by k is reached the print-out includes in addition to the above, a table of observed and predicted values of Y and a plot of Y against X showing both the observed and the predicted values.

The limitations are that $k \leqslant 10$; n the number of cases (pairs of values of X and Y) must be greater than $(k+1)$ and less than 500; the number of variable FORMAT cards must be between one and ten.

In the card deck, the first section up to the first end of record card gives the job number and the instructions for calling the Biomedical programmes, the next section specifies the programme required, BMD05R; these are all job control cards and the remainder of the deck consists of ordinary data cards.

The first of these data cards is required to be PROBLM on which details of the problem to be solved are given. Columns 1 to 6 contain the word PROBLM; columns 7 and 8 give the user's number for the problem; 9 to 11 state the number of pairs of X,Y values; 12 and 13 give k, the degree of the regression required; 14 and 15 contain 01 if a table of residuals and a plot is required, 00 otherwise; 16 and 17 have 01 if transgeneration, that is, mathematical transformation, of Y is required before the regression analysis; if so, special transgeneration cards are added with codes indicating the transformation required (e.g. Y to log Y) if no such transformations are required 00 is punched in columns 16 and 17; in columns 71 and 72, the number of variable FORMAT cards is stated.

The FORMAT card (or cards) does not contain the word, but has the usual type of specification enclosed in brackets; X and Y are read in pairs. The next card(s) are the numerical data punched according to the format and consisting of alternate values of X and Y.

The last card is FINISH followed by the end of file card required by the CDC compiler.

In the publication describing the programmes, exact details of the card punching and deck assembly are given with a test case for each programme.

The numerical output for the data above, which is for the example in § 6 is

```
BMD05R - POLYNOMIAL REGRESSION

PROBLEM NO.     9
SAMPLE SIZE     5

REGRESSION - ONE INDEPENDENT VARIABLE
XMEAN.....               6.00000
YMEAN.....              19.80000
INTERCEPT (A VALUE)...   7.80000
REG. COEFFICIENT......   2.00000
STD. ERROR OF REG. COEF.  .52281
CORRELATION COEF......    .91098

             ANALYSIS OF VARIANCE FOR SIMPLE LINEAR REGRESSION
     SOURCE OF VARIATION      DEGREE OF    SUM OF      MEAN        F
                              FREEDOM      SQUARES     SQUARE      VALUE
DUE TO REGRESSION............     1        160.00000   160.00000   14.63415
DEVIATION ABOUT REGRESSION....    3         32.80000    10.93333
                    TOTAL....     4        192.80000

                    TABLE OF RESIDUALS
NO.    X VALUE     Y VALUE     Y PREDICTED    RESIDUAL
 1      2.0000     15.0000       11.8000       3.2000
 2      4.0000     12.0000       15.8000      -3.8000
 3      6.0000     20.0000       19.8000        .2000
 4      8.0000     22.0000       23.8000      -1.8000
 5     10.0000     30.0000       27.8000       2.2000
```

The equation of the regression line is

$$Y = 7·8 + 2·0X$$

The correlation coefficient is $r = 0·91098$ and the variance about regression is

$$V_v = 10·93.$$

Problems

1. A simple absorptiometer is calibrated for the determination of iron. The reading of the instrument should be a linear function of the iron concentration. The following results were obtained in calibrating the absorptiometer with solutions of known concentrations.

Concentration X	1	2	3	4	5	6	7
Reading Y	8·4	14·0	22·1	32·5	36·6	48·8	52·5

   Calculate the equation of the regression line Y/X. A solution of unknown concentration gave an instrument reading of 35·2; what is the estimate of its concentration from the regression line and what are the limits of error of this estimate ($P = 0·95$)?

2. The efficiency of a chemical batch process over a long period was found to be 88·3 per cent, with a standard deviation of 2·9 per cent. A change in the design of the plant used gave six successive values of efficiency as 89·2, 90·1, 88·0, 89·3, 91·5, 90·6 per cent. Do these figures indicate a significant improvement in efficiency?

3. A series of repeated pH determinations of a buffer solution with two different pH meters gave the following results:

   Meter A 8·01 8·04 8·03 8·07 8·02 7·98 8·05 8·10 8·06 8·04
   Meter B 8·08 8·03 8·17 8·02 8·17 8·19 8·07 8·18 8·04 8·08

   Calculate the means and standard deviations of these results. Is the difference between the two means significant and does the variance ratio indicate that one meter is more reliable than the other as far as random error is concerned?

4. After inspection of 1000 ampoules of a solution for injection, 137 were rejected. A modification of the technique for filling the ampoules reduced the rejection rate to 123 in a batch of 1000. On the basis of this information and the χ^2 test, could the reduction be considered to be significant?

5. A group of ten animals are weighed after regular treatment over a period with a preparation which is reputed to promote growth. These weights, together with those of a control group who have had the same diet except that they have not been given the preparation, are shown below.

Treated group	73	75	69	88	72	74	63	81	79	67
Untreated group	65	69	74	71	77	60	72	56	79	70

   Do these results indicate that the preparation has any significant effect on the mean weight of the animals?

6. Use COMP 8 to calculate t for the data in the example in § 2 (page 194).

7. Modify COMP 8 to solve Problem 3.

8. Use COMP 9 to find the correlation coefficient and regression equation for Problem 1.

9. In Problem 4 of Chapter 5, the initial concentration of reactant was $a = 26·92$ mol/litre; use COMP 9 to find the regression line of log $(a-x)$ against t. From the results, determine the half-life of the reaction.

CHAPTER 14

Some applications of statistics to biological assay and bacteriology

IN THIS CHAPTER some statistical principles are illustrated by their application to quantitative biology, and the values of the variance of a function of variates (see Appendix VII) are frequently used to calculate limits of error for a final result. Since the reproducibility of measurements on biological systems is usually poor, the theory of statistics plays an important part both in the design of experiments so as to obtain the maximum amount of information from them and in the interpretation of the results to give realistic limits of error for the final mean result.

1. Biological assays

Biological assay methods are used to compare the potencies of two preparations of a drug by observing their respective actions on biological systems. One preparation is a standard S whose potency in arbitrary units is known from a comparison with an agreed stable standard, and the other is a test preparation T whose potency is required.

Threshold dose assay. When using this type of assay, the test and standard preparations, or suitable dilutions of them, are injected into animals, and the minimum volume (the 'dose') required to produce a particular effect in each animal is determined. The results are worked out as the volume X of undiluted preparation per unit of body weight of the animal required to produce the effect. The mean values of X for the test and standard preparations, $\bar{X}_T$ and $\bar{X}_S$, respectively, are found. The *mean potency ratio (test/standard)* is then $\bar{X}_S/\bar{X}_T$, the potency of each preparation being inversely proportional to the volume required to produce the effect.

Limits of error (L.E.) for the potency ratio can be calculated from the standard errors of the means and the factor t (see Chap. 12, §§ 3, 7). A probability level for t of $P = 0.95$ is usually adopted, and the final answer then indicates that there is a 95 per cent probability that the true potency ratio of the two preparations lies between (mean−L.E.) and (mean+L.E.). In the *British Pharmacopoeia*, these limits are usually expressed as percentages of the observed mean.

SOME APPLICATIONS OF STATISTICS TO BIOLOGICAL ASSAY 211

In some assays it is found that the distribution of $\log X$ about its mean is more nearly normal than is that of X itself, and in such cases $\log X$ should be used as the variate. The calculations are made as before, but the result is the *mean log potency ratio*. The limits of error are symmetrical about this mean, but when antilogarithms are taken the limits of error of the potency ratio itself are not symmetrical about the mean, the upper limit being wider than the lower.

When the variation is small, both methods of calculation give similar results.

EXAMPLE. A preparation of digitalis tincture was assayed by comparing the dose required to arrest the heart of a guinea-pig with the dose of a standard preparation (potency, 1·316 units per ml) required to produce this effect in a similar animal. Each tincture was diluted with normal saline and injected continuously at a steady rate until the heart of the animal was arrested; a total of eight comparisons were made. The results are summarised in the table below, X being the volume in millilitres of the undiluted tincture per kilogramme of animal body weight required to produce this effect.

Test		Standard	
X_T	$X_T{}^2$	X_S	$X_S{}^2$
0·88	0·7744	1·51	2·2801
0·62	0·3844	1·50	2·2500
0·81	0·6561	1·55	2·4025
0·77	0·5929	1·47	2·1609
0·74	0·5476	1·29	1·6641
0·84	0·7056	1·28	1·6384
0·71	0·5041	1·15	1·3225
0·83	0·6889	1·36	1·8496
$\Sigma(X_T) = 6\cdot20$	$\Sigma(X_T{}^2) = 4\cdot8540$	$\Sigma(X_S) = 11\cdot11$	$\Sigma(X_S{}^2) = 15\cdot5681$

$\bar{X}_T = \Sigma(X_T)/N = 6\cdot20/8$
$\quad = 0\cdot775$

$V_T = \dfrac{\Sigma(X_T{}^2) - \Sigma^2(X_T)/N}{N-1}$

$\quad = \dfrac{4\cdot8540 - (6\cdot20)^2/8}{7}$

$\quad = 0\cdot00700$

$s_T = \sqrt{V_T} = 0\cdot0837$

Standard error of $\bar{X}_T$:

$e_T = s_T/\sqrt{N} = 0\cdot0837/\sqrt{8}$
$\quad = 0\cdot0296$

Coefficient of variation of $\bar{X}_T$:

$E_T = 100(e_T/\bar{X}_T)$ per cent
$\quad = 3\cdot819$ per cent

$\bar{X}_S = \Sigma(X_S)/N = 11\cdot11/8$
$\quad = 1\cdot389$

$V_S = \dfrac{\Sigma(X_S{}^2) - \Sigma^2(X_S)/N}{N-1}$

$\quad = \dfrac{15\cdot5681 - (11\cdot11)^2/8}{7}$

$\quad = 0\cdot01987$

$s_S = \sqrt{V_S} = 0\cdot1409$

Standard error of $\bar{X}_S$:

$e_S = s_S/\sqrt{N} = 0\cdot1409/\sqrt{8}$
$\quad = 0\cdot0498$

Coefficient of variation of $\bar{X}_S$:

$E_S = 100(e_S/\bar{X}_S)$ per cent
$\quad = 3\cdot585$ per cent

From Appendix VII (iv), the coefficient of variation E of the ratio $\bar{X}_S/\bar{X}_T$ is given by

$$E = \sqrt{(E_S{}^2 + E_T{}^2)} = \sqrt{27\cdot437} = 5\cdot24 \text{ per cent.}$$

The mean value of the potency ratio $\bar{X}_S/\bar{X}_T$ is $1\cdot389/0\cdot775$, i.e. $1\cdot792$. Limits of error ($P = 0\cdot95$) are found from the value of t at this probability level with the number of degrees of freedom ϕ equal to 14 (a total of 16 results are used to find E, but two means have been calculated, leaving 14 degrees of freedom). The appropriate value of t (see Appendix XII, Table 2) is $2\cdot14$. Hence

$$\text{L.E.} = \pm 2\cdot14 \times 5\cdot24 = \pm 11\cdot2 \text{ per cent} \quad (P = 0\cdot95).$$

Using the known potency of the standard preparation ($1\cdot316$ units per ml), the potency of the test preparation is

$$1\cdot316 \times 1\cdot792 = 2\cdot358 \text{ units per ml} \pm 11\cdot2 \text{ per cent.}$$

There is thus a probability of $0\cdot95$ that the true potency of the test preparation lies between $2\cdot10$ and $2\cdot62$ units per ml.

If $\log X$ is taken as the variate, the final result is slightly different. The calculations may be summarised as

mean $\log X_T = -0\cdot1130$ mean $\log X_S = 0\cdot1406$
standard error $= 0\cdot0173$ standard error $= 0\cdot0160$
mean log potency ratio = mean $\log X_S$ − mean $\log X_T$
$= 0\cdot1460 - (-0\cdot1130)$
$= 0\cdot2536.$

The standard error of the mean log potency ratio is found from Appendix VII (ii) to be

$$\sqrt{\{(0\cdot0173)^2 + (0\cdot0160)^2\}} = 0\cdot0236.$$

The limits of error ($P = 0\cdot95$ and $\phi = 14$) of the log potency ratio are

$$\pm t(0\cdot0236) = \pm 2\cdot14 \times 0\cdot0236 = \pm 0\cdot0505.$$

This means that the true log potency ratio lies between $(0\cdot2536 - 0\cdot0505)$ and $(0\cdot2536 + 0\cdot0505)$, i.e. between $0\cdot2031$ and $0\cdot3041$. Taking antilogarithms, it is found that the mean potency ratio is $1\cdot79$, with limiting values of $1\cdot59$ and $2\cdot01$. Multiplying these values by the potency of the standard preparation gives a mean potency for the test preparation of $2\cdot36$ units per ml, with limiting values of $2\cdot09$ and $2\cdot65$ units per ml; these limits are not symmetrical about the mean, and when expressed as percentages, they are $-11\cdot4$ and $+12\cdot3$ per cent of the mean value.

Quantitative response (2×2 assay). If a quantitative response to a drug can be developed in a biological system, then the potencies can be compared by measuring the responses produced by the two preparations, or suitable dilutions of them. In the 2×2 assay (or 'four point' assay), test animals are divided into four equal groups; those in Group 1 all receive a relatively low dose of the test preparation, giving a mean response T_1; those in Group 2 receive a higher dose of the test preparation, giving a mean response T_2; those in Groups 3 and 4 receive low and high doses respectively of the standard preparation, giving respective mean responses, S_1 and S_2. The ratio of the two doses of the standard preparation should be the same as that of the two doses of the test preparation.

The quantitative response that is measured is often the gain in weight

SOME APPLICATIONS OF STATISTICS TO BIOLOGICAL ASSAY

by a whole animal during a period on a standard diet, as in a vitamin assay, or the gain in weight of a particular organ, as in a hormone assay.

The most accurate results are obtained if the responses to all the doses can be measured on a single animal or tissue. If this is possible, it is important to arrange that the order in which the preparations are tested is as random as possible. This is ensured by using a *Latin square* design for determining the order in which the doses are administered. A Latin square design is an arrangement of n symbols in n lines and n columns in such a way that no symbol occurs twice in the same column or in the same row.

If the responses to two doses, $^Sd_1$ and $^Sd_2$, of the standard preparation are to be compared on a single tissue with those to two doses, $^Td_1$ and $^Td_2$, of the test preparation, a suitable sequence for the order of injection is given by the following design:

$$\begin{array}{cccc} ^Sd_1 & ^Sd_2 & ^Td_1 & ^Td_2 \\ ^Sd_2 & ^Sd_1 & ^Td_2 & ^Td_1 \\ ^Td_2 & ^Td_1 & ^Sd_1 & ^Sd_2 \\ ^Td_1 & ^Td_2 & ^Sd_2 & ^Sd_1 \end{array}$$

With this scheme, four results are obtained for each dose of both preparations.

To interpret the results of a 2×2 assay (i.e. 2 doses of test preparation and 2 doses of standard preparation), the values of the four mean responses are first calculated, together with the standard errors of the means. It has been found by extensive experimentation that in most cases quantitative responses are approximately a linear function of the logarithm of the amount of drug given. If an observed response is plotted against log dose, where dose means the volume of preparation used, a straight line should result. The position of the straight line for one preparation will differ from that of another preparation unless the potencies, i.e. amount of active material per ml, of the two preparations are identical; if not coincident, the two lines should, however, have similar slopes.

For simplicity in calculation, equal doses of the test and standard preparations may be given, the test solution being diluted, if necessary, so that its potency is of the same order as that of the standard solution.

$$^Td_1 = {^Sd_1} = d_1, \quad \text{and} \quad ^Td_2 = {^Sd_2} = d_2.$$

If the four mean responses are plotted against log dose, ideally two parallel lines should result (see Fig. 14a). Assuming that the mean response Y to a given amount X of drug is a linear function of $\log X$, then

$$Y = b \log X + a,$$

where b is the slope of the line and a is a constant. If z is the concentration of drug per ml in the standard preparation and r is the potency ratio, then the relation between response and dose of the standard preparation, putting

$Y = S$ and $X = zd$, where d is the volume (the 'dose') of the preparation, is given by
$$S = b \log (zd) + a = b \log d + (a + b \log z);$$
this is the equation of the lower straight line in Fig. 14a. Similarly, the equation for the line relating the response and dose of the test preparation, putting $Y = T$ and $X = zrd$, is given by
$$T = b \log (zrd) + a = b \log d + (a + b \log z + b \log r).$$

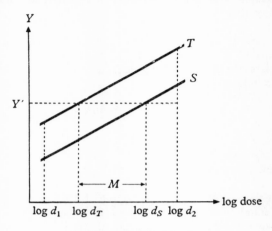

FIG. 14a. An ideal 2×2 assay

The lines are parallel, as the slope b is the same for both equations. Substituting the observed mean responses and doses in these equations gives

$$S_1 = b \log d_1 + (a + b \log z), \qquad \text{(i)}$$
$$S_2 = b \log d_2 + (a + b \log z), \qquad \text{(ii)}$$
$$T_1 = b \log d_1 + (a + b \log z + b \log r), \qquad \text{(iii)}$$
$$T_2 = b \log d_2 + (a + b \log z + b \log r). \qquad \text{(iv)}$$

Subtracting (i) from (ii) gives
$$S_2 - S_1 = b(\log d_2 - \log d_1)$$
or
$$b = (S_2 - S_1)/\log (d_2/d_1).$$
In the appendix to the *British Pharmacopoeia* dealing with this type of assay, the symbol I is used to represent $\log (d_2/d_1)$. Hence
$$b = (S_2 - S_1)/I.$$
Similarly, subtracting (iii) from (iv) gives
$$T_2 - T_1 = b(\log d_2 - \log d_1),$$
and hence
$$b = (T_2 - T_1)/I.$$

SOME APPLICATIONS OF STATISTICS TO BIOLOGICAL ASSAY

In practice, the two values of b are seldom identical, and the best estimate of b from all the results is taken as the arithmetic mean, i.e.

$$b = \tfrac{1}{2}\left(\frac{S_2-S_1}{I}+\frac{T_2-T_1}{I}\right).$$

Hence

$$b = (T_2-T_1+S_2-S_1)/2I. \qquad (14.1)$$

The object of the measurement is to calculate the potency ratio r
Subtracting (i) from (iii) gives

$$T_1-S_1 = b \log r.$$

Hence

$$\log r = (T_1-S_1)/b.$$

Subtracting (ii) from (iv) gives

$$T_2-S_2 = b \log r.$$

Hence

$$\log r = (T_2-S_2)/b.$$

Again, these two values are unlikely to agree exactly, and the best estimate of $\log r$ (written as M in the notation of the *British Pharmacopoeia*) is given by

$$\log r = \tfrac{1}{2}\left(\frac{T_1-S_1}{b}+\frac{T_2-S_2}{b}\right).$$

Hence

$$M = (T_1+T_2-S_1-S_2)/2b. \qquad (14.2)$$

If the test and standard lines are parallel, as in Fig. 14a, then M is equal to the horizontal distance between the lines, since if the broken horizontal line in the figure represents a definite response, Y', then the abscissae ($\log d_T$ and $\log d_S$) at the points of intersection of this horizontal line with the test and standard lines respectively correspond to doses d_T and d_S of test and standard preparations respectively which produce the same response. But by definition, the potency ratio is equal to d_S/d_T. Hence

$$M = \log r = \log (d_S/d_T) = \log d_S - \log d_T,$$

and $\log d_S - \log d_T$ is the horizontal distance between the two parallel lines.

TEST FOR PARALLELISM. It is important to determine whether the slopes of the test and standard lines diverge more than can be expected from chance error in the mean responses; if so, the assay is not valid. If not, then the mean value of M with appropriate limits of error can be accepted as significant. The two slopes are compared by means of a t test of significance.

The slope of the test line is given by $b = (T_2-T_1)/I$, and that of the standard line by $b = (S_2-S_1)/I$. If the null hypothesis that the

216   CHAPTER 14

difference between the two values of b is not significant is adopted, then t can be calculated from the equation

$$t = \frac{\text{difference between the values of } b}{\text{standard error of this difference}}.$$

The difference between the values of b is $[(T_2-T_1)-(S_2-S_1)]/I = (T_2-T_1-S_2+S_1)/I$, and the standard error of this difference is

(standard error of $T_2-T_1-S_2+S_1$)/I,

the error in I being negligible. From Appendix VII (i) and (ii), the standard error of $(T_2-T_1-S_2+S_1)$ is calculated to be

$$\sqrt{(e_1{}^2+e_2{}^2+e_3{}^2+e_4{}^2)},$$

where e_1, e_2, e_3 and e_4 are the standard errors of the mean responses T_1, T_2, S_1 and S_2 respectively. In the *British Pharmacopoeia* this sum of the standard errors squared is put equal to $4V$, where V is the average variance of the mean responses. Hence

$$(e_1{}^2+e_2{}^2+e_3{}^2+e_4{}^2) = 4V.$$

The value of t in the test for parallelism is therefore given by

$$t = \frac{(T_2-T_1-S_2+S_1)/I}{2\sqrt{(V)}/I}$$
$$= (T_2-T_1-S_2+S_1)/2\sqrt{V}.$$

The value of t is compared with theoretical values for the normal distribution with ϕ equal to $4n-4$.

STANDARD ERROR OF THE POTENCY RATIO. It follows from (14.2) and Appendix VII (iv) that

$$\frac{\text{standard error of } M}{M} = \sqrt{\left\{\left(\frac{\text{standard error of } T_1+T_2-S_1-S_2}{T_1+T_2-S_1-S_2}\right)^2 + \left(\frac{\text{standard error of } 2b}{2b}\right)^2\right\}}.$$

Now,

standard error of $T_1+T_2-S_1-S_2$
$$= \sqrt{(e_1{}^2+e_2{}^2+e_3{}^2+e_4{}^2)} \quad \text{[see Appendix VII (ii)]}$$
$$= 2\sqrt{V}.$$

From (14.2)
$$T_1+T_2-S_1-S_2 = 2bM.$$

Further, it follows from (14.1) and Appendix VII (i), (ii) and (iv) that

standard error of $2b = \sqrt{(e_1{}^2+e_2{}^2+e_3{}^2+e_4{}^2)}/I = 2\sqrt{V}/I.$

Therefore,
$$\frac{\text{standard error of } M}{M} = \sqrt{\left(\frac{4V}{4b^2M^2} + \frac{4V/I^2}{4b^2}\right)}.$$

Hence
$$\text{standard error of } M = \frac{1}{b}\sqrt{\left(V + \frac{VM^2}{I^2}\right)},$$

and the limits of error for M are therefore

$$\pm \frac{t}{b}\sqrt{\left(V + \frac{VM^2}{I^2}\right)}, \tag{14.3}$$

where t is the theoretical value for $P = 0.95$ and $\phi = 4n-4$. This equation should be corrected if the variance V is large; according to the *British Pharmacopoeia*, an index of significance, called g and equal to Vt^2/I^2b^2, should be calculated (t is again the theoretical value for $P = 0.95$ and $\phi = 4n-4$). This index is the square of the ratio of the limits of error of b to the value of b. If g exceeds 0.1, a more complicated formula, involving g (see page 224), should be used to calculate the limits of error. If the above limits are written as $\pm k$, then the log (limits per cent) $= 2 \pm k$. These latter limits are the upper (r_2) and lower (r_1) values of the potency ratio expressed as a percentage of the mean $\bar{r}$. Thus

$$\log r_2 = \log (\bar{r}) + k, \text{ and } \log r_1 = \log (\bar{r}) - k.$$

Hence
$$\log (r_2/\bar{r}) = k, \text{ and } \log (r_1/\bar{r}) = -k,$$

or
$$\log (100 r_2/\bar{r}) = 2 + k, \text{ and } \log (100 r_1/\bar{r}) = 2 - k.$$

EXAMPLE. *In the assay of posterior pituitary extracts for oxytocic activity based on contraction produced in uterine muscle, doses of the test and standard preparations are added to the organ bath, two doses of the test and two of the standard being selected so as to produce contractions (the measured quantitative response) of between 30 and 70 per cent of the maximum contraction possible; the ratio of the test doses is the same as that of the standard doses. The solutions are added successively, and the contractions for the different doses are measured by recording the contractions on a smoked drum.*

In a particular assay, the test sample was diluted 550 times with normal saline; the standard preparation contained 0.02 unit per ml. The following results were obtained:

Preparation		Dose (in ml)			Height of contraction (in mm)			
Standard	S_2	0.7	39	39	37	39	38	36
	S_1	0.5	25	27	21	21	16	13
Test	T_2	0.7	38	35	31	30	34	31
	T_1	0.5	24	12	14	11	11	12

What is the potency of the test preparation?

The calculation of mean responses and standard errors of the means is made in the usual way; thus

$T_1 = 14{\cdot}00$, $T_2 = 33{\cdot}17$, $S_1 = 20{\cdot}50$ and $S_2 = 38{\cdot}00$.
$e_1{}^2 = 4{\cdot}200$, $e_2{}^2 = 1{\cdot}561$, $e_3{}^2 = 4{\cdot}650$ and $e_4{}^2 = 0{\cdot}267$.
$$I = \log(7/5) = \log(1{\cdot}4) = 0{\cdot}1461.$$
$$\begin{aligned}b &= (T_2 - T_1 + S_2 - S_1)/2I \\ &= (33{\cdot}17 - 14{\cdot}00 + 38{\cdot}00 - 20{\cdot}50)/0{\cdot}2922 \\ &= 125{\cdot}5.\end{aligned}$$
$$V = (e_1{}^2 + e_2{}^2 + e_3{}^2 + e_4{}^2)/4 = 10{\cdot}678/4 = 2{\cdot}67.$$

The significance of b can be checked by the t test; thus
$$t = (T_2 - T_1 - S_2 + S_1)/2\sqrt{V} = 1{\cdot}67/3{\cdot}268 = 0{\cdot}511.$$

From Appendix XII, Table 2, the value of t for $P = 0{\cdot}95$ and $\phi = 20$ (i.e. $24-4$) is $2{\cdot}09$. The calculated value of t is much smaller than this, so that the slopes of the two lines are not significantly different and the calculation of M will be valid.

The index of significance of b is given by
$$g = \frac{Vt^2}{I^2 b^2} = \frac{2{\cdot}67 \times (2{\cdot}09)^2}{(0{\cdot}1461)^2 (125{\cdot}5)^2}$$
$$= 0{\cdot}0347.$$

Therefore, since g is less than $0{\cdot}1$, (14.3) can be used to calculate the limits of error of M; thus
$$M = (T_1 + T_2 - S_1 - S_2)/2b = -11{\cdot}33/251 = -0{\cdot}0452.$$

The limits of error for M are given by
$$\pm \frac{t}{b}\sqrt{\left(V + \frac{VM^2}{I^2}\right)} = \pm \frac{2{\cdot}09}{125{\cdot}5}\sqrt{\left\{2{\cdot}67 + \frac{2{\cdot}67 \times (-0{\cdot}0452)^2}{(0{\cdot}1461)^2}\right\}}$$
$$= \pm 0{\cdot}0285 \quad (P = 0{\cdot}95)$$

The mean potency ratio r is therefore given by antilog M, i.e.
$$\text{antilog}(-0{\cdot}0452) = 0{\cdot}901,$$
with limits of
$$\text{antilog}(-0{\cdot}0452 - 0{\cdot}0285) \quad \text{and} \quad \text{antilog}(-0{\cdot}0452 + 0{\cdot}0285),$$
i.e. $0{\cdot}844$ and $0{\cdot}962$. The original test preparation contains therefore a mean of
$$550 \times 0{\cdot}02 \times 0{\cdot}901 = 9{\cdot}91 \text{ units per ml},$$
with limiting values ($P = 0{\cdot}95$) of $9{\cdot}28$ and $10{\cdot}58$ units per ml. In percentages, these limits are $-6{\cdot}4$ per cent and $+6{\cdot}7$ per cent, or $93{\cdot}6$ and $106{\cdot}7$ per cent of the mean value.

COMP 10: Fortran programme for bioassay. COMP 10 below gives the details of a programme for working out the example above by computation.

```
C      COMP 10
C      BIOASSAY WITH P=0.95 ERROR LIMITS
       DIMENSION S1(6),S2(6),T1(6),T2(6)
       READ(5,1) N,DR,TTAB,S1,S2,T1,T2
1      FORMAT(I2,2F5.0/12F5.0/12F5.0)
```

SOME APPLICATIONS OF STATISTICS TO BIOLOGICAL ASSAY

```
C       SUM OF NUMBERS AND SQUARES
        SUMS1 = 0.
        SUMS2 = 0.
        SUMT1 = 0.
        SUMT2 = 0.
        S1SQ = 0.
        S2SQ = 0.
        T1SQ = 0.
        T2SQ = 0.
        DO 2 I = 1,N
        SUMS1 = SUMS1 + S1(I)
        SUMS2 = SUMS2 + S2(I)
        SUMT1 = SUMT1 + T1(I)
        SUMT2 = SUMT2 + T2(I)
        S1SQ = S1SQ + S1(I)**2
        S2SQ = S2SQ + S2(I)**2
        T1SQ = T1SQ + T1(I)**2
    2   T2SQ = T2SQ + T2(I)**2
C       MEANS
        EN = N
        S1M = SUMS1/EN
        S2M = SUMS2/EN
        T1M = SUMT1/EN
        T2M = SUMT2/EN
C       VARIANCE
        VS1 = (S1SQ - (SUMS1**2)/EN)/(EN*(EN - 1.))
        VS2 = (S2SQ - (SUMS2**2)/EN)/(EN*(EN - 1.))
        VT1 = (T1SQ - (SUMT1**2)/EN)/(EN*(EN - 1.))
        VT2 = (T2SQ - (SUMT2**2)/EN)/(EN*(EN - 1.))
        V = (VS1 + VS2 + VT1 + VT2)/4.
C       MEAN SLOPE AND SIGNIFICANCE
        DI = ALOG10(DR)
        B = (T2M - T1M + S2M - S1M) / (2.*DI)
        T = (T2M - T1M - S2M + S1M) / (2.*SQRT(V))
        IF (T-TTAB) 6,6,3
    3   WRITE (6,4)
    4   FORMAT(29H1THE SLOPE IS NOT SIGNIFICANT)
        WRITE(6,5) S1M,S2M,T1M,T2M,VS1,VS2,VT1,VT2,V,T
        STOP
    6   G =(V*(TTAB**2)) / ((DI*B)**2)
        IF (G-0.1) 9,9,7
    7   WRITE(6,8) G
    8   FORMAT (4 H1INDEX OF SIGNIFICANCE IS GREATER THAN 0.1
       1 //4H G =,F6.4)
        WRITE(6,5) S1M,S2M,T1M,T2M,VS1,VS2,VT1,VT2,V,T
        STOP
C       POTENCY RATIO AND LIMITS OF ERROR
    9   EM = (T1M + T2M - S1M - S2M) / (2.*B)
        EEM = (TTAB/B) * SQRT(V*(1. + (EM/DI)**2))
```

```
      AL10 = ALOG(10.)
      R = EXP(AL10 * EM)
      R1 = EXP(AL10 * (EM - EEM))
      R2 = EXP(AL10 * (EM + EEM))
      WRITE(6,5) S1M,S2M,T1M,T2M,VS1,VS2,VT1,VT2,V,T
5     FORMAT(6H1MEANS//6H S1M =,F10.4//6H S2M =,F10.4//
     1 6H T1M =,F10.4//6H T2M =,F10.4////10H VARIANCES
     2 //6H VS1 =,F10.4//6H VS2 =,F10.4//6H VT1 =,F10.4
     3 //6H VT2 =,F10.4//21H AVERAGE VARIANCE V =,F10.4
     4 ////14H T FOR SLOPE =,F10.4//)
      WRITE(6,10) R1,R,R2
10    FORMAT(17HORESULTS OF ASSAY//14H LOWER LIMIT =,F10.4//
     1 7H MEAN =,F10.4//14H UPPER LIMIT =,F10.4)
      STOP
      END
```

The individual responses are read in as elements of the dimensioned variables S1, S2, T1, and T2, together with N the number of responses in each group, which is the same for each group, DR the dose ratio within the groups and TTAB the tabulated value of t for $P = 0.95$ and $\phi = 20(6 \times 4 - 4)$. The value of TTAB in this problem is 2·09. The reading FORMAT calls for three data cards, the first containing N, DR and TTAB, the second having the data for the standard preparation, and the third the data for the test preparation.

The sums of numbers and squares in each group are all set to zero and then evaluated in a single DO loop, this is possible since N is the same for all four groups.

For the calculation of group means, N is given a real name EN. The variances are variances of the means, that is, the standard errors squared, called VS1, etc. in the programme and e_1^2 etc. in the previous working of the problem.

The log dose ratio, previously called I, is called DI in the programme so that it has a real name; it is calculated from DR by the supplied function ALOG10( ). The mean slope of the log dose, response lines, B, is calculated by equation (14.1) and to carry out the test for parallelism, t for the difference of test and standard slopes is computed.

The significance of the difference in slope of the two lines is assessed by comparing the calculated t with the tabulated value TTAB, and the result is used to decide how the execution of the programme will proceed by using an arithmetic IF statement. This type of IF statement is more compact than the logical IF which was used in COMP 3 (page 90). IF is followed by brackets containing one quantity minus another, the brackets are closed, and then there are three statement numbers separated by commas. When the quantity in the brackets is negative, execution goes to the first statement number ignoring the programme between IF and this statement; when the quantity is zero, execution goes to the second statement number; when it is positive, execution goes to the third statement.

In the test for parallelism, if t for the slope difference is less than or equal to the tabulated value, the difference of slopes is not significant and the calculation continues, going to statement 6. If t is greater than TTAB, that is (T−TTAB) is positive, then the lines cannot be considered to be parallel and the assay is not valid. Execution goes to statement 3 which writes out a message and prints out the information already computed, as this might be useful in assessing the reason for failure, and then stops the execution.

If the results have survived the test for parallelism they proceed to statement 6 where they are subjected to another arithmetic IF to determine whether the index of significance, g, is greater than 0·1; if it is, execution goes to statement 7 where a message is written out together with the computed information. Alternatively, the calculation could have been continued using the more complicated equation for limits of error.

If g equals or is less than 0·1, the logarithm of potency ratio is calculated by equation (14.2). This quantity is called EM rather than M so that it has a real name. Limits of error for EM, called EEM, are calculated by equation (14.3).

The mean potency ratio R is the antilogarithm of EM and this is found by multiplying by $\ln(10) = 2·3026$; in the programme this is done by the supplied function ALOG( ) with 10· in the brackets. This gives the natural logarithm of the potency ratio and the ratio itself is obtained from the supplied function EXP( ).

The $P = 0·95$ limits R1 and R2 are then calculated by finding the antilogarithms of (EM−EEM) and (EM+EEM).

The final print-out gives all the computed information concluding with the mean potency ratio and its limits.

```
MEANS
S1M =     20.5000
S2M =     38.0000
T1M =     14.0000
T2M =     33.1667

VARIANCES
VS1 =      4.6500
VS2 =       .2667
VT1 =      4.2000
VT2 =      1.5611
AVERAGE VARIANCE V =  2.6694

T FOR SLOPE =      .5100

RESULTS OF ASSAY
LOWER LIMIT =      .8440
MEAN =      .9012
UPPER LIMIT =      .9623
```

CHAPTER 14

Quantal response assay. In this assay, the mean response of a group of test animals to a given quantity of drug is determined by the percentage of positive effects observed in the group, each animal giving either a positive or a negative (all or none) response.

The assay can be carried out in the 2×2 form by dividing the animals into four equal groups and giving each group doses of test or standard preparations. In order to be able to use the method of the previous section to calculate the potency ratio, it is necessary to convert the percentage response of each group into some quantity which is a linear function of the logarithm of the amount of drug given to each member of the group. A suitable quantity for this purpose can be derived from normal distribution theory and is called a *probit*. Gaddum (*Nature*, 1945, **156**, 463) has pointed

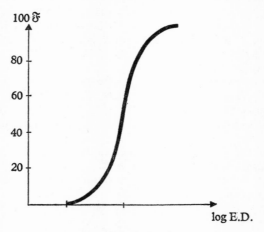

FIG. 14b. Cumulative frequency curve

out that the logarithms of the sensitivities to drugs among individual animals of the same species, as measured by the dose required to cause some definite effect, are normally distributed in practically every case. In the interpretation of all quantal response assays (assays in which a percentage of positive effects is observed), the assumption is made that log effective dose (log E.D.) for each individual animal is normally distributed about the average log E.D. for a large number of such animals.

The percentage cumulative frequency $100\mathfrak{F}$ (see page 187) is the percentage of values of the variate, log E.D., in the whole distribution equal to or less than a given value of the variate. If $100\mathfrak{F}$ is plotted against log E.D., a sigmoid curve is obtained as for any normal distribution (see Fig. 14b). If an amount X_1 of the drug is given to each of a group of animals, all those for whom the effective dose is less than X_1, i.e. log E.D. is less than log X_1, will show a positive effect, while all those for whom the

SOME APPLICATIONS OF STATISTICS TO BIOLOGICAL ASSAY 223

effective dose is greater than X_1 will show a negative effect. The cumulative frequency $100\mathfrak{F}$ (as a percentage) at this point gives the expected percentage of animals showing positive effects.

The cumulative frequency is related to the corresponding value of X_1 by the normal distribution equation (12.10), i.e.

$$\mathfrak{F} = \frac{1}{\sqrt{(2\pi)}} \int_{-\infty}^{u_1} \exp\left(-\tfrac{1}{2}u^2\right) du, \qquad (14.4)$$

where u_1 is the normal deviate corresponding to X_1. Hence

$$u_1 = \frac{\log X_1 - \text{mean} \log X}{\sigma}, \qquad (14.5)$$

where $\log X$ is the variate; σ, the standard deviation of the distribution of log E.D., and mean $\log X$ (logarithm of the value of X giving a 50 per cent response) are both constants for a large number of animals of a given species and for a given drug. Hence u_1 is a linear function of $\log X_1$.

Equation (14.4) means that for each value of u_1 there is a definite value of $\mathfrak{F}$, which can be found by computation. Conversely, for each value of $\mathfrak{F}$ there is a definite value of u_1, which from (14.5) is seen to be a linear function of $\log X_1$. Hence u_1 is a function of the observed percentage response, which is, theoretically, also a linear function of log dose.

For practical purposes, $5+u_1$ is the quantity used to calculate the results of a 2×2 quantal assay. By adding 5 to u_1, negative numbers, which pharmacologists appear to dislike, are eliminated; $5+u_1$ is called the *probit of a percentage response*.

The observed percentage response is the value of $100\mathfrak{F}$. From values of the integral in (14.4) obtained from tables, the corresponding value of $5+u_1$ can be found, and this is taken as the mean response of each group of animals. The relation between percentage response and probit is seen to be that between the value of an integral and the upper limit of the integral, the lower limit being fixed.

The calculation of potency ratio is carried out exactly as in the previous section, using probits in place of T_1, T_2, S_1 and S_2. The calculation of limits of error is rather complicated (see J. H. Burn, *Biological Standardisation*, Oxford University Press, London, 1950, p. 142), but a rough estimate can be made by taking V as being $4/\{\Sigma(wn)\}$, where w is the weight factor (see below) of each probit and n is the number of animals in the group; t is the value for the given probability level, with $\phi = \infty$.

Probits corresponding to different responses do not have equal reliabilities. The probit of a 50 per cent response has the maximum reliability but, owing to chance variations, the reliability of probits of other responses decreases as the response diverges either positively or negatively from 50 per cent (see Appendix XII, Table 5). The reliability of a probit can be estimated from normal distribution theory and is expressed in the form of a weight factor w; tables of weight factors for given probits have been calculated (see Appendix XII, Table 6). The weight factors of probits of 0 per cent and 100 per cent responses are zero, and so responses of 0 per cent and 100 per cent should be avoided in designing an assay.

A test for the parallelism of the S and T probit, log dose lines should be made. This usually takes the form of a χ^2 test (see J. H. Burn, *loc. cit.*).

EXAMPLE. *The biological assay of an œstrogen cream was carried out by comparing it with Œstrone, B.P.C., dissolved in arachis oil. Four groups of 10 rats were used in a 2×2 assay. Vaginal smears were taken 48 hours after injection of the test and the standard solutions, and were scored as positive when a predominance of non-nucleated cornified cells were present. The following results were obtained.*

		Dose	No. of animals n	No. of positive effects	Percentage response	Probit*	w†
Standard	S_2	14 i.u.	10	7	70	5·52	0·58
	S_1	8 i.u.	10	1	10	3·72	0·34
Test	T_2	66·15 mg	10	8	80	5·84	0·49
	T_1	37·80 mg	10	3	30	4·48	0·58

What is the potency of the cream?

The following calculations are made.
$$d_2/d_1 = 14/8 = 1·75.$$
Therefore
$$I = \log 1·75 = 0·2430.$$
$$b = (T_2 - T_1 + S_2 - S_1)/2I = (5·84 - 4·48 + 5·52 - 3·72)/0·4860$$
$$= 3·16/0·486 = 6·50.$$
$$M = (T_1 + T_2 - S_1 - S_2)/2b = (4·48 + 5·84 - 3·72 - 5·52)/13·00$$
$$= 1·08/13·00 = 0·0831.$$
$$V = \frac{4}{\Sigma(wn)} = \frac{4}{5·8 + 3·4 + 4·9 + 5·8} = 0·201.$$
$$g = Vt^2/I^2b^2$$
$$= \frac{0·201(1·96)^2}{(0·243)^2(6·50)^2} = \frac{0·7722}{2·495} = 0·309$$

(the value of t for $P = 0·95$ and for infinite degrees of freedom is 1·96). The value of g, being greater than 0·1, is too large to permit calculation of limits of error of M by (14.3). The potency ratio r is antilog $(0·0831) = 1·21$.

Larger groups of animals and more observations are required to reduce the value of V so that reasonably narrow limits of error can be specified.

If the formula given in the *British Pharmacopoeia* is used to calculate the limits of error for M with $g > 0·1$, the limits of M are given by

$$\frac{gM}{1-g} \pm \frac{t}{b(1-g)} \sqrt{\left\{V(1-g) + \frac{VM^2}{I^2}\right\}} \quad (P = 0·95)$$
$$= \frac{0·309 \times 0·0831}{0·691} \pm \frac{1·96}{6·50 \times 0·691} \sqrt{\left\{0·201 \times 0·691 + \frac{0·201(0·0831)^2}{(0·243)^2}\right\}}$$
$$= 0·0372 \pm 0·175$$
$$= -0·138 \quad \text{or} \quad +0·212.$$

* Obtained from Table 5, Appendix XII.
† Obtained from Table 6, Appendix XII.

SOME APPLICATIONS OF STATISTICS TO BIOLOGICAL ASSAY

The limits of M are therefore $(0\cdot083-0\cdot138)$, i.e. $-0\cdot055$ and $(0\cdot083+0\cdot212) = 0\cdot295$. Taking antilogarithms, the potency ratio is therefore between 0·88 and 1·97, with a mean of 1·21; these limits correspond to −27 per cent and +63 per cent, or 73 per cent and 163 per cent of the mean value. They illustrate the uncertainty of an assay of this type using only small numbers of animals in each group. The final result is that there is a probability of 0·95 that the true potency ratio lies between 0·88 and 1·97.

2. Toxicity tests

These are not assays, since only one preparation is used. Suitably graded doses of the drug under test are administered to groups of animals, one group for each dose level. The percentage mortality in each group after a given period is noted, and the probits of these quantal responses are plotted against log dose. The best straight line is drawn by eye through the points, and the log dose on this line which corresponds to a 50 per cent mortality in a group, i.e. to a probit of 5, is read off. The antilog of this is called the median lethal dose (M.L.D.) and gives a quantitative measure of the toxicity of the drug.

If the probit, log dose points are scattered, and the most reliable value of M.L.D. is required, then the equation of the probit, log dose regression line is calculated (see page 199), taking into account the weight factors of the probits. Limits of error for M.L.D. are then those of prediction of any dose from a regression line, and they can be evaluated from the variance of the points about the regression line (see Chap. 13 and Appendix IX).

EXAMPLE. *The acute toxicity of trimethylcyclohexanol was determined by administering a solution in arachis oil containing 200 mg of the substance per ml orally and subcutaneously to mice, weighing 18 to 24 g, using 10 mice for each dose group. The following results were obtained.*

Route of administration	Dose (mg per 20 g body weight)	Log dose	No. of dead animals out of 10	Percentage response	Probit
Oral	90	1·954	3	30	4·48
	100	2·000	4	40	4·75
	110	2·041	8	80	5·84
Subcutaneous	60	1·778	1	10	3·72
	70	1·845	6	60	5·25
	90	1·954	9	90	6·28

What are the oral and subcutaneous M.L.D.?

For the oral tests, it is seen from Fig. 14c that

$$\log \text{M.L.D.} = 1\cdot990.$$

Hence

$$\text{oral M.L.D.} = 97\cdot7 \text{ mg per 20 g of body weight}$$
$$= 4\cdot88 \text{ g per kg of body weight.}$$

For the subcutaneous tests,
$$\log \text{M.L.D.} = 1{\cdot}848 \text{ (see Fig. 14c)}.$$
Hence
subcutaneous M.L.D. = 70·5 mg per 20 g of body weight
= 3·52 g per kg of body weight.

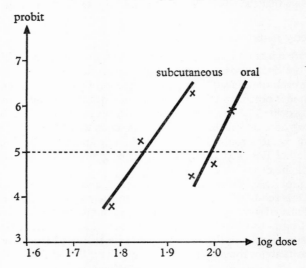

FIG. 14c. Median lethal dose determination

3. Analysis of variance

Analysis of variance is sometimes used for interpreting the results of biological assays, and it is particularly valuable for examining the reliability of a new type of assay.

When a biological assay process is being devised, doses of increasing magnitude are given to groups of animals, and the responses are noted. In an ideal situation, the response of each animal to a given dose would always be the same; moreover, the responses at the different dose levels would be a linear function of the logarithm of the dose. This ideal situation is never found in practice, and the object of the analysis of variance is to compare the variation of response due to increasing dosage with that due to other causes, particularly animal variation within a dose group. It is then possible to assess whether the variation due to increasing dosage is significant when viewed against the background of animal variation. If the latter contributes significantly to the total variation, the selected biological assay process is unsatisfactory, and the responses do not truly represent the effect of increasing dosage. From the analysis of variance it is also possible to decide whether the response, log dose relation is, in fact, linear within the limits imposed by animal variation.

As an example of the analysis of variance, the following data are taken from the work of Lou and Fairbairn (*J. Pharm. Pharmacol.*, 1951, 3, 295) on the development of a biological assay for cascara. For this purpose, increasing doses of dry extract of cascara were given to groups of mice, each dose-group containing five pairs of mice, and the response observed was the number of wet faeces per pair of mice. The results are given in Table I.

TABLE 1

Pair	Number of wet faeces per pair of mice					
	Dose (in mg per pair)				
	80	120	180	270	405
A	1	4	19	25	27
B	5	1	15	25	26
C	1	5	12	24	20
D	1	10	17	17	37
E	2	8	19	24	29
Totals	10	28	82	115	139
Average (Z)	2·0	5·6	16·4	23·0	27·8

In the analysis of these results the following symbols are used:

X is the logarithm of the dose for each group.
Y represents any individual response, i.e. the number of wet faeces.
Z is the average response for a dose-group.
N is the total number of results.
n is the number of dose-groups.

First, the results are all treated as repeats of the same measurement and their grand mean m is calculated; thus

$$m = \Sigma(Y)/N.$$

The sum of the squares of the deviations of all the results from m is then calculated; this is called the *total sum of squares* S_1. Thus

$$S_1 = \Sigma(Y-m)^2 = \Sigma(Y^2) - \Sigma^2(Y)/N.$$

S_1 has $N-1$ degrees of freedom, one degree of freedom having been used to calculate the mean m. This total sum of squares is now divided into two main components: (i) S_2, which measures the squared deviations caused by the dose differences, and (ii) S_3, which measures the squared deviations due to animal variation within the dose groups, i.e.

$$S_1 = S_2 + S_3.$$

S_2 is called *the sum of squares between dose-groups*; it is calculated by finding the total sum of squares when the mean response Z for each group is taken to represent each result in the group. Thus, for the purpose of this calculation, the results summarised in Table 1 are rewritten in the form shown in Table 2.

TABLE 2

Pair	Number of wet faeces per pair of mice				
	Dose (in mg per pair)				
	80	120	180	270	405
A	2·0	5·6	16·4	23·0	27·8
B	2·0	5·6	16·4	23·0	27·8
C	2·0	5·6	16·4	23·0	27·8
D	2·0	5·6	16·4	23·0	27·8
E	2·0	5·6	16·4	23·0	27·8

The grand mean of these results will still have the same value m, but the sum of squared deviations from m will be free from effects due to variations within the dose-groups. If each response in this table is denoted by Y_1, then

$$S_2 = \Sigma(Y_1-m)^2 = \Sigma(Y_1^2) - Nm^2$$
$$= \Sigma(Y_1^2) - \Sigma^2(Y_1)/N \quad \text{(see Chap. 12, § 1).}$$

The values of Y_1 are the dose-group means Z. $\Sigma(Y_1^2)$ is $\Sigma(Z^2)$ multiplied by the number of rows in the table; in the general case of N results arranged in n columns, there will be $N/n\ (=l)$ rows. Hence

$$S_2 = l\Sigma(Z^2) - \frac{\Sigma^2(Y)}{N}.$$

The number of degrees of freedom of S_2 are $n-1$, since one degree of freedom has been used to calculate the grand mean of n groups.

The sum of squares within dose-groups S_3 may be found by considering the results in Table 1 again and calculating for each dose-group the sum of squared deviations from the group mean Z and adding these values together. However, S_3 is more usually evaluated by subtracting S_2 from S_1, i.e.

$$S_3 = S_1 - S_2.$$

The number of degrees of freedom of S_3 are $N-n$, since the sum is obtained from N results after calculating n group means.

The values of S_1, S_2 and S_3 are now computed for this example. Thus from Table 1, $\Sigma(Y) = 374$, $\Sigma(Y^2) = 8324$, $\Sigma(Z^2) = 1606·16$, $N = 25$, $n = 5$, and $l\ (= N/n) = 5$. Hence

$$S_1 = 8324 - (374)^2/25 = 8324 - 5595·04$$
$$= 2728·96 \text{ (degrees of freedom } \phi_1 = 24).$$

Further,
$$S_2 = 5 \times 1606 \cdot 16 - 5595 \cdot 04 = 2435 \cdot 76 \quad (\phi_2 = 4),$$
and
$$S_3 = 2728 \cdot 96 - 2435 \cdot 76 = 293 \cdot 20 \quad (\phi_3 = 20).$$

The variances within doses and between doses, V_2 and V_3 respectively, are found by dividing the sums, S_2 and S_3, by their degrees of freedom. Hence
$$V_2 = S_2/\phi_2 = 608 \cdot 94, \quad \text{and} \quad V_3 = S_3/\phi_3 = 14 \cdot 66.$$

The variance due to dose differences is very much greater than that due to animal variation, and therefore the response does give a reasonable measure of the dose level and can be used for assay purposes. To confirm that the two variances are significantly different, the ratio (larger variance)/(smaller variance) is calculated, i.e. $608 \cdot 94/14 \cdot 66 = 41 \cdot 5$, and compared with the theoretical value of the variance ratio F for $P' = 0 \cdot 05$ and the appropriate degrees of freedom, ϕ_2 and ϕ_3. From Appendix XII, Table 4, the theoretical value of F is interpolated as $2 \cdot 9$ approximately; this is very much smaller than the observed ratio of $41 \cdot 5$ and so it is very unlikely that V_2 and V_3 are sample variances from the same distribution, i.e. the difference between them is significant.

Further information about the suitability of the process for biological assay purposes can be derived by carrying out a regression analysis on the results. For this purpose it is assumed that the mean group-response Z is a linear function of log dose X. The sum of squares due to doses S_2 is subdivided into two components: (i) S_4, which is the sum of the squared deviations of the X, Z points from the regression line, and (ii) S_5, which is the sum of squared deviations from the mean of the theoretical responses calculated from this line. S_4 is called the *sum of squares about regression* and simply measures the scatter of the X, Z points about the line. S_5 is called the *sum of squares due to regression* and measures the squared deviations from the grand mean m due to the fact that there is a Z/X regression line with finite slope.

A formula (13.8) for calculating the sum of squares about regression has already been deduced. Since each value of Z is taken as representing l individual responses, the value for the X, Z scatter is multiplied by l; thus, from (13.8)
$$S_4 = l[S(Z^2) - n\bar{Z}^2 - b^2\{\Sigma(X^2) - n\bar{X}^2\}]. \qquad (14.6)$$

Since there are n dose groups, there are n pairs of values of X and Z.

The mean of the group means $\bar{Z}$ is the grand mean m; hence
$$l\{\Sigma(Z^2) - n\bar{Z}^2\} = l\{\Sigma(Z^2)\} - Nm^2 = l\{\Sigma(Z^2)\} - \Sigma^2(Y)/N = S_2.$$

Substituting this value in (14.6) gives
$$S_4 = S_2 - b^2 l\{\Sigma(X^2) - n\bar{X}^2\}.$$

CHAPTER 14

Now, $S_4 + S_5 = S_2$, i.e. $S_4 = S_2 - S_5$. Therefore

$$S_5 = b^2 l \{\Sigma(X^2) - \Sigma^2(X)/n\},$$

writing $\bar{X} = \Sigma(X)/n$.

It is usual to compute S_5 from the data, and to find S_4 by subtracting S_5 from S_2. To avoid calculating the value of b, its value from (13.5) [note there are n pairs of values in this regression calculation instead of the N pairs in (13.5)] is substituted in the expression for S_5, giving

$$S_5 = l\left[\frac{\{\Sigma(XZ) - \Sigma(X)\,\Sigma(Z)/n\}^2}{\Sigma(X^2) - \Sigma^2(X)/n}\right]. \tag{14.7}$$

S_5 is then calculated directly from the group means Z and the log dose values X.

The degrees of freedom of S_4 are $n-2$, since the regression is calculated from n pairs of X, Z values from which the two means, $\bar{X}$ and $\bar{Z}$, have already been computed. The degrees of freedom of S_5 are equal to the difference between those for S_2 and S_4, i.e. $(n-1) - (n-2) = 1$.

In this example, S_5 is found from the values of X and Z in Table 3.

TABLE 3

Group dose	Log dose X	X^2	Mean response Z	XZ
80	1·9031	3·6218	2·0	3·8062
120	2·0792	4·3231	5·6	11·6435
180	2·2553	5·0864	16·4	36·9869
270	2·4314	5·9117	23·0	55·9222
405	2·6075	6·7990	27·8	72·4885
	$\Sigma(X) = 11·2765$	$\Sigma(X^2) = 25·7420$	$\Sigma(Z) = 74·8$	$\Sigma(XZ) = 180·8473$

$$N = 25, \quad n = 5, \quad l = 5$$

$$\Sigma^2(X)/n = (11·2765)^2/5 = 25·4318$$

$$\Sigma(X)\Sigma(Z)/n = 11·2765 \times 74·8/5 = 168·6964$$

Substituting these values in (14.6) gives

$$S_5 = \frac{5(180·8473 - 168·6964)^2}{25·7420 - 25·4318} = 2379·80 \quad (\phi = 1).$$

Since $S_2 = 2435·76$ (see page 229), it follows that

$$S_4 = S_2 - S_5 = 2435·76 - 2379·80 = 55·96 \quad (\phi = 3).$$

SOME APPLICATIONS OF STATISTICS TO BIOLOGICAL ASSAY

The variances about regression and due to regression, V_4 and V_5 respectively, are found by dividing the sums of squares by their degrees of freedom, and the ratios of these variances to the variance within groups V_3 are calculated and compared with theoretical values of F.

The complete results for the analysis of variance are summarised in Table 4.

TABLE 4

	Sum of squares		Degree of freedom ϕ	Variance	Variance ratio	
	Symbol	Value			Calculated	F ($P' = 0.05$)
Total	S_1	2728·96	24	—	—	—
Between doses	S_2	2435·76	4	608·94	41·54	2·9
Within doses	S_3	293·20	20	14·66	—	—
About regression	S_4	55·96	3	18·65	1·27	3·1
Due to regression	S_5	2379·80	1	2379·80	162	4·4

The variance within doses is taken as the yardstick against which the other variances are measured. The variance ratio about regression is equal to V_4/V_3, and although V_4 is greater than V_3, comparison of this ratio with the appropriate theoretical value of F in the last column of the table shows that $V_4/V_3 < F$ for $P' = 0.05$, and it is therefore reasonable to assume that V_4 and V_3 are sample variances from the same distribution, i.e. the difference between them is not significant. This means that the scatter of the X, Z points about their regression is only what might be expected from animal variation, and it is reasonable to assume that a linear response–log dose relation exists. If V_4/V_3 had been greater than F, it would have meant that the scatter about regression was greater than could be ascribed to animal variation, suggesting that the true response–log dose relation was not linear.

The variance ratio due to regression is V_5/V_3. This value is seen from the table to be considerably larger than the theoretical value of F, and so the difference between V_5 and V_3 is significant. This result confirms the conclusion reached from the ratio V_2/V_3 that there is a significant change of response with dose. Had the reverse result been obtained, the method would have been unsuitable for assay purposes.

In the present case it has been shown that the variation of response with dose is highly significant, while the scatter about regression is only what might be expected from animal variability. The method is therefore suitable for biological assay work based on a linear response–log dose regression line.

To conclude this section, the calculation of analysis of variance in a problem of this type with a total of N results arranged in n columns, each containing l results, is summarised in Table 5; Y is an individual response, X is the quantity which is varied from one column to another, and Z is the column mean.

TABLE 5

Sum of squares	Symbol	Formula	Degree of freedom ϕ	Variance
Total	S_1	$\Sigma(Y^2)-\Sigma^2(Y)/N$	$N-1$	$S_1/(N-1)$
Between columns	S_2	$l\{\Sigma(Z^2)\}-\Sigma^2(Y)/N$	$n-1$	$S_2/(n-1)$
Within columns	S_3	S_1-S_2	$N-n$	$S_3/(N-n)$
About regression	S_4	S_2-S_5	$n-2$	$S_4/(n-2)$
Due to regression	S_5	$l\left[\dfrac{\{\Sigma(XZ)-\Sigma(X)\Sigma(Z)/n\}^2}{\Sigma(X^2)-\Sigma^2(X)/n}\right]$	1	S_5

4. Bacteriological counts

Repeated observations of the number of bacteria in different samples of a liquid give a set of results which form a Poisson distribution about their mean, the circumstance of a bacterium being counted being the rare event which gives the low value of p in the binomial generating function (see page 168).

Bacteria can either be counted directly by means of a microscope or, more conveniently, they can be counted by plating out a small sample of the medium containing them on to a nutrient gel. The inoculated plate is incubated for some time, when each bacterium in the gel gives rise to a visible colony. The number of colonies can be counted directly, so giving an estimate of the number present in the original medium.

If the organisms are randomly distributed in the medium, the frequencies with which different numbers of organisms are counted will approximate to the frequencies given by the Poisson distribution, in which the variance about the mean is equal to the mean itself. In fact, the Poisson distribution only applies to rare events whereas these counts may give quite large values. It is perhaps best to consider the χ^2 test described below as being made to check with a binomial distribution of maximum variance rather than with a Poisson distribution. When the variate for a binomial distribution consists of positive integers such as counts, the maximum variance is the mean (Appendix V). If, however, the mixing is poor, or the organisms tend to aggregate or to show mutual

SOME APPLICATIONS OF STATISTICS TO BIOLOGICAL ASSAY

repulsion, this result will not be obtained, and the variance of repeated results about the mean will differ significantly from the mean.

In order to test whether a plating technique is satisfactory, a number of repeated counts are made and the hypothesis that they form a binomial distribution of maximum variance is tested by means of a χ^2 test of significance.

EXAMPLE. *The following counts were obtained for the numbers of colonies of* Escherichia coli *in a medium, using the surface viable method of Miles and Misra with 30 replicate plates:*

14	22	20	15	28	15	19	17	20	20
20	19	22	16	21	26	23	20	21	16
20	26	20	22	24	15	18	22	18	23

Is this a suitable plating technique?

The following calculations are made:

$$\Sigma(X) = 602 \quad \text{and} \quad \Sigma(X^2) = 12{,}430.$$

To test whether these results do fit the binomial distribution of maximum variance with reasonable accuracy, the null hypothesis that they do in fact agree with this distribution, i.e. the variance V is equal to the mean m, is assumed, i.e. that

$$s^2 = V = m,$$

where s is the standard deviation. Dividing the sum of the squared deviations from the mean by the mean gives

$$\frac{\Sigma(X-m)^2}{m} = \frac{\Sigma(X-m)^2}{s^2} = \frac{(X_1-m)^2}{s^2} + \frac{(X_2-m)^2}{s^2} + \ldots,$$

i.e. equal to a sum of squared normal deviates $u_1^2 + u_2^2 + \ldots$, where $u_1 = (X_1-m)/s$, $u_2 = (X_2-m)/s, \ldots$.

If m is moderately large, the binomial distribution becomes approximately a normal one (see Appendix VIII, 3). This sum of squared normal deviates can be compared with the tabulated value of χ^2 for an equal number of degrees of freedom. In this case

$$\chi^2 = \frac{\Sigma(X-m)^2}{m} = \frac{\Sigma(X^2) - \Sigma^2(X)/N}{\Sigma(X)/N}$$

$$= \frac{12{,}430 - (602)^2/30}{602/30} = \frac{12{,}430 - 12{,}080 \cdot 1}{20 \cdot 07}$$

$$= \frac{349 \cdot 9}{20 \cdot 07} = 17 \cdot 43.$$

The number of degrees of freedom ϕ of χ^2 are 29, since one mean has been calculated from the 30 results. From Appendix XII, Table 3, it is seen that the theoretical value of χ^2 for $P' = 0 \cdot 05$ is 42·6; since this is very much greater than the observed value, the hypothesis that the variance of the results is equal to their mean is reasonable, and the plating technique can be considered to be satisfactory.

Mean single survivor time. In extinction methods for determining bactericidal activity, the bactericide solution is inoculated with the organism under investigation. After varying intervals a set of small samples of the solution are withdrawn, plated out, and the number of surviving organisms in each sample is estimated. The results for the samples in a set will vary among one another, but the mean number of survivors λ at each time will decrease logarithmically as the time of exposure of the organism to the bactericide increases (an experimental law); $\ln \lambda$ is therefore a linear function of time.

In practice, instead of observing the mean number of survivors in each set of samples, the proportion p of samples in each set which show no growth is noted. A relation between p and λ can be deduced from the theory of the Poisson distribution; the survival of organisms after the longer exposure times is a sufficiently rare event for the theory of this distribution to apply. Thus, each time a set of samples is taken, the number of survivors in single samples will be distributed about the mean value λ for the set according to the Poisson distribution law. The fraction of samples in the set which show no growth will be given by the first term of the Poisson series, i.e. $\exp(-\mu)$, while the fraction which shows one survivor will be the second term of this series, i.e. $\mu \exp(-\mu)$, and so on; in this case, $\mu = \lambda$, the mean number of survivors at each time. From this it is seen that the proportion of any set of samples which shows no survivors is $\exp(-\lambda)$, i.e. $p = \exp(-\lambda)$, or $\ln p = -\lambda$. Hence

$$\lambda = -\ln p.$$

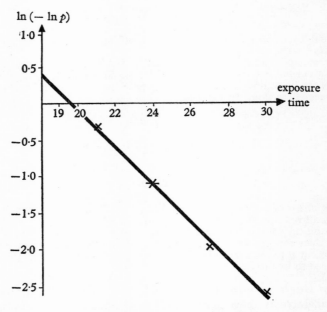

Fig. 14d. Mean single survivor time

SOME APPLICATIONS OF STATISTICS TO BIOLOGICAL ASSAY 235

This equation gives a method for finding λ from the proportion p of samples in a set which show no growth.

This is the basis of the log–log analysis of extinction time described by Mather (*Biometrics*, 1949, **5**, 127), to which the reader is referred for more detail of the method. In practice, since $\ln \lambda = \ln(-\ln p)$, $\ln(-\ln p)$ is plotted against time, and the regression line is drawn through the points. The time at which $\ln(-\ln p)$ is zero is then found; this is called the *mean single survivor time*, since at this point

$$\ln(-\ln p) = 0 = \ln \lambda, \quad \text{i.e.} \quad \lambda = 1,$$

and 1 is the mean number of survivors at this time.

EXAMPLE. *A 1·10 per cent phenol solution at 20° was inoculated with a standardised volume of a suspension of* Escherichia coli. *Uniform samples were removed into sterile tubes and after exposures varying from 15 to 33 minutes, a nutrient medium was added to each tube. The determinations were repeated 15 times. After incubation, the following numbers of negative tubes were found for the various exposure times.*

Exposure time (minutes)	15	18	21	24	27	30	33
Number of negative samples	0	3	8	11	13	14	15
p	0	0·200	0·533	0·733	0·867	0·933	1·000
$\ln(-\ln p)$	—	+0·476	−0·463	−1·169	−1·947	−2·668	—

What is the mean single survivor time?

The values of $\ln(-\ln p)$ give a fairly straight line when plotted against time. The mean single survivor time corresponding to $\ln(-\ln p) = 0$ can be read off from the graph as 19·6 minutes (see Fig. 14d).

CHAPTER 15

Some applications of statistics in pharmacy

1. Sampling

The problem of selecting from a batch of material a sample that will give some idea of the mean properties of the batch is an important one in pharmacy. With liquid preparations the problem does not arise unless they are grossly heterogeneous, the smallest sample of a homogeneous liquid being identical in composition with the bulk from which it is drawn. With batches of solid material this is not so, and the composition of a small sample may differ appreciably from the bulk.

Two types of sampling procedure are commonly used for solid material:

(i) Random sampling, which is used when the batch is considered to be reasonably uniform. Individual units, e.g. tablets, are chosen from the batch in a way as random as possible to form the sample.
(ii) Representative sampling, which is used when the batch is known to be heterogeneous. For example, with a batch of commercial opium the individual units are chosen by dividing the batch into sections and making up the sample with one unit from each section.

If the batch is large and the sample fairly small, the mean value m of a property of the sample will usually differ from the true mean value μ of the whole batch. With random sampling, the distribution of m about μ will approximate to that of a sample mean about the universe mean in a normal distribution, i.e. it will have a variance of σ^2/N, where σ is the standard deviation of individual results about μ and N is the number of results in the sample. The difference between m and μ is called the sampling error; the limits of this error for a given probability level are $\pm u\sigma/\sqrt{N}$, or $\pm ts_1/\sqrt{N}$ it s_1, the standard deviation of individual results in the sample about m, is used (see Chap. 12, § 6, 7); u and t are the theoretical values (see Appendix XII, Tables 1 and 2) for the probability level chosen.

In addition to the sampling error there will usually be an appreciable experimental error in the determination of the property. The total error in m will be made up of both these sources of random error.

If the property under consideration Z is the content of active ingredient in a tablet, then Z will be the product of the weight X of the tablet and the proportion Y of active ingredient in the tablet, i.e. $Z = X \times Y$. The variation in Y is the assay error, which will be independent of X over the normal range of tablet weights in a sample. If E_Z is the coefficient of

SOME APPLICATIONS OF STATISTICS IN PHARMACY

variation, i.e. standard deviation expressed as a percentage, of Z, while E_X is that of the tablet weights and E_Y is that due to the error of the assay method, then, from Appendix VII (iii),

$$E_Z = \sqrt{(E_X^2 + E_Y^2)}.$$

The mean content of active ingredient m in the sample is given by

$$m = \bar{X} \times \bar{Y},$$

where $\bar{X}$ and $\bar{Y}$ are the sample means of X and Y. If there are N tablets in the sample, the coefficient of variation of $\bar{X}$ is the standard error of $\bar{X}$ expressed as a percentage, i.e.

$$E_{\bar{X}} = E_X/\sqrt{N}.$$

Similarly,

$$E_{\bar{Y}} = E_Y/\sqrt{N}.$$

Hence

$$E_m = \sqrt{(E_X^2/N + E_Y^2/N)}.$$

The limits of error for the mean result m can be calculated from the factor t (see Chap. 12, § 7), providing N is not less than 4 and are equal to $= \pm tE_m$.

If, however, the property is determined on the whole sample, as in the assays of the *British Pharmacopoeia* for the determination of the content of active ingredient in tablets, then

$$m = \bar{X} \times Y,$$

where Y is a single determination, or perhaps the mean of two determinations, and

$$E_m = \sqrt{(E_X^2/N + E_Y^2)}.$$

The coefficient of variation of the analytical method E_Y can be found from repeated analyses, preferably by a number of different workers, on a uniform batch of powder containing a known amount of the active ingredient.

In a case where the analytical result Y is obtained from a single determination, or is the mean of a duplicate determination, the use of t for calculating limits of error, with the assumption of normal distribution of the results about the batch mean, is not justifiable; t should only be used to give limits for a mean of at least four results, unless a normal distribution of single results has been proved. If limits for single results, or for means of two or three repeated results, are required, the Camp–Meidell expression (see Chap. 12, § 10) should be used. In using this expression, normal distribution is not assumed, but a standard deviation derived theoretically, or known from a large number of previous results, is required.

The maximum values of u, the factor for limits for a single result, derived from the Camp–Meidell expression are compared with the values of u for normal distribution in Table 1.

TABLE 1
Values of u for limits for a single result

	Probability level			
	0·95	0·99	0·995	0·999
Normal distribution	1·96	2·58	2·81	3·29
Camp–Meidell expression	2·98	6·67	9·43	21·08

The differences in the values of u become very marked at high probability levels, and, in fact, it is usually of little value to try to express high probability limits for single results.

2. Tests for uniformity of weight of tablets

Uniformity of weight of tablets is controlled in the *British Pharmacopoeia* by a test in which 20 tablets are weighed singly. The requirement is that not more than two tablets in the sample should differ from the mean weight by more than x per cent and no tablet should differ by more than $2x$ per cent. The value of x depends on the weight of the tablets; for tablets having an average weight of less than 120 mg, x is 10; for those of more than 120 mg and less than 300 mg, x is 7·5; and for those of more than 300 mg, x is 5. This test is a qualitative one, and although the tablets are weighed individually the information so obtained is not used fully. The test is, in fact, a limit on the proportion of defectives, or perhaps outsiders is a better word, in the sample.

It is possible to calculate limits for the proportion of outsiders, i.e. tablets differing from the mean weight by more than $\pm x$ per cent, in a batch from which a sample which just passes this test is drawn. In such a sample, two tablets have weights that differ from the mean weight by more than $\pm x$ per cent, and the probability p that any single tablet in the sample will have a weight outside these limits is therefore 2/20, i.e. 0·1. In the whole batch this probability will be $\mathfrak{p}$, the value of which will usually differ from p, which is the best estimate of $\mathfrak{p}$ that can be derived from the sample. The proportion of outsiders in the whole batch is $\mathfrak{p}$, and the problem is to determine the limits between which $\mathfrak{p}$ may be expected to lie when p is equal 0·1 for a sample of 20.

From considerations of binomial probability (see Chap. 11, § 3), the probability of obtaining two outsiders, or less, in a sample of 20 tablets is the sum of the first three terms of the expansion of $(\mathfrak{q}+\mathfrak{p})^{20}$, where $\mathfrak{q} = 1-\mathfrak{p}$. Thus, the probability p_1 is given by

$$p_1 = \mathfrak{q}^{20} + 20\mathfrak{q}^{19}\mathfrak{p} + \frac{20.19}{1.2}\mathfrak{q}^{18}\mathfrak{p}^2.$$

If a definite value, say 0·025, is given to this probability, then, since $q = 1-p$, there is only one unknown quantity, i.e. p, in the above equation, and, with much arithmetic labour, the equation can be solved to give a value of p which is just large enough, at the 0·025 probability level, to give two outsiders in a sample of 20; a greater value of p would tend to give more than two outsiders. This value of p can therefore be regarded as an upper limit, and the probability that it will be exceeded in the batch is 0·025.

The probability p_2 of there being more than two outsiders in a sample of 20 is the sum of the remaining 18 terms of the binomial, i.e.

$$p_2 = \frac{20.19.18}{1.2.3} q^{17} p^3 + \ldots + p^{20}.$$

Again, if the value 0·025 is given to this probability, the equation can be solved to give a value of p which is just sufficient to give two outsiders in a sample of 20; a lower value of p would tend to give less than two outsiders at the chosen probability level. This value of p can therefore be regarded as a lower limit, and the probability that p will be less than this is 0·025.

Therefore, there is a probability of $(1-0·025-0·025)$, i.e. 0·95, that the value of p for a batch will lie between these two limits if a sample of 20 gives two outsiders. These values are therefore the required limits for the expected proportion of outsiders in the batch. Fortunately, there is no need to work out a set of binomial terms every time a calculation of this type has to be made. Tables are available giving limiting values of p for different values of a, the number of outsiders in a sample of n individuals, at various probability levels. Such tables are computed from the binomial distribution. The probability level given in them is the probability of exceeding a single limit. The probability P that p will lie between a pair of limits is given by

$$P = 1 - (2 \times \text{tabulated probability}).$$

An adaptation of such a table is given in Table 2, which shows limits for p corresponding to different values of a in a sample of 20 at $P = 0·95$.

It is seen that the information about the proportion of outsiders in a batch that can be derived from the test in the *British Pharmacopoeia* is vague. Even if there are no outsiders in a sample of 20, there may be up to 16·8 per cent of them in the batch; if there is one in the sample, there may be from 0·1 to 24·8 per cent in the batch, and in a sample that just passes the test, i.e. has two outsiders, there may be from 1·2 to 31·7 per cent of them in the batch.

Closer limits for p can be obtained by examining larger samples, but it is doubtful whether this is worth the extra work involved. A sample of 50 tablets with five outsiders would give limits for the percentage of outsiders in the batch ($P = 0·95$) of 3·3 to 21·8, while a sample of 100 tablets with 10 outsiders would give limits of 4·9 to 17·6.

TABLE 2
Proportion of outsiders in a batch
(Test sample of 20; $P = 0.95$)

a	p ($= a/20$)	Limits for expected number in sample of 20	Limits for proportion in batch ($\mathfrak{p}$)	Limits for percentage of outsiders in batch ($= 100\mathfrak{p}$)
0	0	0 to 3.37	0 to 0.168	0 to 16.8
1	0.05	0.025 to 4.97	0.001 to 0.248	0.1 to 24.8
2	0.1	0.247 to 6.34	0.012 to 0.317	1.2 to 31.7
3	0.15	0.642 to 7.58	0.032 to 0.379	3.2 to 37.9
4	0.2	1.15 to 8.73	0.057 to 0.436	5.7 to 43.6
5	0.25	1.73 to 9.82	0.086 to 0.491	8.6 to 49.1

The second part of the *British Pharmacopoeia* test, which requires that no tablet should deviate from the mean by more than $2x$ per cent, is designed only to prevent gross discrepancies. However, it may be seen from Table 2 that a sample that passes this test may have been drawn from a batch that contains up to 16.8 per cent of such tablets.

A test for uniformity of weight of tablets based on sounder principles is that of the *Swedish Pharmacopoeia*, 1946, in which the relative standard deviation of individual tablets about the mean is required, for uncoated tablets, not to exceed 4.5 per cent. The calculation of standard deviation utilises all the information obtained from the weighing of the individual tablets. This type of test has the following advantages over that of the *British Pharmacopoeia*:

(i) By establishing a clear maximum value for the sampling error, it permits statistically better limits to be set for the content of active ingredient in the tablets.
(ii) It gives a better indication of the variation in the tablet weights in the batch.
(iii) It gives better protection from gross variations in tablet weights.

The first advantage is discussed in § 4; the other two points can be illustrated by a few calculations. Suppose a sample of 20 tablets gives a standard deviation at the limit of the Swedish test, i.e. 4.5 per cent. Then the maximum value for the batch variance can be estimated from the variance ratio F. From Appendix XII, Table 4, it is seen that, for infinite degrees of freedom in the higher variance (the batch variance) and 19 degrees of freedom in the other variance (the sample variance), F is, by interpolation, 1.9. The sample variance is $(4.5)^2$. Therefore, the maximum estimated batch variance is $1.9 \times (4.5)^2$, and the maximum estimated batch standard deviation is the square root of this, i.e. $4.5\sqrt{1.9}$, or 6.2.

If a normal distribution of tablet weights in the batch is assumed, then

SOME APPLICATIONS OF STATISTICS IN PHARMACY

the normal deviate u corresponding to a difference from the mean of 10 per cent will be $10/6 \cdot 2$, i.e. $1 \cdot 61$, and the normal deviate for a difference of 20 per cent will be $20/6 \cdot 2$, i.e. $3 \cdot 22$. From the properties of the normal distribution (see Chap. 12, § 5, and Appendix XII, Table 1), it is found that the proportion $(1-P)$ of the distribution having these values of u, or higher, are $0 \cdot 107$ and $0 \cdot 001$, respectively, and these are expected maximum proportions of tablets in the batch having 10 and 20 per cent deviations from the mean. If the distribution is not normal, higher proportions than these may result.

In a normal distribution, one-tenth of the values ($P = 0 \cdot 9$) will differ from the mean by an amount exceeding $1 \cdot 645\sigma$ (see Appendix XII, Table 1). If this difference is 10 per cent, then $1 \cdot 645\sigma = 10$, i.e. $\sigma = 6 \cdot 1$. Thus, a batch with normal distribution which is on the borderline of the *British Pharmacopoeia* requirements and which contains 10 per cent of outsiders will have a standard deviation of $6 \cdot 1$, which is in good agreement with the Swedish requirement. The intention of the two types of test therefore appears to be similar, but the use of standard deviation is more effective. For tablets of mean weight greater than 120 mg the Swedish test becomes rather less stringent than the British one, but it could be improved by fixing limits for the standard deviation which vary with the weight of the tablets. Suitable limits can be calculated as follows. In a batch of tablets in which the weights are normally distributed and the weights of one-tenth of the tablets differ from the mean weight by more than $7 \cdot 5$ per cent, then $1 \cdot 645\sigma = 7 \cdot 5$, and hence $\sigma = 4 \cdot 56$; if one-tenth of the tablets differ from the mean by more than 5 per cent, $\sigma = 5/1 \cdot 645$, i.e. $3 \cdot 04$. If these batch standard deviations are divided by the factor $1 \cdot 37$, i.e. $\sqrt{1 \cdot 9}$, derived from the variance ratio (see page 240), the results are $3 \cdot 3$ and $2 \cdot 2$ respectively, and these would be suitable percentage standard deviation limits for samples of 20 tablets taken from batches of mean weights between 120 mg and 300 mg and over 300 mg respectively.

A disadvantage of a test based on standard deviation is that it involves some computation. However, this can be done rapidly by means of a suitable calculating machine. Alternatively, if the weighings are made to one place of decimals, an abridged calculation can be made without the use of a machine, as is illustrated in the following example.

EXAMPLE. The weights w of twenty 120-mg tablets were

122·4	120·6	116·4	121·8	124·2
117·0	117·0	127·2	118·2	109·2
122·4	124·2	119·4	124·8	117·6
123·6	116·4	120·6	115·2	122·4

In order to determine their standard deviation without a calculating machine, the variate is transformed so as to give a set of small integers. If the weights are given to one decimal place, a suitable transformation is given by

$X = 10(w - \text{stated weight})$, i.e. $X = 10(w - 120)$.

Hence
$$w = X/10 + 120.$$
If m is the mean value of w, then
$$m = \Sigma(w)/N = \Sigma(X/10+120)/N = \Sigma(X)/10N + 120 = \bar{X}/10 + 120.$$
If V_w is the variance of w and V_X is that of X, then the relation between V_w and V_X may be found, since
$$V_w = \Sigma(w-m)^2/(N-1) = \Sigma\{(X/10+120)-(\bar{X}/10+120)\}^2/(N-1)$$
$$= \Sigma(X-\bar{X})^2/10^2(N-1) = V_X/10^2.$$
If a machine is not available, it is much easier to calculate V_X and $\bar{X}$ than V_w and m.

w	X $\{=10(w-120)\}$	X^2
122·4	+24	576
120·6	+6	36
116·4	−36	1296
121·8	+18	324
124·2	+42	1764
117·0	−30	900
117·0	−30	900
127·2	+72	5184
118·2	−18	324
109·2	−108	11664
122·4	+24	576
124·2	+42	1764
119·4	−6	36
124·8	+48	2304
117·6	−24	576
123·6	+36	1296
116·4	−36	1296
120·6	+6	36
115·2	−48	2304
122·4	+24	576
	$\Sigma(X) = +6$	$\Sigma(X^2) = 33{,}732$

Now
$$\bar{X} = 6/20 = 0{\cdot}3, \quad \text{and} \quad m = 0{\cdot}3/10 + 120 = 120{\cdot}0.$$
From (12.1)
$$V_X = \{\Sigma(X^2) - \Sigma^2(X)/N\}/(N-1)$$
$$= (33{,}732 - 36/20)/19 = 1775.$$
$$V_w = 1775/10^2 = 17{\cdot}75.$$
The standard deviation of w is
$$\sqrt{V_w} = 4{\cdot}213.$$
The relative standard deviation is
$$\frac{4{\cdot}213}{120} \times 100 = 3{\cdot}5 \text{ per cent}.$$
The sample therefore passes the standard deviation test of the *Swedish Pharmacopoeia*, and also complies with the test of the *British Pharmacopoeia*,

SOME APPLICATIONS OF STATISTICS IN PHARMACY 243

since there are only two tablets that deviate from the mean weight by more than 10 per cent and none that deviate by 20 per cent or more. The large contribution made to the value of $\Sigma(X^2)$ by the two tablets that deviate by more than 10 per cent from the mean should be noted. This effect gives protection from large discrepancies without the need for a special test for this purpose.

The above calculation is facilitated by using tables of squares and square roots.

3. Qualitative tests on samples

From Table 2 in §2 it is apparent that qualitative tests on samples of 20 tablets give meagre information about the properties of the batch from which the samples are drawn. To ensure that a batch contains no defectives it is necessary to examine every unit in the batch. For example, if none of a sample of 20 ampoules from a batch shows visible defects, then all that can be concluded is that the proportion of defectives in the batch is less than 16·8 per cent ($P = 0.95$). To ensure that the batch has no defectives, all the ampoules must be examined.

4. Limits for content of active ingredient in tablets

In the monographs on tablets in the *British Pharmacopoeia* the amount of active ingredient in each tablet of average weight is required to be within certain fixed percentage limits of the amount prescribed or stated. 'These ranges have been framed to allow for all permissible variations, including that of the active ingredient itself and that due to the process of manufacturing the tablets.' In the assay of a batch, a sample of 20 tablets is taken, weighed individually and powdered, and the powder is assayed by a suitable method. The weight m of active ingredient in a tablet of average weight is the product of the mean tablet weight $\bar{X}$ of the sample and the percentage Y of active ingredient in the powder. Both $\bar{X}$ and Y are subject to independent random errors; that in $\bar{X}$ is due to sampling error, while that in Y is due to inherent experimental error of the assay method used. It would be extremely useful if the percentage standard deviations of the different assay methods of the *British Pharmacopoeia* could be published in the monographs.

As has already been shown

$$E_m = \sqrt{(E_X^2/N + E_Y^2)},$$

where E is the coefficient of variation of the quantity indicated by the subscript. E_X can be calculated from the standard deviation of individual tablet weights about their mean. For 120-mg tablets that just comply with the British and Swedish tests for uniformity, this has been shown to be about 6 per cent. Hence, $E_X/\sqrt{N} = 6/\sqrt{20}$, i.e. 1·34. E_Y can be calculated from the results of repeated assays, carried out preferably by a number of different analysts, on the same batch of powder. In addition, a third source of error, due to permitted variation in the purity of the drug from which the tablets were made, may have to be included.

Combination of errors from various sources. If the percentage standard deviations due to the different independent sources of error are accurately known, there is no difficulty in calculating the coefficient of variation of the final result; the E values are squared, added together and the square root taken of the sum. Direct addition of the E values may seriously overestimate the standard error of the final result.

In some cases one of the errors may be expressed as, say, ± 2 per cent. An approximate standard deviation can be calculated by using the factor 3 (see Chap. 12, § 10) as relating the limits of error of single results to their standard deviation. The coefficient of variation then equals 2/3, i.e. 0·67 per cent, and this value can then be included in the estimate for E for the final result.

The coefficients of variation which are combined in this way should be either theoretical ones or known from a large number of repeated measurements. If they are determined from a few results, the corresponding maximum values of the universe standard deviation should be calculated using the variance ratio (see § 2) and these used in the expression for the E value for the final result.

Limits of error for the final result can be calculated using values of u for the normal distribution if each term involved in the final result is the mean of at least four determinations. If this is not so, then the maximum value of the Camp–Meidell expression should be used.

It is sometimes necessary to estimate the error in a result which is the product of several quantities subject to independent random errors whose standard deviations are not known, but whose errors are stated as, say, $\pm x$ per cent and $\pm y$ per cent, without reference to probability limits. The best estimate of the error of the final result is then $\pm \sqrt{(x^2+y^2)}$.

Limits for content of active ingredient. The standard deviation E_m of the result of an assay of a sample of tablets by the method of the *British Pharmacopoeia* is given by

$$E_m = \sqrt{(E_X^2/N + E_Y^2 + E_P^2)},$$

the term E_P being introduced to allow for variation in the purity of the active ingredient; it is calculated as outlined above. If the permitted limits for purity are not symmetrical, as for example in the case of calcium gluconate in the *British Pharmacopoeia* (-1 to $+3$ per cent), it is best to take the wider limit for the calculation of E_P.

The limits of error for m will be $\pm uE_m$, and the content of active ingredient in any sample whose value of m falls outside these limits of error can be deemed to differ significantly from the stated value. If m is within these limits of error, then the difference between the observed and stated values may well be due to chance. The limits for content of active ingredient, therefore, should never be less than $\pm uE_m$.

SOME APPLICATIONS OF STATISTICS IN PHARMACY 245

The value of u depends on the level of significance chosen, usually $P = 0.95$, and also on the distribution of m. In the *British Pharmacopoeia* assays the percentage of active ingredient in the powder may depend on the result of a single determination, or may be the mean of only two results, so it cannot be assumed that the distribution of Y, and hence of m, is normal. The Camp–Meidell value of u should therefore be used. For $P = 0.95$, this is approximately 3, so that the limits for m should be $\pm 3E_m$. In practice, it is probably advisable to specify slightly wider limits than this, but they should never be less than $\pm 3E_m$.

Limits for active ingredients in small tablets (less than 120 mg). To illustrate this section, phenobarbitone tablets B.P. will be considered. The value of E_m for these tablets can be computed as follows. If the sample of 20 tablets just passes the test for uniformity of weight, the batch standard deviation is likely to be up to 6 per cent, and this will cause a sampling standard error, i.e. $E_X/\sqrt{20}$, of 1·34 per cent. Forty-two repeated determinations of phenobarbitone by a number of different analysts on a single uniform batch of powder gave a standard deviation of 1·5 per cent for the assay method. There is no permitted variation in the purity of phenobarbitone B.P., and so E_m is given by

$$E_m = \sqrt{\{(1.34)^2 + (1.50)^2\}} = 2.01 \text{ per cent.}$$

The limits for the standard ($P = 0.95$) should therefore be at least ± 6 per cent; the actual limits given in the *British Pharmacopoeia* are ± 7.5 per cent.

Even if there were no analytical error, the permitted weight variation for 120-mg tablets may cause a value of E_Z of 1·34, and with $P = 0.95$, this alone would require limits of ± 4 per cent.

Borderline samples. The great problem in any quality sampling scheme is to decide when a batch of material should be rejected because a sample from it fails a test. Providing the limits for the test are statistically sound, i.e. they are greater than $\pm 3E_m$, the smallest deviation of a result from the limit indicates a significant difference between the observed and required content of active ingredient in a batch of tablets. If the discrepancy is small, further samples should be examined, and if another sample also gives a result outside the limits, the batch must be presumed to be faulty. If two samples out of three give results that are outside the limits, it is fairly certain that the batch is unsatisfactory or so close to the limit that other samples drawn from it are likely to fail the test.

EXAMPLE. *The following results were obtained by different analysts in the assay of three separate samples of a batch of 60-mg phenobarbitone tablets; $\bar{X}$ is the mean weight, in milligrams, of a sample of 20 tablets, Y is the percentage of active ingredient found in the powder from 20 tablets, and m, which is equal to $\bar{X}Y$, is the mean weight, in milligrams, of active ingredient in each tablet of average weight.*

Analyst	$\bar{X}$	Y		m	
A	87·36	62·50,	62·29	54·60,	54·42
B	87·84	65·10,	64·69	57·18,	56·82
C	88·62	62·36,	62·29	55·26,	55·20

The lower limit in the British Pharmacopoeia *for these tablets requires that at least* 92·5 *per cent of the stated weight should be phenobarbitone, i.e. they should contain at least* 55·50 *mg of the drug.* Analyst A reported that the tablets failed to comply with the standard, B claimed that this was not so, while C confirmed the finding of A. *Who is correct?*

An examination of all the results as repeated values of the same measurement gives a mean value of $\bar{X}$ of 87·96 mg and $E_{\bar{X}}$ equal to 0·724 per cent, a value smaller than the likely maximum value (1·34 per cent) calculated from the test for uniformity of weight, and so the variation between the different results for $\bar{X}$ is only what is to be expected from sampling error.

The values of Y have a mean of 63·20, and E_Y is 2·08, i.e. rather larger than the known experimental value of 1·50. However, the variance ratio $(2·08/1·50)^2$, i.e. 1·92, is appreciably less than the theoretical value of 2·2, interpolated from Appendix XII, Table 4, for F with $\phi_1 = 5$ and $\phi_2 = \infty$, and so there is no reason to suppose that the variation in Y is more than is to be expected from the experimental error of the assay. Further confirmation of this point can be obtained by finding the ratio of the maximum deviation of a single value of Y from the mean to the standard deviation of the assay method; this ratio is

$$(65·10 - 63·20)/1·50 = 1·27.$$

This value is much less than 3, the value of u ($P = 0·95$) derived from the Camp–Meidell expression.

When the mean of the values of m is calculated, it is found to be 55·56 mg, which suggests that the batch might be within the specification. However, a better criterion is the fact that two samples out of three have been rejected. All the analysts agree that the phenobarbitone content of the tablets is well below the stated value of 60 mg, and it should be realised that if control samples from a batch are near the permitted limits, there is a possibility that subsequent samples may fail the test. In fact, the product should be controlled to within $\pm 2E_m$ or better in order to ensure that a large proportion of samples drawn from the batch will pass the test.

Probability of a batch passing or failing to pass a test. If the mean weights m of active ingredient per tablet for samples of 20 tablets are assumed to be normally distributed about the batch mean μ with a standard error of E_m per cent, it is possible to calculate the probability that a given sample will pass or fail to pass the test. This probability will vary with the value of μ.

If A is the lower limit of the test, then the probability that a given sample mean chosen at random from the distribution will exceed A, i.e. the sample will pass the test, is represented by the unshaded area to the right of the ordinate at A (see Fig. 15a) divided by the total area under the curve. A is fixed, but as μ increases, the curve moves to the right and the ratio of unshaded to total area also increases. This ratio is equal to 1 minus the fraction of the whole area under the curve lying to the left of A, i.e. to $1-\mathfrak{F}$, where $\mathfrak{F}$ is the cumulative frequency at A. The normal deviate at A is u and is equal to $(A-\mu)/E_m$. It should be noted that since E_m is expressed as a percentage of μ, $A-\mu$ should be similarly expressed and μ itself then becomes 100 in the above expression. Theoretical values of $\mathfrak{F}$ for different values of u are known from the theory of normal distribution (see Chap. 12, § 11).

For 60-mg phenobarbitone tablets B.P., A is 55·5 mg. In Table 3, values of u for various values of the mean weight μ of active ingredient in each tablet in the batch are calculated, and in the final column the probability that a sample of 20 tablets from each batch will comply with the

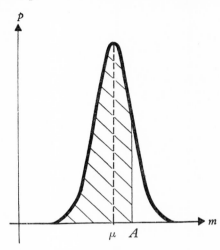

FIG. 15a

specification is shown. The value of E_m is taken as the batch value just complying with the test for uniformity of weight, i.e. 2 per cent (see page 245).

TABLE 3

Probability of acceptance of a sample of 20 tablets drawn from a normally distributed batch of mean weight μ

μ (mg)	A (as percentage of μ)	u $\{=(A-100)/E_m\}$	Probability of acceptance of sample $(=1-\mathfrak{F})$
48·0	115·6	7·8	extremely small
51·0	108·8	4·4	5×10^{-6}
52·8	105·1	2·55	0·005
54·0	102·8	1·40	0·08
54·6	101·65	0·82	0·21
55·2	100·54	0·27	0·39
55·8	99·46	−0·27	0·61
56·4	98·40	−0·80	0·79
57·0	97·37	−1·31	0·90
57·6	96·35	−1·82	0·97
58·2	95·36	−2·32	0·99
59·4	93·4	−3·3	0·9995

From this table it is seen that if the mean weight of active ingredient in each tablet in a batch is between 54·6 and 55·2 mg, the probability of a sample passing the test is about one-third, i.e. only one sample in three would comply with the standard. Reversing this argument, if the result given by only one sample out of three is above the lower limit, then the batch mean is likely to be between 54·6 and 55·2 mg, and the sample does not comply with the *British Pharmacopoeia* standard.

If the batch mean is between 55·8 and 56·4 mg, the probability of acceptance is about two-thirds, i.e. samples drawn from such a batch would be expected to fail the test once in three tests. In order to reduce the chance of a sample failing to pass the test to one in 20 $(1-\mathfrak{F} = 0.95)$, it is necessary for the batch mean to be about 57·6 mg, differing from the stated amount of 60 mg by -4 per cent, or $-2E_m$. This calculation illustrates the statement already made, that products should be controlled so that samples give results for the weight of active ingredient in each tablet of average weight within $\pm 2E_m$ of the stated value.

Similar considerations apply to the upper limits of the *British Pharmacopoeia* standards.

5. Dispensing errors

Under normal conditions, dispensing errors are relatively large compared with analytical errors.

Dispensing of liquids. The paper by Capper, Cowell and Thomas (*Pharm. J.*, 1955, *175*, 241) on the measurement of liquids indicates a considerable systematic error which may arise in the dispensing of liquids in an ordinary pharmaceutical measure due to alternative methods of illuminating the meniscus. One type of background made a false meniscus more prominent than the real one, and half the dispensers co-operating in the work were found to be reading this false meniscus, which caused a systematic error of from -1 to -3 per cent in the volume dispensed.

In addition to this systematic error, other random personal errors in the measurement of liquids were found. A further error of about 2 per cent might arise from the permitted Board of Trade tolerances for the graduations on the measure.

Taking 60 minims of liquefied phenol as the test volume, Capper *et al* found that the standard deviation for the personal error ($\phi = 9$) was 1·4 per cent of the mean result, which was 59·15 minims, all the measurements being made to the true meniscus. The variance ratio for $\phi_1 = \infty$ and $\phi_2 = 9$ is 2·7 (see Appendix XII, Table 4), so that an estimate of the maximum universe standard deviation due to the personal error is $1.4\sqrt{2.7}$, i.e. 2·3 per cent (see § 2).

The error due to permitted tolerances for the measure graduations was 2·3 per cent, corresponding to an estimated standard deviation due to this cause of 2·3/3, i.e. 0·77 (see § 4). The combined standard deviation due to

SOME APPLICATIONS OF STATISTICS IN PHARMACY 249

personal error and location of graduations is therefore estimated as $\sqrt{\{(2\cdot3)^2+(0\cdot77)^2\}}$, i.e. 2·42 per cent. Using the Camp–Meidell value of u, this gives limits of error ($P = 0·95$) for a further single dispensing as $\pm3\times2\cdot42$, i.e. $\pm7\cdot26$ per cent. If the possible systematic error of 1·4 per cent due to reading the false meniscus is taken into account, the lower limit must be widened to $-(7\cdot3+1\cdot4)$, i.e. $-8\cdot7$; the upper limit is unaffected by this possibility. The final limits are therefore $+7\cdot3$ and $-8\cdot7$.

The variations in dispensed volume obtained under ordinary dispensing conditions when 60 minims of liquefied phenol is prescribed can be estimated from the published results of the Ministry of Health's Drug Testing Scheme (*Pharm. J.*, 1956, *176*, 47) as is shown in Table 4.

TABLE 4

Estimate of standard deviation in dispensing 60 minims of liquefied phenol

Range of deviations from 60 min., as percentage (both + and −)	Central deviation in group (X)	Number of results in group (f)	X^2	fX^2
0 to 1	0·5	51	0·25	12·75
1·01 to 2	1·505	3	2·265	6·795
2·01 to 5	3·505	26	12·285	319·4
5·01 to 10	7·505	9	56·325	506·9
11 to 20	15·5	8	240·25	1922·0
21 to 30	25·5	3	650·25	1950·75
31 to 50	40·5	1	1640·25	1640·25
51 to 75	63·0	1	3969·00	3969·00
		$\Sigma(f) = \overline{102}$		$\Sigma(fX^2) = \overline{10327\cdot845}$

The deviation ranges include both positive and negative deviations, and so the mean $\Sigma(fX)$ cannot be calculated. However, assuming that the positive and negative deviations cancel each other out, $\Sigma(fX) = 0$. The variance V of the results about the prescribed volume of 60 minims is then given by

$$V = \Sigma(fX^2)/N = 10327\cdot845/102 = 101\cdot3.$$

It should be noted that N is used in the denominator instead of the usual $N-1$, because a mean has not been calculated from the data, and so N degrees of freedom remain for calculating the variance.

The percentage standard deviation, or coefficient of variation, is $\sqrt{V}$, i.e. 10·1 per cent. This high standard deviation includes errors arising in the adjustment of the preparation to its final volume, errors due to faulty sampling after dispensing, and errors in the analysis for phenol, none of which are present in the results of Capper *et al.*, which represent only variations in measuring liquefied phenol. In addition, the grouping of the data leads to a higher variance than the true one obtained from ungrouped results. Examination of the final column of Table 4 shows that the small number of results that deviate widely from the prescribed amount make a very large contribution to the sum of squares, and hence to the standard deviation. These large deviations are probably due to mistakes in calculation rather than to random dispensing errors, but it does not seem justifiable to ignore them altogether, since this introduces a selection of results for the statistical calculation. The straightforward calculation of standard deviation does, however, appear to give undue weight to these widely deviating results, and a better estimate of random dispensing error as is required for fixing tolerance limits for the future can be made by considering the proportion of all the results within certain limits of the prescribed value. Such a calculation can only be made when, as in the present case, a large number of results are available. The limits should be chosen so as to include about 95 per cent of the results, corresponding to the usual probability level for limits of error. The limits of error for single results about the mean are given by $\pm u\sigma$, where u is the maximum value from the Camp-Meidell expression corresponding to a value of P equal to the proportion of results within the limits chosen. P should not exceed 0·98, since above this probability the values of u become large and uncertain.

For the results given in Table 4, suitable limits are ± 20 per cent, a proportion of 97/102, i.e. 0·951, of the results being within these limits. The maximum value of u ($P = 0.951$) is 3·0. The estimate of σ is given by $\pm u\sigma = \pm 20$ per cent. Hence $\sigma = 20/3$, i.e. 6·7 per cent. This calculation applies the ideas of limits of error in reverse, using the limits to estimate the standard deviation, but the estimate so obtained is very approximate.

An alternative method for calculating the standard deviation of the normal distribution which fits most closely the results near the mean is to use probability graph paper. To use this method it is necessary to know the signs of the deviations of the results from the mean. The central results are classified into groups of suitable width, and the cumulative frequency for each group is calculated as in the example in Chap. 12, § 11, taking into account the diverging results. Values of $100\mathfrak{F}$ are plotted against the upper values of the variate for each group, the best straight line is drawn through the central points on the plot near the mean, and the mean and standard deviation are calculated (see Chap. 12, § 12). In this method the widely diverging results are not ignored, since they are taken into account in the calculation of $\mathfrak{F}$, but they are not given undue weight since the straight line drawn on the probability paper plot, as always, is not influenced to any great extent by the points at the extremities of the distribution.

In the data summarised in Table 4, only the magnitudes of the deviations are given, and it is not known whether they are positive or negative. However, if it is assumed that the distribution is symmetrical about the mean, this method of calculation can still be applied. First, the fractional

SOME APPLICATIONS OF STATISTICS IN PHARMACY 251

frequencies for each group are calculated, which are then halved, and one-half are assumed to have positive deviations and the other half negative (see Table 5). The cumulative frequency and percentage cumulative frequency (100$\mathfrak{F}$) are calculated as in the example in Chap. 12, § 11, and the values of 100$\mathfrak{F}$ are plotted against the upper value of the variate for the group.

TABLE 5

Range of deviations from 60 min., as percentage	Number of results in group (f)	f/N	f/2N
0 to 1	51	0·500	0·250
1 to 2	3	0·029	0·0145
2 to 5	26	0·255	0·1275
5 to 10	9	0·088	0·044
11 to 20	8	0·078	0·039
21 to 30	3	0·029	0·0145
31 to 50	1	0·010	0·005
51 to 75	1	0·010	0·005

The last three groups are classified together as widely diverging results with a total value of $f/2N$ of 0·025. If the deviations are now divided equally into positive and negative values, each group having a fractional frequency of $f/2N$, the cumulative frequency table (see Table 6) is obtained. It should be noted that in this table the upper group limit is taken as the value of the variate for each group, since the cumulative frequency includes all values of the variate up to and including this upper limit.

TABLE 6

Range of deviations from 60 min., as percentage	X	Fractional frequency of group (f/2N)	Fractional cumulative frequency ($\mathfrak{F}$)	100$\mathfrak{F}$
Below −20	−20	0·0245	0·0245	2·5
−20 to −10	−10	0·039	0·0635	6·4
−10 to −5	−5	0·044	0·1075	10·8
−5 to −2	−2	0·1275	0·235	23·5
−2 to −1	−1	0·0145	0·2495	25·0
−1 to 0	0	0·250	0·4995	50·0
0 to 1	1	0·250	0·7495	75·0
1 to 2	2	0·0145	0·764	76·4
2 to 5	5	0·1275	0·8915	89·2
5 to 10	10	0·044	0·9355	93·6
11 to 20	20	0·039	0·9745	97·5
above 20	—	0·0245	0·999	99·9

The plot on probability paper shows considerable deviations from linearity, but the best straight line through the central points of the plot gives a standard deviation of 3·4 per cent, and this is probably the best estimate of standard deviation due to random dispensing and analytical errors.

Dispensing of solids. Theoretically the error in dispensing a quantity of a solid such as 240 grains of sodium bicarbonate should be small. However, when the report already quoted on the drug-testing scheme is examined, the variations for this amount of sodium bicarbonate are seen to be quite considerable. As the preparation being tested is liquid it is probable, however, that the variations are largely due to errors in adjusting the prescription to the final volume rather than in weighing the sodium bicarbonate. The effect of a small proportion of widely diverging results is considerable, and these results are more likely to represent failure to ensure complete dissolution of the sodium bicarbonate before dividing the official sample into the requisite three parts or to be due to errors in calculation rather than to faulty weighing. In Table 7 a direct estimate of the standard deviation of the published results is made.

TABLE 7
Estimate of standard deviation in dispensing 240 grains of sodium bicarbonate

Range of deviations from 240 gr., as percentage (both + and −)	Central deviation in group (X)	Number of results in group (f)	X^2	fX^2
0 to 1	0·5	350	0·25	87·5
1·01 to 2	1·505	109	2·265	246·9
2·01 to 5	3·505	174	12·285	2137·6
5·01 to 10	7·505	46	56·325	2591·0
11 to 20	15·5	30	240·25	7207·5
21 to 30	25·5	4	650·25	2601·0
31 to 50	40·5	5	1640·25	8201·25
51 to 75	63·0	1	3969·0	3969·0
		$\Sigma(f) = 719$		$\Sigma(fX^2) = 27041·7$

One of the results given in the report has been left out; this result is shown as having a deviation in excess of 100 per cent and a footnote indicates that the sodium bicarbonate had been omitted.

The variance V of the results about the prescribed weight of 240 grains is given by

$$V = \Sigma(fX^2)/N = 27041·7/719 = 37·61.$$

SOME APPLICATIONS OF STATISTICS IN PHARMACY   253

The percentage standard deviation is $\sqrt{V}$, i.e. 6·13. As in the dispensing of liquefied phenol, this estimate of the standard deviation is higher than might be anticipated, due to the inclusion of a small proportion of widely diverging results. In the 'reversed limits of error' calculation, limits of ±10 per cent are appropriate; they include a proportion of 679/719, i.e. 0·944, of the results. Limits of ±20 per cent would give a proportion of 0·986, which is too high for a reliable estimation of u. The maximum value of u ($P = 0·944$) is 2·8 (see § 1). The estimate of σ is given by $\pm u\sigma = 10$. Hence $\sigma = 10/2·8$, i.e. 3·6 per cent.

Estimation of standard deviation by the cumulative frequency method, using a plot on probability paper, gives a value of σ of 2·7 per cent, and this is probably the best estimate of the standard deviation due to random dispensing and analytical errors.

6. Analysis of variance

As was shown in Chap. 14, analysis of variance is a useful method for studying the results obtained in the development of an assay. To illustrate this further with a chemical assay, the following results are used. They summarise the values for percentage w/v of ferric chloride found in a single batch of gargle of ferric chloride B.P.C. by five different laboratories, each using four different assay methods. (The assay is complicated by the fact that potassium chlorate is present in the gargle.)

TABLE 8

Analysis of variance for percentage of ferric chloride in gargle of ferric chloride B.P.C.

Method	Laboratory					Method mean
	A	B	C	D	E	
Gravimetric	0·423	0·437	0·424	0·439	0·423	0·4292
Titrimetric (a) with titanous chloride	0·423	0·423	0·421	0·453	0·416	0·4272
(b) with sodium edetate	0·413	0·383	0·386	0·390	0·405	0·3954
Spectrophotometric	0·401	0·395	0·389	0·385	0·409	0·3958
Column mean (Z)	0·415	0·4095	0·405	0·41675	0·41325	—

This table differs from the analysis of variance table in Chap. 14, § 3, in that all the results are, in fact, repeats of the same measurement Y. In this case no regression is involved, and the first step is to resolve the total sum of squares into components between columns and within columns. The former component is due to variation between laboratories and the latter to variation between the different assay methods.

The total sum of squares S_1 is given by
$$S_1 = \Sigma(Y^2) - \Sigma^2(Y)/N \quad \text{(see Chap. 14, Table 5).}$$
The sum of squares between columns S_2 is given by
$$S_2 = l\,\Sigma(Z^2) - \frac{\Sigma^2(Y)}{N},$$
where N is the number of results and l is the number of means. The sum of squares within columns S_3 is given by
$$S_3 = S_1 - S_2.$$
The degrees of freedom ϕ of these sums are $N-1$, $n-1$, and $N-n$ respectively. The corresponding variances are the appropriate sums divided by their degrees of freedom.

From Table 8,
$$N = 20, \quad n = 5, \quad \text{and} \quad N/n = 4.$$
$\Sigma(Y) = 8\cdot238$, grand mean $m = 0\cdot4119$, $\Sigma(Y^2) = 3\cdot400720$
and
$$\Sigma(Z)^2 = 0\cdot848396.$$
Hence
$S_1 = 74\cdot88 \times 10^{-4}, \quad S_2 = 3\cdot53 \times 10^{-4}, \quad \text{and} \quad S_3 = 71\cdot35 \times 10^{-4},$
and
$V_1 = 3\cdot940 \times 10^{-4}, \quad V_2 = 0\cdot882 \times 10^{-4}, \quad \text{and} \quad V_3 = 4\cdot757 \times 10^{-4}.$

The variance V_3 due to different analytical methods is seen to be much greater than the variance V_2 between laboratories. To test whether this difference is significant, the variance ratio F (larger/smaller) is calculated and compared with the theoretical value. The calculated value of F is $4\cdot757/0\cdot882$, i.e. $5\cdot39$. The theoretical value of F ($P' = 0\cdot05$) for $\phi_1 = 15$ and $\phi_2 = 4$ is $5\cdot86$ (see Appendix XII, Table 4). The difference between the two values of F is small, suggesting that there may be a significant difference between V_2 and V_3, i.e. that the different methods give results which differ more than would be expected from the variation of results between laboratories. This possibility can be further explored by calculating individual means and variances for the different analytical results, i.e. means and variances for each row in Table 8. If this is done the results are summarised in Table 9.

TABLE 9

Method		Mean		Variance
Gravimetric	m_g	0·4292	V_g	$0\cdot6525 \times 10^{-4}$
Titrimetric				
(a) with titanous chloride	m_t	0·4272	V_t	$2\cdot1625 \times 10^{-4}$
(b) with sodium edetate	m_e	0·3954	V_e	$1\cdot6825 \times 10^{-4}$
Spectrophotometric	m_s	0·3958	V_s	$0\cdot9125 \times 10^{-4}$

In each case, $\phi = 4$

SOME APPLICATIONS OF STATISTICS IN PHARMACY 255

The variances show no large differences between the methods, the greatest variance ratio being V_t/V_g, which equals 3·31. The theoretical value for $F(P' = 0·05)$ for $\phi_1 = \phi_2 = 4$ is 6·4, so that there are no grounds for rejecting any of the four methods on the basis of having too great a random variation.

When the means are examined, it is seen that the differences are comparatively large. To test the significance of the largest difference $m_g - m_e$, a t test (see Chap. 13, § 2) can be used, taking V_2, i.e. $0·882 \times 10^{-4}$ ($\phi = 4$), as a measure of the overall variance between laboratories. Now

$$t = \frac{\text{difference between means}}{\text{standard error}}.$$

Standard error, with five results to each mean, is equal to $\sqrt{(V_2/5)}$, and

$$\text{standard error} = \frac{\sqrt{V_2}}{\sqrt{5}} = \frac{0·939 \times 10^{-2}}{2·24} = 0·419 \times 10^{-2}.$$

Therefore

$$t = \frac{0·4292 - 0·3954}{0·00419} = \frac{0·0338}{0·00419} = 8·1.$$

The theoretical value of t ($P' = 0·05$) for $\phi = 4$ is 2·78 (see Appendix XII, Table 2), so that there is a highly significant difference between these two mean results, suggesting that one of them includes a systematic error.

Instead of using $\sqrt{(V_2/5)}$ as the denominator in the expression for t, a value of t can be calculated from the results of the gravimetric and edetate methods using the formula given in Chap. 13, § 2. Thus

$$V = \frac{\Sigma d_g^2 + \Sigma d_e^2}{N_g + N_e - 2} = 1·167 \times 10^{-4}.$$

Hence

$$s = \sqrt{V} = 0·0108.$$

Therefore, from (13·2),

$$t = \frac{m_g - m_e}{s\sqrt{(1/N_g + 1/N_e)}} = \frac{0·0338}{0·0108\sqrt{0·4}} = 5·0.$$

The theoretical value of t ($P' = 0·05$) for $\phi = 8$ is 2·3, and so again a significant difference between means is established.

The values of t calculated for other pairs of means, using $\sqrt{(V_2/5)}$ as the denominator in the expression for t, are given in Table 10, the theoretical value of t being 2·78 in each case. The negative value in the last result does not affect the test, which is a comparison of the magnitudes of observed and calculated values of t.

The results of this analysis show that although there is no significant difference between the random errors of the four methods, there is a significant difference between the means. The gravimetric and titanous chloride titration methods agree with one another, but their means differ

TABLE 10

Difference between means		t		
$m_g - m_t$	0·0020	0·0020/0·00419 =	0·48	not significant
$m_g - m_s$	0·0334	0·0334/0·00419 =	8·0	highly significant
$m_t - m_e$	0·0318	0·0318/0·00419 =	7·6	highly significant
$m_t - m_s$	0·0314	0·0314/0·00419 =	7·5	highly significant
$m_e - m_s$	−0·0004	−0·0004/0·00419 =	−0·1	not significant

significantly from those obtained with the edetate titration and spectrophotometric methods.

In order to decide between the four methods with the data available in Table 8, the variances should first be examined. The tendency would be to give preference to the method with the least variance. The gravimetric and spectrophotometric methods appear to give the lowest variances and hence the least random errors, but since there is a significant difference between their means, one of the methods introduces a systematic error. The statistical analysis of these results does not show which is the most accurate method, and a decision can be made only by carrying out further work with mixtures of accurately known composition.

APPENDIX I

Fundamental constants, approximations and conversion factors

Fundamental constants

Avogadro number $\quad N_A = 6 \cdot 02 \times 10^{26}$ kmol^{-1}
Gas constant $\quad R = 8314 \cdot 3$ J kmol^{-1} K^{-1}
Absolute zero $\quad = -273 \cdot 15°$C
Boltzmann's constant $\quad k = R/N_A$
$\quad = 1 \cdot 38 \times 10^{-23}$ J K^{-1}

Molar volume of ideal gas at 0°C and 760 mm pressure $\quad = 22 \cdot 414$ litres mol^{-1} = $22 \cdot 414$ m^3 kmol^{-1}
Acceleration due to gravity at Greenwich $\quad g = 9 \cdot 81$ m s^{-2}
Charge on electron $\quad -e = 1 \cdot 602 \times 10^{-19}$ coulomb (C)
Mass of proton $\quad = 1 \cdot 6723 \times 10^{-27}$ kg
Planck's constant $\quad h = 6 \cdot 6256 \times 10^{-34}$ J s
$\quad = 4 \cdot 1356 \times 10^{-15}$ eV s

Approximations

If x is very small, i.e. $x \ll 1$, then

$$(1 \pm x)^n \simeq 1 \pm nx$$
$$\ln(1 \pm x) \simeq \pm x$$
$$\sin x \simeq x \text{ (in radians)}$$
$$\cos x \simeq 1$$
$$e^x \simeq 1 + x$$
$$e^{-x} \simeq 1 - x$$
$$a^x \simeq 1 + x \ln a$$

If N is very large, i.e. $N \to \infty$, then

$$1/N \to 0$$
$$\tan^{-1} N \to \tfrac{1}{2}\pi$$
$$\int_0^N \exp(-\tfrac{1}{2}u^2)\, du \to \sqrt{(\tfrac{1}{2}\pi)}$$
$$\ln(N!) \to N \ln(N) - N \quad \text{(Stirling's approximation)}$$
$$(1 - m/N)^N \to e^{-m}$$

Conversion factors

Length

1 metre (m) $\quad = 39 \cdot 3701$ inches
1 inch (in) $\quad = 2 \cdot 5400$ centimetres

APPENDIX I

Volume

1 litre (l)	= 1·00 cubic decimetre = 0·001 m³
	= 1000·0 cubic centimetres
	= 1·7598 pints = 35·196 fluid ounces
1 millilitre (ml)	= 16·894 minims
1 pint (pt)	= 0·568245 litres
1 fluid ounce (fl oz)	= 28·412 millilitres
1 minim (min)	= 0·059192 millilitres

Mass

1 kilogramme (kg)	= 35·274 ounces = 2·20462 pounds
1 gramme (g)	= 15·432 grains
1 pound (avoirdupois) (lb)	= 453·592 grammes
1 ounce (avoirdupois) (oz)	= 28·3495 grammes = 437·5 grains
1 grain (gr)	= 64·7989 milligrammes

APPENDIX II

Triangles, lengths, areas, volumes and analytical geometry

Triangles

The area of any triangle is equal to half the length of the base multiplied by the perpendicular height; thus in the figure, area ABC $= \tfrac{1}{2} a . h$.

In any triangle

$$\frac{a}{\sin A} = \frac{b}{\sin B} = \frac{c}{\sin C},$$

and

$$a^2 = b^2 + c^2 - 2bc \cos A.$$

In a right-angled triangle, in which $\widehat{A} = 90°$,

$$a^2 = b^2 + c^2.$$

This is the statement of Pythagoras' theorem. Since this theorem is the basis of relations between trigonometric functions, a geometric proof of it is given below.

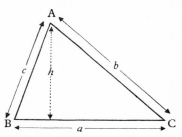

FIG. IIa. Sides and angles of a triangle

Pythagoras' theorem. The right-angled triangle is ABC. Construct the squares ADEC, BGFC and AIHB on the sides AC, BC and AB, respectively. Draw a perpendicular from A to FG meeting FG at J. Join A and F, B and E, C and H, and A and G.

Consider the triangles AFC and BEC.

CF = BC, since both are sides of the square BGFC.

Also

$\widehat{ACF} = \widehat{ACB} + \widehat{BCF}$

$= \widehat{ACB} + 90°$, since $\widehat{BCF}$ is an angle of a square,

and

$\widehat{BCE} = \widehat{ACB} + \widehat{ACE}$

$= \widehat{ACB} + 90°$, since $\widehat{ACE}$ is also an angle of a square.

Therefore

$\widehat{ACF} = \widehat{BCE}.$

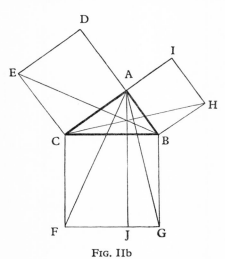

FIG. IIb

APPENDIX II

The triangles AFC and BEC therefore have two corresponding sides and the angle between them equal, and hence the triangles are identical and have the same area. Thus

$$\text{area AFC} = \tfrac{1}{2}(\text{base})(\text{perpendicular height}) = \tfrac{1}{2}CF \times FJ$$
$$= \tfrac{1}{2}BC \times FJ, \quad \text{since } CF = BC, \text{ both being sides of the same square.}$$

Further,

$$\text{area BEC} = \tfrac{1}{2}CE \times AC = \tfrac{1}{2}AC^2, \quad \text{since } CE = AC, \text{ both being sides of the same square.}$$

In an exactly similar manner, it can be shown that the triangles BAG and BHC are identical and have the same area. Thus

$$\text{area BAG} = \tfrac{1}{2}BG \times GJ = \tfrac{1}{2}BC \times GJ, \quad \text{since } BG = BC,$$

and

$$\text{area BHC} = \tfrac{1}{2}BH \times AB = \tfrac{1}{2}AB^2, \quad \text{since } BH = AB.$$

Now

$$\text{area BHC} + \text{area BEC} = \text{area BAG} + \text{area AFC}.$$

Therefore

$$\tfrac{1}{2}AB^2 + \tfrac{1}{2}AC^2 = \tfrac{1}{2}BC \times GJ + \tfrac{1}{2}BC \times FJ$$
$$= \tfrac{1}{2}BC(GJ + FJ)$$
$$= \tfrac{1}{2}BC^2, \quad \text{since } GJ + FJ = FG = BC.$$

Hence

$$BC^2 = AB^2 + AC^2,$$

or

$$a^2 = b^2 + c^2.$$

Lengths

Circumference of a circle of radius $a = 2\pi a$ $(\pi = 3 \cdot 1416)$.
Circumference of an ellipse of semi-axes a and $b = \pi \sqrt{(2a^2 + 2b^2)}$.

Areas

Area of a rectangle = (base)(height).
Area of a parallelogram = (base)(perpendicular height).
Area of a triangle = $\tfrac{1}{2}$(base)(perpendicular height).
Area of a circle of radius $a = \pi a^2$.
Area of an ellipse of semi-axes a and $b = \pi ab$.
Surface of a sphere of radius $a = 4\pi a^2$.
Area of a trapezium of sides a and b and width $c = \tfrac{1}{2}c(a+b)$, i.e., it is the mean height of the trapezium multiplied by the width.

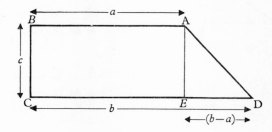

Fig. IIc. Area of a trapezium

This result is used in Chap. 7, § 1, to determine the area under a curve. It can be proved as follows. If AE is the perpendicular from A to CD, then

area of trapezium ABCD = area of rectangle ABCE+area of
triangle ADE
$= ac+\frac{1}{2}(b-a)c$, since DE $= b-a$
$= \frac{1}{2}c(a+b)$.

Volumes

Volume of a sphere of radius $a = \frac{4}{3}\pi a^3$.

Volume of a prolate ellipsoid $= \frac{4}{3}\pi ab^2$

Volume of an oblate ellipsoid $= \frac{4}{3}\pi a^2 b$

(a is the semi-major axis and b is the semi-minor axis. Prolate obtained by rotating an ellipse about the major axis; oblate by rotating about the minor axis).

Analytical geometry

Curved lengths and areas and volumes of curved shapes can be evaluated by means of analytical geometry, which employs the techniques of calculus to solve geometrical problems. An outline of some of the general methods of analytical geometry is given in the remainder of this appendix.

Curved lengths. In order to find the curved length AB of any part of the curve $y = f(x)$ between ordinates at $x = a$ and $x = b$, the curve within

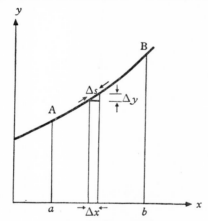

Fig. IId. Curved length

these limits is divided into a number of elements, each of length Δs. The length of each of these elements is approximately equal to the length of the hypotenuse of a right-angled triangle whose other sides are Δx and Δy (see Fig. IId). Hence

$$(\Delta s)^2 \simeq (\Delta x)^2+(\Delta y)^2.$$

If Δs is made very small, the curvature of each element becomes negligible and the above relation becomes exact, i.e. as $\Delta s \to 0$

$$(ds)^2 = (dx)^2+(dy)^2$$
$$= (dx)^2\{1+(dy/dx)^2\}$$

Therefore

$$ds = \sqrt{\{1+(dy/dx)^2\}}dx.$$

The curved length AB is then the sum of all such elements between $x = a$ and $x = b$. Hence

$$s = \int_a^b \sqrt{\{1+(dy/dx)^2\}}\, dx;$$

dy/dx is evaluated from the equation of the curve.

If, for example, the curve is the arc of a quadrant of a circle of radius a, the equation of the curve is

$$x^2 + y^2 = a^2$$

and hence $dy/dx = -x/y$; the limits are $x = 0$ and $x = a$. By substituting these values in the above formula and eliminating y, the result that $s = \frac{1}{2}\pi a$ is obtained; the total circumference of a circle is four times the length of the arc of the quadrant, i.e. $2\pi a$.

Plane area. The plane area of certain figures can usually be found by using integration to determine the area between the curve $y = f(x)$, the

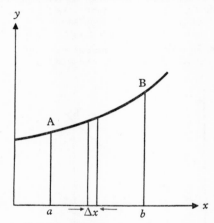

FIG. IIe.   Plane area under a curve

vertical lines at $x = a$ and $x = b$, and the x-axis (see Fig. IIe). This area A is given by

$$A = \int_a^b y\, dx \quad \text{(see Chap. 7, § 1)}$$

In the case of a quadrant of a circle, for example, the integration is carried out between the limits $x = 0$ and $x = a$ for the equation $x^2 + y^2 = a^2$. Converting to polar co-ordinates greatly facilitates the integration, i.e. $x = a \cos \theta$, and $y = a \sin \theta$.

Surface of solid of revolution. A solid of revolution is generated by rotating a curve about an axis. For example, if the curve $y = f(x)$ between ordinates at $x = a$ and $x = b$ is rotated about the x-axis, a solid figure is generated. The surface area of the solid is found by first considering the area ΔA resulting from the rotation about the axis of a single strip of curved length Δs. This area will be approximately equal to the width of the strip multiplied by the mean circumference of the circle formed on rotation.

The radius of the circle will be equal to y, the ordinate at the point considered, and the circumference will be $2\pi y$. Hence

$$\Delta A \simeq 2\pi y\, \Delta s.$$

If Δs is made very small, the approximation becomes exact, and

$$dA = 2\pi y\, ds$$
$$= 2\pi y \sqrt{\{1+(dy/dx)^2\}}\, dx.$$

The total area A is then obtained by summing all such area elements between the limits $x = a$ and $x = b$; thus

$$A = 2\pi \int_a^b y\sqrt{\{1+(dy/dx)^2\}}\, dx.$$

The surface area of a hemisphere, for example, is found by rotating the circle $x^2+y^2 = a^2$ about the x-axis between the limits $x = 0$ and $x = a$.

Volume of a solid of revolution. This is found by applying the same principles as were used in the previous section. The volume of the solid figure is divided into infinitesimal segments or slices. The volume of each

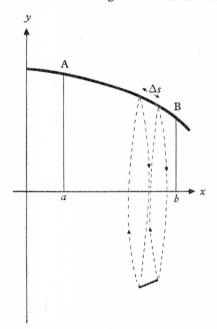

Fig. IIf. Solid of revolution

segment is the depth of the segment Δx (instead of Δs) multiplied by the mean area of the two faces of the segment (see Fig. IIf). If Δx is small, this mean area is approximately πy^2. By reducing Δx to an infinitesimal length and summing between the limits, the volume v is given by

$$v = \int_a^b \pi y^2\, dx.$$

The volume of a hemisphere, for example, is found by rotating the circle $x^2+y^2 = a^2$ about the x-axis, between the limits $x = 0$ and $x = a$.

The reader should determine the circumference of a circle, the area of a circle, the surface area of a sphere and the volume of a sphere by the methods outlined above. All the integrals are greatly simplified by converting from rectangular co-ordinates x and y to polar co-ordinates $a \cos \theta$ and $a \sin \theta$.

Multiple integrals. Volumes of solids which are not solids of revolution can be found by dividing the solid into elements in three dimensions. The total volume is then a triple integral of an appropriate function of three rectangular co-ordinates x, y and z. Limits corresponding to the shape of the solid are imposed after each integration.

A volume integral is written as

$$\int_e^f \int_c^d \int_a^b f(x, y, z) \, dx \, dy \, dz.$$

It is evaluated by integrating firstly with respect to z, treating x and y as constants; the limits e and f are then inserted and the result is integrated with respect to y, treating x as a constant; limits c and d are inserted into this result, which is then finally integrated with respect to x and the limits a and b are applied.

Areas of solids which are not solids of revolution can often be evaluated by double integration. An area integral is written as

$$\int_c^d \int_a^b f(x, y) \, dx \, dy.$$

APPENDIX III

Standard integrals

The integration constants C have been omitted; a is a constant.

$\int x^n \, dx = \dfrac{x^{(n+1)}}{n+1}$, except when $n = -1$.

$\int x^{-1} \, dx = \ln x$.

$\int \dfrac{dx}{x^2 + a^2} = \dfrac{1}{a} \tan^{-1} \dfrac{x}{a}$.

$\int \dfrac{dx}{x^2 - a^2} = \dfrac{1}{2a} \ln \dfrac{x-a}{x+a}$.

$\int \dfrac{dx}{a^2 - x^2} = \dfrac{1}{2a} \ln \dfrac{a+x}{a-x}$.

$\int \exp x \, dx = \exp x$.

$\int a^x \, dx = \dfrac{a^x}{\ln a}$.

$\int \sin x \, dx = -\cos x$.

$\int \cos x \, dx = \sin x$.

$\int \tan x \, dx = -\ln(\cos x)$.

$\int \cot x \, dx = \ln(\sin x)$.

$\int \ln x \, dx = x \ln(x) - x$, or $x(\ln x - 1)$.

$\int a\, f(x)\, dx = a \int f(x)\, dx$, where $f(x)$ is some function of x.

$\int (u+v)\, dx = \int u\, dx + \int v\, dx$, where u and v are functions of x.

$\int u\, dv = uv - \int v\, du$.

$\int \dfrac{dx}{\sqrt{(a^2 - x^2)}} = \sin^{-1} \dfrac{x}{a}$.

$\int \dfrac{dx}{\sqrt{(a^2 + x^2)}} = \ln\{x + \sqrt{(x^2 + a^2)}\}$.

$\int \dfrac{dx}{\sqrt{(x^2 - a^2)}} = \ln\{x + \sqrt{(x^2 - a^2)}\}$.

APPENDIX IV

Stirling's approximation

Stirling's approximation gives a value for ln $(N!)$ when N is large, i.e.
$$\ln (N!) \simeq N \ln (N) - N.$$
This approximation can be shown by plotting ln x against x for a series of whole number values of x from 1 upwards, as is shown in the figure.

x	1	2	3	4	5	6	7	8
ln x	0·00	0·69	1·10	1·39	1·61	1·79	1·95	2·08

The area between this curve, the x-axis and any ordinate can be measured by dividing it up in vertical strips of unit width. Considering each strip as a

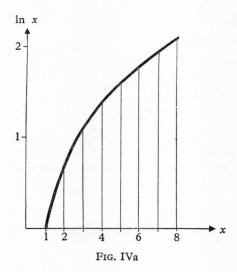

Fig. IVa

trapezium, its area is approximately equal to the mean of the heights of the ordinates forming the sides of the strip multiplied by the strip width (unity).

The total area A between the curve, the x-axis and the ordinates at $x = 1$ and $x = N$ is therefore given by

$A \simeq \frac{1}{2}(\ln 1 + \ln 2) + \frac{1}{2}(\ln 2 + \ln 3) + \frac{1}{2}(\ln 3 + \ln 4) + \ldots$
$\quad + \frac{1}{2}\{\ln (N-2) + \ln (N-1)\} + \frac{1}{2}\{\ln (N-1) + \ln N\}$
$\simeq \frac{1}{2} \ln 1 + \{\ln 2 + \ln 3 + \ln 4 + \ldots + \ln (N-1)\} + \frac{1}{2} \ln N$
$\simeq \{\ln 2 + \ln 3 + \ln 4 + \ldots + \ln (N-1) + \ln N\} - \frac{1}{2} \ln N$, since ln 1 = 0
$\simeq \ln \{2 \times 3 \times 4 \times \ldots \times (N-1) \times N\} - \frac{1}{2} \ln N$
$\simeq \ln (N!) - \frac{1}{2} \ln N.$

STIRLING'S APPROXIMATION

This area can be found exactly by integration; thus

$$A = \int_1^N \ln x \, dx = \left[x \ln(x) - x \right]_1^N = \{N \ln(N) - N\} - (-1).$$

Equating the two values of A gives

$$\ln(N!) - \tfrac{1}{2} \ln N \simeq N \ln(N) - N + 1.$$

Hence

$$\ln(N!) \simeq (N + \tfrac{1}{2}) \ln(N) - N + 1.$$

If N is very large, then $\tfrac{1}{2} \ln N$ and 1 become negligible compared with $N \ln N$ and N, so that

$$\ln(N!) \simeq N \ln(N) - N.$$

APPENDIX V

Mean and variance of the binomial and Poisson distributions

These quantities can be deduced theoretically by using *moment generating functions*.

The rth moment μ_r of a distribution of N values of a variate X about zero is defined as the mean of the sum of all the rth powers of X, i.e.

$$\mu_r = \Sigma(fX^r)/N,$$

where f is the frequency of occurrence of each value of X. [*Note*: moments about zero are usually denoted by μ', but in this treatment the prime has been omitted.] If f is replaced by the fractional frequency p, where $p = f/N$, then

$$\mu_r = \Sigma(pX^r).$$

The first moment μ_1 is the arithmetic mean, i.e.

$$\mu_1 = \Sigma(pX).$$

The second moment μ_2 is related to the variance, since

$$\mu_2 = \Sigma(pX^2).$$

The variance V is the sum of squared deviations from the mean, i.e.

$$V = \Sigma p(X-\mu_1)^2 = \Sigma(pX^2) - \mu_1^2 \text{ [see Chap. 12, § 1]}.$$

Hence
$$V = \mu_2 - \mu_1^2.$$

Any moment of a given theoretical distribution can be found by means of the *moment generating function* $M(\theta)$, this is defined by the equation

$$M(\theta) = \Sigma(pe^{\theta X}),$$

where θ is an undefined constant. Expanding $e^{\theta X}$ gives

$$M(\theta) = \Sigma p\left(1 + \theta X + \frac{\theta^2 X^2}{2!} + \frac{\theta^3 X^3}{3!} + \ldots\right)$$

$$= \Sigma p + \theta\Sigma(pX) + \frac{\theta^2}{2!}\Sigma(pX^2) + \frac{\theta^3}{3!}\Sigma(pX^3) + \ldots$$

$$= 1 + \theta\mu_1 + \frac{\theta^2}{2!}\mu_2 + \frac{\theta^3}{3!}\mu_3 + \ldots \quad (\Sigma p = 1, \text{ by definition}).$$

The moments of the distribution of X therefore appear as coefficients in the above power series in θ. It is also possible to expand $M(\theta)$ as a power series in θ by means of Maclaurin's theorem (see Chap. 6, § 2); thus

$$M(\theta) = 1 + \theta\frac{d}{d\theta}\Big[M(\theta)\Big]_{\theta=0} + \frac{\theta^2}{2!} \cdot \frac{d^2}{d\theta^2}\Big[M(\theta)\Big]_{\theta=0} + \ldots$$

These two power series express the same result for all values of θ, i.e.

MEAN OF THE BINOMIAL AND POISSON DISTRIBUTIONS

they are identities in θ, and the coefficients of like powers of θ are equal in the two expressions. Therefore

$$\mu_1 = \frac{d}{d\theta}\Big[M(\theta)\Big]_{\theta=0}, \quad \text{and} \quad \mu_2 = \frac{d^2}{d\theta^2}\Big[M(\theta)\Big]_{\theta=0}.$$

These two equations can be used to evaluate μ_1 and μ_2 for a theoretical distribution, and so giving values for the mean and variance.

The binomial distribution

In the binomial distribution, the value of p for any whole number value of X is the $(X+1)$th term of the binomial $(q+p)^n$ (see Chap. 11, § 3),

$$p = {}^nC_X p^X q^{n-X}.$$

The moment generating function is therefore given by

$$M(\theta) = \Sigma e^{\theta X} \times {}^nC_X p^X q^{n-X} = \Sigma^n C_X (pe^\theta)^X q^{n-X}.$$

This summation is the expansion of the binomial $(q+pe^\theta)^n$. Therefore

$$M(\theta) = (q+pe^\theta)^n = (1-p+pe^\theta)^n \quad \text{(since } q = 1-p\text{)}.$$

Differentiating this with respect to θ, np and q being constant for a given distribution, gives

$$\frac{d}{d\theta}\Big[M(\theta)\Big] = n(1-p+pe^\theta)^{n-1}e^\theta p.$$

Putting $\theta = 0$, i.e. $e^\theta = 1$, gives

$$\frac{d}{d\theta}\Big[M(\theta)\Big]_{\theta=0} = np = \mu_1.$$

The mean of the binomial distribution is therefore np.

The second moment is found by differentiating $M(\theta)$ again; thus

$$\frac{d^2}{d\theta^2}\Big[M(\theta)\Big] = n(1-p+pe^\theta)^{n-1}pe^\theta + n(n-1)(1-p+pe^\theta)^{n-2}(pe^\theta)^2.$$

Putting $\theta = 0$, i.e. $e^\theta = 1$, gives

$$\frac{d^2}{d\theta^2}\Big[M(\theta)\Big]_{\theta=0} = np+n(n-1)p^2 = np+n^2p^2-np^2.$$

This is the value of the second moment μ_2. The variance V is given by

$$V = \mu_2 - \mu_1^2 = np + n^2p^2 - np^2 - n^2p^2$$
$$= np - np^2 = np(1-p) = npq.$$

The variance of the binomial distribution is therefore npq, the maximum value of which is $np = \mu$, which occurs when $q \to 1$; the maximum variance of a binomial distribution of positive integers, such as frequencies, is therefore the mean.

Poisson distribution

This is a binomial distribution in which p is very small, i.e. $p \ll 1$ and hence $q \simeq 1$, since $q = 1-p$. Therefore

$$V = npq \simeq np.$$

In a Poisson distribution, the mean and variance are both equal to np.

APPENDIX VI

The normal distribution

The normal distribution equation is
$$p_X/p_\mu = \exp\{-\tfrac{1}{2}(X-\mu)^2/V\},$$
where p_X and p_μ are the fractional frequencies of occurrence of any value of the variate X and of the mean μ respectively. The normal distribution is a limiting case of the binomial distribution in which n, the index of the generating function $(q+p\theta)^n$, is very large, and both np and nq are large numbers.

The equation of the normal distribution can be derived by considering the ratio p_X/p_μ for any binomial; thus
$$\frac{p_X}{p_\mu} = \frac{{}^nC_X p^X q^{n-X}}{{}^nC_\mu p^\mu q^{n-\mu}} = \frac{\mu!(n-\mu)!}{X!(n-X)!}\left(\frac{p}{q}\right)^{X-\mu}.$$

In any binomial distribution, $\mu = np$, and hence $n-\mu = n(1-p) = nq$. Substituting $np+x$ for X, where x is the deviation of X from the mean μ, i.e. np, gives
$$\frac{p_X}{p_\mu} = \frac{(np)!(nq)!}{(pn+x)!(nq-x)!}\left(\frac{p}{q}\right)^x \quad \text{(since } n-X = n-np-x = nq-x,$$
$$\text{and } X-\mu = x).$$

If x is moderately small, all the factorials are those of large numbers, and they can be simplified by Stirling's approximation (see Appendix IV); thus
$$\ln(N!) \simeq N\ln(N) - N.$$
On taking natural logarithms, the non-logarithmic terms from the approximation cancel out with the result that
$$\ln(p_X/p_\mu) = np\ln np + nq\ln nq - (np+x)\ln(np+x) - $$
$$- (nq-x)\ln(nq-x) + x\ln p - x\ln q.$$
Rearranging this expression gives
$$\ln\left(\frac{p_X}{p_\mu}\right) = -np\ln\left(\frac{np+x}{np}\right) - nq\ln\left(\frac{nq-x}{nq}\right) - x\ln\left(\frac{np+x}{p}\right) + x\ln\left(\frac{nq-x}{q}\right).$$
Now
$$\ln\left(\frac{np+x}{p}\right) = \ln\left\{n\left(1+\frac{x}{np}\right)\right\} = \ln n + \ln\left(1+\frac{x}{np}\right).$$
Similarly,
$$\ln\left(\frac{nq-x}{q}\right) = \ln n + \ln\left(1-\frac{x}{nq}\right).$$
Substituting these in the logarithmic equation for p_X/p_μ and rearranging all the logarithms in the form $\ln(1+y)$ gives
$$\ln\left(\frac{p_X}{p_\mu}\right) = -np\ln\left(1+\frac{x}{np}\right) - nq\ln\left(1-\frac{x}{nq}\right) - x\ln\left(1+\frac{x}{np}\right) + x\ln\left(1-\frac{x}{nq}\right).$$

THE NORMAL DISTRIBUTION

If x is moderately small, these logarithms can all be expressed approximately by the first two terms of the logarithmic series; thus
$$\ln(1+y) \simeq y - \tfrac{1}{2}y^2.$$
Making this substitution and multiplying out the result gives
$$\ln\left(\frac{p_X}{p_\mu}\right) = -\frac{x^2}{2np} - \frac{x^2}{2nq} + \frac{x^3}{2(np)^2} - \frac{x^3}{2(nq)^2}.$$
Since x is small and np and nq are large, the last two terms are very small compared with the first two; hence
$$\ln\left(\frac{p_X}{p_\mu}\right) = -\frac{x^2(p+q)}{2npq}$$
$$= -\frac{x^2}{2npq}, \quad \text{since } (p+q) = 1.$$
The variance V of any binomial distribution is npq. Therefore
$$\ln(p_X/p_\mu) = -\tfrac{1}{2}x^2/V = -\tfrac{1}{2}(X-\mu)^2/V,$$
and hence
$$p_X/p_\mu = \exp\{-\tfrac{1}{2}(X-\mu)^2/V\}.$$

Evaluation of $\int_{-\infty}^{\infty} \exp(-\tfrac{1}{2}u^2)\,du$

This quantity is the denominator of the important probability integral and it has a finite value.

Since $\exp(-\tfrac{1}{2}u^2)$ is an even function of u, then
$$\int_{-\infty}^{\infty} \exp(-\tfrac{1}{2}u^2)\,du = 2\int_0^{\infty} \exp(-\tfrac{1}{2}u^2)\,du.$$
If it is assumed that $\int_0^\infty \exp(-\tfrac{1}{2}u^2)\,du$ has a finite value A, then A depends only on the limits of the integral and will not be changed by altering the letter used to represent the variable, for example
$$A = \int_0^\infty \exp(-\tfrac{1}{2}u^2)\,du = \int_0^\infty \exp(-\tfrac{1}{2}z^2)\,dz = \int_0^\infty \exp\{-\tfrac{1}{2}(zu)^2\}\,d(zu).$$
For any particular value of z, $d(zu) = z\,du$. Therefore for any particular value of z it follows that
$$A = \int_0^\infty \exp\{-\tfrac{1}{2}(zu)^2\}z\,du.$$
If both sides of this equation are multiplied by $\exp(-\tfrac{1}{2}z^2)$ and this factor taken into the integral (a legitimate procedure for a particular value of z), the result is
$$A\exp(-\tfrac{1}{2}z^2) = \int_0^\infty \exp\{-\tfrac{1}{2}(zu)^2\}\exp(-\tfrac{1}{2}z^2)z\,du$$
$$= \int_0^\infty \exp\{-\tfrac{1}{2}z^2(u^2+1)\}z\,du.$$
If both sides of the equation are now multiplied by dz and integrated over all positive values of z, i.e. 0 to ∞, then
$$A\int_0^\infty \exp(-\tfrac{1}{2}z^2)\,dz = \int_0^\infty \int_0^\infty \exp\{-\tfrac{1}{2}z^2(u^2+1)\}z\,du\,dz.$$

APPENDIX VI

The integral on the left-hand side of the equation is equal to A; that on the right-hand side is solved by integrating with respect to z (treating u as a constant), substituting the limits for z, and integrating the result with respect to u. The first integration is made by putting $y = -\tfrac{1}{2}z^2(u^2+1)$, i.e.

$$dy = -z(u^2+1)\, dz \quad (u \text{ being treated as a constant}).$$

When $z = 0$, then $y = 0$; when $z = \infty$, then $y = -\infty$. Therefore

$$\int_0^\infty \exp\{-\tfrac{1}{2}z^2(u^2+1)\}z\, dz = \int_0^{-\infty} \frac{\exp(y)\, dy}{-(u^2+1)}.$$

Since u is treated as a constant in this integration, it follows that

$$\int_0^{-\infty} \frac{\exp(y)\, dy}{-(u^2+1)} = -\left[\frac{1}{u^2+1}\exp(y)\right]_0^{-\infty}$$

$$= -\frac{1}{u^2+1}\{\exp(-\infty) - \exp(0)\}.$$

Now, $\exp(-\infty) = 1/e^\infty = 0$, and $\exp(0) = e^0 = 1$. Therefore the integral is equal to $1/(u^2+1)$.

In the second stage of the double integration, this result has to be integrated with respect to u; this can be done directly as it is in a standard form. Thus

$$\int_0^\infty \frac{du}{u^2+1} = \left[\tan^{-1}(u)\right]_0^\infty = \tan^{-1}(\infty) - \tan^{-1}(0).$$

The principal value of the angles whose tangent is ∞ is $\tfrac{1}{2}\pi$, and that of the angles whose tangent is zero is 0. Therefore the final result of the double integration is

$$A^2 = \int_0^\infty \int_0^\infty \exp\{-\tfrac{1}{2}z^2(u^2+1)\}z\, du\, dz = \tfrac{1}{2}\pi.$$

Hence
$$A = \sqrt{(\tfrac{1}{2}\pi)}.$$
Therefore
$$\int_{-\infty}^\infty \exp(-\tfrac{1}{2}u^2)\, du = 2A = \sqrt{(2\pi)}.$$

APPENDIX VII

Variance of a function of variates

General case

Suppose Z is a function of two variates, X and Y, i.e. $Z = f(X, Y)$. Let m_x, m_y and m_z be the mean values of X, Y and Z respectively; then $m_z = f(m_x, m_y)$.

Suppose two values of X and Y are taken which deviate from their means by the small amounts ΔX and ΔY respectively; then the deviation ΔZ of the corresponding value of Z from its mean is given by

$$\Delta Z \simeq \frac{\partial Z}{\partial X}\Delta X + \frac{\partial Z}{\partial Y}\Delta Y.$$

If a large number N of pairs of values of X and Y are considered, the variance V_z of the resulting values of Z about their mean will be $\Sigma(\Delta Z)^2/N$. But

$$(\Delta Z)^2 = \left(\frac{\partial Z}{\partial X}\right)^2 (\Delta X)^2 + \left(\frac{\partial Z}{\partial Y}\right)^2 (\Delta Y)^2 + 2\left(\frac{\partial Z}{\partial X}\right)\left(\frac{\partial Z}{\partial Y}\right)(\Delta X)(\Delta Y).$$

Therefore

$$V_z = \left(\frac{\partial Z}{\partial X}\right)^2 \frac{\Sigma(\Delta X)^2}{N} + \left(\frac{\partial Z}{\partial Y}\right)^2 \frac{\Sigma(\Delta Y)^2}{N} + 2\left(\frac{\partial Z}{\partial X}\right)\left(\frac{\partial Z}{\partial Y}\right)\frac{\Sigma\{(\Delta X)(\Delta Y)\}}{N}.$$

If X and Y are independent, i.e. each chosen at random from their distributions, then $\Sigma\{(\Delta X)(\Delta Y)\}$ is zero (see page 203). Now $\Sigma(\Delta X)^2/N$ and $\Sigma(\Delta Y)^2/N$ are the variances of X and Y respectively, i.e. V_x and V_y respectively. Therefore

$$V_z = \left(\frac{\partial Z}{\partial X}\right)^2 V_x + \left(\frac{\partial Z}{\partial Y}\right)^2 V_y.$$

The equation can be extended in a similar form if Z is a function of more than two variates.

(i) **Sum of variates.** If $Z = X + Y$, then

$$V_z = V_x + V_y, \quad \text{since } \partial Z/\partial X = \partial Z/\partial Y = 1.$$

The relation between the standard deviations, σ_z, σ_x and σ_y is given by

$$\sigma_z = \sqrt{(\sigma_x^2 + \sigma_y^2)}.$$

(ii) **Difference of variates.** If $Z = X - Y$, then again

$$V_z = V_x + V_y, \quad \text{since } (\partial Z/\partial Y)^2 = (-1)^2 = 1,$$

and

$$\sigma_z = \sqrt{(\sigma_x^2 + \sigma_y^2)}.$$

(iii) **Product of variates.** If $Z = XY$, then $(\partial Z/\partial X) = Y$ and $(\partial Z/\partial Y) = X$. Hence

$$V_z = Y^2 V_x + X^2 V_y.$$

Dividing by $(XY)^2$ gives

$$\frac{V_z}{(XY)^2} = \frac{V_x}{X^2}+\frac{V_y}{Y^2},$$

$$\frac{V_z}{Z^2} = \frac{V_x}{X^2}+\frac{V_y}{Y^2}.$$

This result can be expressed more conveniently in terms of the coefficients of variation of Z, X and Y, i.e. E_z, E_x and E_y respectively.
Now,

$$E_z = \frac{100\sigma_z}{m_z} = \frac{100\sqrt{V_z}}{Z} \quad (Z \text{ being the mean value}).$$

Therefore

$$E_z{}^2 = \frac{100^2 V_z}{Z^2}.$$

Similarly

$$E_x{}^2 = \frac{100^2 V_x}{X^2} \quad \text{and} \quad E_y{}^2 = \frac{100^2 V_y}{Y^2}.$$

Substituting these values in the equation for V_z/Z^2 gives

$$E_z{}^2 = E_x{}^2 + E_y{}^2,$$

i.e.

$$E_z = \sqrt{(E_x{}^2+E_y{}^2)}.$$

(iv) *Quotient of variates.* If $Z = Y/X$, then $(\partial Z/\partial X) = -Y/X^2$, and $(\partial Z/\partial Y) = 1/X$. Hence

$$V_z = \frac{Y^2}{X^4}V_x + \frac{1}{X^2}V_y.$$

Dividing throughout by $(Y/X)^2$ gives

$$\frac{V_z}{(Y/X)^2} = \frac{V_x}{X^2}+\frac{V_y}{Y^2},$$

or

$$\frac{V_z}{Z^2} = \frac{V_x}{X^2}+\frac{V_y}{Y^2}.$$

In terms of coefficients of variation this becomes

$$E_z{}^2 = E_x{}^2 + E_y{}^2.$$

Hence

$$E_z = \sqrt{(E_x{}^2+E_y{}^2)}.$$

(v) *Linear function of variates.* If $Z = aX+bY$, then $\partial Z/\partial X = a$, and $\partial Z/\partial Y = b$. Hence

$$V_z = a^2 V_x + b^2 V_y.$$

Therefore

$$\sigma_z = \sqrt{(a^2\sigma_x{}^2+b^2\sigma_y{}^2)}.$$

If Z is a function of 3 or more variates, i.e.

$$Z = aU+bW+cX+dY+\ldots,$$

then

$$V_z = a^2 V_u + b^2 V_w + c^2 V_x + d^2 V_y + \ldots.$$

APPENDIX VIII

Some theorems in statistics

1. The variance of a sample mean is V/N.

Consider a very large distribution of values of a variate X. Then if m is the mean of a sample of N of these values,

$$m = \frac{X_1 + X_2 + X_3 + \ldots}{N}.$$

As $X_1, X_2, X_3 \ldots$ are drawn at random from the whole distribution, they can therefore be regarded as independent of one another; m is then a linear function of these independent quantities and so the variance (V_m) of m can be deduced from Appendix VII (v); thus

$$V_m = V(X_1)/N^2 + V(X_2)/N^2 + \ldots.$$

$V(X_1), V(X_2), \ldots$ are each equal to the distribution variance V, and as there are N terms in the expression for V_m, it follows that

$$V_m = NV/N^2 = V/N.$$

As V_m is the variance of the sample mean about the universe mean μ, the square root of V_m is the standard deviation of m about μ, i.e. the standard error (s_m) of m; thus

$$s_m = \sqrt{V_m} = \sqrt{(V/N)} = \sigma/\sqrt{N},$$

where σ is the distribution standard deviation, and $\sigma^2 = V$.

2. χ^2 for data in two classes

Data in two classes means here a set of observations of the number of occasions on which an event either did or did not occur. If n is the total number of observations, X is the number of times the event occurred and Y is the number of times it did not occur, then $X + Y = n$. If p and q are the probabilities of occurrence and failure of occurrence respectively in a single trial, then $\mathrm{p} + \mathrm{q} = 1$. The values of X and Y will be binomially distributed about their expected values $n\mathrm{p}$ and $n\mathrm{q}$, each with a variance of $n\mathrm{pq}$.

If χ^2 is defined as the sum of the squared deviations from expectation divided by the expected value, then

$$\chi^2 = \frac{(X - n\mathrm{p})^2}{n\mathrm{p}} + \frac{(Y - n\mathrm{q})^2}{n\mathrm{q}}.$$

From the relations $X + Y = n$, and $\mathrm{p} + \mathrm{q} = 1$, it follows that

$$X - n\mathrm{p} = (n - Y) - n(1 - \mathrm{q}) = -(Y - n\mathrm{q}).$$

Therefore

$$\chi^2 = \frac{(X - n\mathrm{p})^2}{n}\left(\frac{1}{\mathrm{p}} + \frac{1}{\mathrm{q}}\right)$$

$$= \frac{(X - n\mathrm{p})^2}{n\mathrm{pq}} = \frac{(X - \mu)^2}{V} = \left(\frac{X - \mu}{\sigma}\right)^2 = u^2,$$

where μ is the mean of the binomial distribution of X, and V and σ are the variance and standard deviation of this distribution, respectively. From the above equation, it is seen that χ^2 has the form of a squared normal deviate u^2; this is the basis for comparing values of χ^2 calculated from the formula

$$\chi^2 = \frac{(O-E)^2}{E},$$

with theoretical values of χ^2 calculated from the properties of the normal distribution as the sum of a number of squared normal deviates drawn at random from the distribution, i.e.

$$\chi^2 = u_1^2 + u_2^2 + u_3^2 + \ldots$$

In the case of data in two classes, there is only one degree of freedom in χ^2, and $\chi^2 = u_1^2$. Similar results can be deduced for data in more than two classes, the number of degrees of freedom in χ^2 being the number of squared normal deviates which are summed in obtaining the theoretical value of χ^2.

3. A Poisson distribution becomes approximately normal when μ is large

From the properties of the Poisson distribution (see Chap. 11, § 6)

$$\frac{p_X}{p_\mu} = \frac{\mu^X \exp(-\mu)}{X!} \cdot \frac{\mu!}{\mu^\mu \exp(-\mu)} = \mu^{X-\mu} \frac{\mu!}{X!}.$$

If μ is large and the deviation x of the value of X considered (X is a value of the variate in the distribution) is relatively small, then the factorials can both be simplified by applying Stirling's approximation; thus writing $X = \mu + x$, and taking natural logarithms gives

$$\ln(p_X/p_\mu) = x \ln \mu + \mu \ln \mu - \mu - (\mu+x) \ln(\mu+x) + (\mu+x)$$
$$= -(\mu+x) \ln(1+x/\mu) + x.$$

If x/μ is small, the logarithm $\ln(1+x/\mu)$ can be written approximately as

$$x/\mu - x^2/(2\mu^2).$$

Making this substitution in the equation for $\ln(p_X/p_\mu)$ and multiplying out gives

$$\ln\left(\frac{p_X}{p_\mu}\right) = -\frac{x^2}{2\mu} + \frac{x^3}{2\mu^2}.$$

when x is small and μ is large, the second term can be neglected compared with the first giving

$$\ln\left(\frac{p_X}{p_\mu}\right) = -\frac{x^2}{2\mu} = -\frac{(X-\mu)^2}{2\mu}.$$

This is the equation of a normal distribution whose variance is equal to μ.

APPENDIX IX

Regression variance

Variance of regression coefficient

The variance of the regression coefficient b of Y upon X, where Y is subject to random error and X is not, can be calculated by expressing b in the form

$$b = \frac{\Sigma\{(X-\bar{X})(Y-\bar{Y})\}}{\Sigma(X-\bar{X})^2}$$

$$= \frac{\Sigma\{(X-\bar{X})Y\}}{\Sigma(X-\bar{X})^2}. \quad \text{[see equation (13.6)].}$$

Expanding the numerator gives

$$b = \frac{(X_1-\bar{X})}{\Sigma(X-\bar{X})^2}Y_1 + \frac{(X_2-\bar{X})}{\Sigma(X-\bar{X})^2}Y_2 + \ldots$$

The variance of b will arise entirely from the variances of $Y_1, Y_2, \ldots$ about their true values, which are the regression-line values corresponding to $X_1, X_2, \ldots$. Each value of Y will have the variance V_Y, which has been evaluated in equation (13.8). The variance of b, i.e. V_b, is then that of a linear function of the variates [see Appendix VII (v)]; thus

$$V_b = \frac{(X_1-\bar{X})^2}{\{\Sigma(X-\bar{X})^2\}^2}V_Y + \frac{(X_2-\bar{X})^2}{\{\Sigma(X-\bar{X})^2\}^2}V_Y + \ldots$$

$$= \frac{V_Y}{\{\Sigma(X-\bar{X})^2\}^2}\{(X_1-\bar{X})^2 + (X_2-\bar{X})^2 + \ldots\}.$$

But since $(X_1-\bar{X})^2 + (X_2-\bar{X})^2 + \ldots = \Sigma(X-\bar{X})^2$, then

$$V_b = \frac{V_Y}{\Sigma(X-\bar{X})^2}.$$

If s_b and s_Y are the corresponding standard deviations, then

$$s_b = s_Y/\sqrt{\{\Sigma(X-\bar{X})^2\}}.$$

Limits of error for b are $\pm t s_b$, where t is the value for the chosen probability level with the number of degrees of freedom ϕ equal to $N-2$ (N is the number of pairs of values of Y and X, and two means are computed from these in the calculation of b).

Variance of a value of X estimated from a single value of Y and the regression line

The equation of the regression line is

$$Y - \bar{Y} = b(X - \bar{X}),$$

where X is the predicted value for a given value of Y. Hence

$$X = \bar{X} + Y/b - \bar{Y}/b.$$

APPENDIX IX

Of the quantities on the right-hand side of the equation, $\bar{X}$ is not subject to random variation, while Y, $\bar{Y}$ and b are all subject to random errors caused by the variance of the values of Y. Hence

$$\frac{\partial X}{\partial Y} = \frac{1}{b}, \quad \frac{\partial X}{\partial \bar{Y}} = -\frac{1}{b}, \quad \text{and} \quad \frac{\partial X}{\partial b} = -\frac{Y-\bar{Y}}{b^2}.$$

Using the general equation of Appendix VII extended to a function of three variates gives

$$V_X = \frac{1}{b^2}V_Y + \frac{1}{b^2}V_{\bar{Y}} + \frac{(Y-\bar{Y})^2}{b^4}V_b.$$

Now $V_{\bar{Y}} = V_Y/N$, where N is the number of values of Y in the regression calculation, and

$$V_b = V_Y/\Sigma(X-\bar{X})^2 \quad \text{(see previous section)}.$$

Therefore

$$V_X = \frac{V_Y}{b^2}\left[1 + \frac{1}{N} + \frac{(Y-\bar{Y})^2}{b^2\{\Sigma(X-\bar{X})^2\}}\right].$$

Limits of error for X are $\pm t\sqrt{V_X}$, where t has $N-2$ degrees of freedom.

If m repeats are made to determine the value of Y, V_Y is reduced to V_Y/m. Therefore

$$V_X = \frac{V_Y}{b^2}\left[\frac{1}{m} + \frac{1}{N} + \frac{(Y-\bar{Y})^2}{b^2\{\Sigma(X-\bar{X})^2\}}\right].$$

APPENDIX X

Solution of the diffusion equation

In order to solve the partial differential equation

$$\left(\frac{\partial c}{\partial t}\right)_x = D\left(\frac{\partial^2 c}{\partial x^2}\right)_t, \tag{X.1}$$

which expresses Fick's laws of diffusion, a quantity y is introduced. This quantity is a function of both x and t, and is defined by the relation

$$y = \frac{x}{2\sqrt{(Dt)}} = \frac{xt^{-1/2}}{2\sqrt{D}}.$$

Both x and t can be completely eliminated from (X.1) by replacing them with y. This is done as follows. If x remains constant, then

$$\frac{\partial c}{\partial t} = \frac{\partial c}{\partial y} \cdot \frac{\partial y}{\partial t}.$$

Now, $\partial y/\partial t$ is found by partial differentiation of the expression which defines y; thus

$$\frac{\partial y}{\partial t} = \frac{x}{2\sqrt{D}}\left(-\frac{1}{2}t^{-3/2}\right) = -\frac{x}{4\sqrt{(Dt^3)}}.$$

Therefore

$$\left(\frac{\partial c}{\partial t}\right)_x = -\frac{x}{4\sqrt{(Dt^3)}}\left(\frac{\partial c}{\partial y}\right)_x.$$

Also

$$\frac{\partial^2 c}{\partial x^2} = \frac{\partial}{\partial x}\left(\frac{\partial c}{\partial x}\right) = \frac{\partial y}{\partial x} \cdot \frac{\partial}{\partial y}\left(\frac{\partial c}{\partial y} \cdot \frac{\partial y}{\partial x}\right).$$

If t remains constant, then from the definition of y, it follows that

$$\frac{\partial y}{\partial x} = \frac{1}{2\sqrt{(Dt)}}.$$

Therefore

$$\frac{\partial^2 c}{\partial x^2} = \frac{1}{2\sqrt{(Dt)}} \cdot \frac{\partial}{\partial y}\left(\frac{\partial c}{\partial y} \cdot \frac{1}{2\sqrt{(Dt)}}\right).$$

As t is constant, $1/2\sqrt{(Dt)}$ is constant and can therefore be taken outside the brackets giving

$$\left(\frac{\partial^2 c}{\partial x^2}\right)_t = \frac{1}{4Dt} \cdot \frac{\partial}{\partial y}\left(\frac{\partial c}{\partial y}\right) = \frac{1}{4Dt}\left(\frac{\partial^2 c}{\partial y^2}\right)_t.$$

If these values for $(\partial c/\partial t)$ and $(\partial^2 c/\partial x^2)$ are substituted into equation (X.1), then

$$-\frac{x}{4\sqrt{(Dt^3)}}\left(\frac{\partial c}{\partial y}\right)_x = \frac{1}{4t}\left(\frac{\partial^2 c}{\partial y^2}\right)_t.$$

Hence

$$-\frac{x}{\sqrt{(Dt)}}\left(\frac{\partial c}{\partial y}\right)_x = \left(\frac{\partial^2 c}{\partial y^2}\right)_t.$$

But

$$\frac{x}{\sqrt{(Dt)}} = 2y,$$

and therefore

$$-2y\left(\frac{\partial c}{\partial y}\right)_x = \left(\frac{\partial^2 c}{\partial y^2}\right)_t.$$

The original partial differential equation has now been reduced to one containing two variables only. A solution to the general equation

$$-2y\frac{dc}{dy} = \frac{d^2c}{dy^2} \qquad (X.2)$$

without restrictions about x and t will hold in all cases and will thus be a solution to the partial differential equation. The general equation (X.2) can be solved by the power series method (see Chap. 9, § 2). Let the solution be

$$c = A + By + Cy^2 + Dy^3 + Ey^4 + Fy^5 + Gy^6 + Hy^7 + \ldots.$$

If this expression is differentiated and the values of dc/dy and d^2c/dy^2 are substituted in equation (X.2), then

$$-2By - 4Cy^2 - 6Dy^3 - 8Ey^4 - 10Fy^5 - 12Gy^6 - 14Hy^7 + \ldots$$
$$= 2C + 6Dy + 12Ey^2 + 20Fy^3 + 30Gy^4 + 42Hy^5 + \ldots.$$

This is an identity in y, and therefore the coefficients of each power of y can be equated; thus

constant term $0 = 2C$;   therefore $C = 0$
coefficients of y $-2B = 6D$;   therefore $D = -\frac{1}{3}B$
coefficients of y^2 $-4C = 12E$;   therefore $E = -\frac{1}{3}C = 0$
coefficients of y^3 $-6D = 20F$;   therefore $F = -3D/10 = \frac{1}{10}B$
coefficients of y^4 $-8E = 30G$;   therefore $G = -4E/15 = 0$
coefficients of y^5 $-10F = 42H$;   therefore $H = -5F/21 = -B/42$

Substituting these values of $A, B, C, \ldots$ in the expression for c gives

$$c = A + By - \frac{By^3}{3} + \frac{By^5}{10} - \frac{By^7}{42} + \ldots.$$

That is

$$c = A + B\left(y - \frac{y^3}{3} + \frac{y^5}{10} - \frac{y^7}{42} + \ldots\right).$$

The series in the brackets is the expansion of $\int_0^y \exp(-\beta^2)\,d\beta$, since

$$\int_0^y \exp(-\beta^2)\,d\beta = \int_0^y \left(1 - \beta^2 + \frac{\beta^4}{2} - \frac{\beta^6}{6} + \ldots\right) d\beta$$

$$= \left[\beta - \frac{\beta^3}{3} + \frac{\beta^5}{10} - \frac{\beta^7}{42} + \ldots\right]_0^y$$

$$= y - \frac{y^3}{3} + \frac{y^5}{10} - \frac{y^7}{42} + \ldots \text{ (since the value of}$$
$$\text{this series for } y = 0 \text{ is zero).}$$

SOLUTION OF THE DIFFUSION EQUATION

Therefore, a solution to (X.2) is

$$c = A + B \int_0^y \exp(-\beta^2)\, d\beta.$$

The values of A and B are found from the boundary conditions for the diffusion; i.e. when $t = 0$ and x is positive, then $c = 0$ (i.e. at the start of diffusion there is only pure solvent above the boundary) and $y = +\infty$; when $t = 0$ and x negative, then $c = c_0$ and $y = -\infty$. Substituting these values in the equation for c gives

$$0 = A + B \int_0^\infty \exp(-\beta^2)\, d\beta,$$

and

$$c_0 = A + B \int_0^{-\infty} \exp(-\beta^2)\, d\beta.$$

The integrals with infinite limits have finite values which can be deduced from Appendix VI, where it is shown that

$$\int_0^\infty \exp(-\tfrac{1}{2}u^2)\, du = \sqrt{(\tfrac{1}{2}\pi)}.$$

Writing $\beta^2 = \tfrac{1}{2}u^2$, i.e. $\beta = u/\sqrt{2}$, it follows that $d\beta = du/\sqrt{2}$ and $du = \sqrt{2}\, d\beta$. Therefore

$$\int_0^\infty \exp(-\tfrac{1}{2}u^2)\, du = \int_0^\infty \exp(-\beta^2) \sqrt{2}\, d\beta = \sqrt{2} \int_0^\infty \exp(-\beta^2)\, d\beta = \sqrt{(\tfrac{1}{2}\pi)}.$$

Hence

$$\int_0^\infty \exp(-\beta^2)\, d\beta = \tfrac{1}{2}\sqrt{\pi}.$$

Since $\exp(-\beta^2)$ is an even function of β, it follows that

$$\int_0^{-\infty} \exp(-\beta^2)\, d\beta = -\tfrac{1}{2}\sqrt{\pi}.$$

Putting these values for the integrals into the equations for A and B gives

$$0 = A + \tfrac{1}{2}B\sqrt{\pi}, \quad \text{and} \quad c_0 = A - \tfrac{1}{2}B\sqrt{\pi}.$$

Adding the equations gives

$$c_0 = 2A, \quad \text{i.e.} \quad A = \tfrac{1}{2}c_0.$$

Subtracting them gives

$$-c_0 = B\sqrt{\pi}, \quad \text{i.e.} \quad B = -c_0/\sqrt{\pi}.$$

The final solution to the differential equation for diffusion with these boundary conditions is therefore

$$c = \tfrac{1}{2}c_0 - \frac{c_0}{\sqrt{\pi}} \int_0^y \exp(-\beta^2)\, d\beta,$$

which is usually written in the form

$$c = \tfrac{1}{2}c_0 \left\{ 1 - \frac{2}{\sqrt{\pi}} \int_0^y \exp(-\beta^2)\, d\beta \right\}.$$

APPENDIX XI

Binomial theorem for any index

The binomial theorem for a positive whole number index has been proved in Chap. 4, § 6, and from this, the rule for differentiation,

$$\frac{d(x^n)}{dx} = nx^{n-1},$$

can be established for positive whole number values of n. To prove the binomial theorem for any index, the above rule for differentiation is first established for any value of n. In order to do this, consider the extreme case in which n is both negative and fractional. Let $n = -j/k$, where j and k are positive integers. If $y = x^n$, then

$$y = x^{-j/k} = \frac{1}{x^{j/k}}, \text{ or } y^k = \frac{1}{x^j}.$$

The two sides of this equation can be differentiated by the rules, since j and k are positive integers. Thus, the right-hand side is treated as a quotient, so that

$$ky^{k-1} \, dy = \frac{x^j \times 0 - 1 \times jx^{j-1} \, dx}{x^{2j}} = -\frac{j \, dx}{x^{j+1}}.$$

Hence

$$\frac{dy}{dx} = -\frac{j}{k} \cdot \frac{1}{x^{j+1} y^{k-1}}.$$

Now $x^{-1/k} = y$, i.e.

$$y^{k-1} = x^{(-1/k)(k-1)}$$
$$= x^{-1+1/k}$$

Substituting this value of y^{k-1} in the equation for dy/dx gives

$$\frac{dy}{dx} = -\frac{j}{k} \cdot \frac{1}{x^{j+1} x^{-1+1/k}}$$

$$= -\frac{j}{k} \cdot \frac{1}{x^{j+1/k}}$$

$$= -\frac{j}{k} x^{-1-1/k}$$

$$= nx^{n-1}.$$

Therefore the rule for differentiating x^n holds for negative non-integral values of n; it can also be shown to hold for positive non-integral values of n.

With this proved, Maclaurin's theorem can be used to give a series for the expansion of $(1+x)^n$ (see Chap. 6, § 2); thus

$$(1+x)^n = 1 + nx + \frac{n(n-1)x^2}{2!} + \frac{n(n-1)(n-2)x^3}{3!} + \dots.$$

If n is a positive integer, one of the factors $(n-1)$, $(n-2)$, $(n-3)$, etc., eventually becomes zero, so that there is a limit to the number of terms of the series. If n is negative or non-integral, none of these terms becomes zero and the series is infinite. In order that its sum shall have a finite value, the series must converge, and to make it convergent it is necessary to limit the values of x to those between $+1$ and -1. The above series is therefore only true for any value of n if $|x| < 1$.

APPENDIX XII

Statistical tables

In all these tables, P is the probability that random choice from a normal distribution will give a value of the quantity less than or equal to the values given in the tables; P', i.e. $1-P$, is the probability that a random choice from a normal distribution will give a value greater in magnitude than the values given in the tables. In tests of significance, P' is the probability that the null hypothesis is correct.

These tables are abridged from tables in Fisher and Yates, 'Statistical Tables for Biological, Agricultural and Medical Research', published by Oliver and Boyd Ltd., Edinburgh, by permission of the authors and publishers.

TABLE 1

Values of the normal deviate u_1

$$P = \sqrt{\frac{2}{\pi}} \int_0^{u_1} \exp\left(-\tfrac{1}{2}u^2\right) du$$

P	u_1	P	u_1
0·05	0·063	0·70	1·037
0·10	0·126	0·80	1·281
0·20	0·253	0·90	1·645
0·30	0·386	0·95	1·960
0·40	0·524	0·99	2·576
0·50	0·674	0·995	2·81
0·60	0·842	0·999	3·29

TABLE 2

Values of t

Degrees of freedom ϕ	Value of t		Degrees of freedom ϕ	Value of t	
	$P = 0·95$ $P' = 0·05$	$P = 0·99$ $P' = 0·01$		$P = 0·95$ $P' = 0·05$	$P = 0·99$ $P' = 0·01$
1	12·71	63·7	11	2·20	3·11
2	4·30	9·92	12	2·18	3·05
3	3·18	5·84	15	2·13	2·95
4	2·78	4·60	20	2·09	2·85
5	2·57	4·03	25	2·06	2·79
6	2·45	3·71	30	2·04	2·75
7	2·37	3·50	40	2·02	2·70
8	2·31	3·36	60	2·00	2·66
9	2·26	3·25	120	1·98	2·62
10	2·23	3·17	∞	1·96	2·58

Table 3

Values of χ^2

Degrees of freedom ϕ	Value of χ^2		Degrees of freedom ϕ	Value of χ^2	
	$P = 0.95$ $P' = 0.05$	$P = 0.99$ $P' = 0.01$		$P = 0.95$ $P' = 0.05$	$P = 0.99$ $P' = 0.01$
1	3.84	6.63	9	16.9	21.7
2	5.99	9.21	10	18.3	23.2
3	7.81	11.3	11	19.7	24.7
4	9.49	13.3	12	21.0	26.2
5	11.1	15.1	15	25.0	30.6
6	12.6	16.8	20	31.4	37.6
7	14.1	18.5	25	37.7	44.3
8	15.5	20.1	30	43.8	50.9

Table 4

Values of the variance ratio at $P = 0.95$ ($P' = 0.05$)

Number of degrees of freedom of the lesser variance ϕ_2	Number of degrees of freedom of the greater variance ϕ_1									
	1	2	3	4	6	8	10	15	30	∞
1	161	199	216	225	239	237	242	246	250	254
2	18.5	19.0	19.2	19.2	19.3	19.4	19.4	19.4	19.5	19.5
3	10.1	9.55	9.28	9.12	8.94	8.85	8.79	8.70	8.62	8.53
4	7.71	6.94	6.59	6.39	6.16	6.04	5.96	5.86	5.75	5.63
6	5.99	5.14	4.76	4.53	4.28	4.15	4.06	3.94	3.81	3.67
8	5.32	4.46	4.07	3.84	3.58	3.44	3.35	3.22	3.08	2.93
10	4.96	4.10	3.71	3.48	3.22	3.07	2.98	2.84	2.70	2.54
15	4.54	3.68	3.29	3.06	2.79	2.64	2.54	2.40	2.25	2.07
30	4.17	3.32	2.92	2.69	2.42	2.27	2.16	2.01	1.84	1.62
∞	3.84	3.00	2.60	2.37	2.10	1.94	1.83	1.67	1.46	1.00

APPENDIX XII

TABLE 5

Probits corresponding to percentages

Percentages	Probits									
	0	1	2	3	4	5	6	7	8	9
0	—	2·67	2·95	3·12	3·25	3·36	3·45	3·52	3·59	3·66
10	3·72	3·77	3·82	3·87	3·92	3·96	4·01	4·05	4·08	4·12
20	4·16	4·19	4·23	4·26	4·29	4·33	4·36	4·39	4·42	4·45
30	4·48	4·50	4·53	4·56	4·59	4·61	4·64	4·67	4·69	4·72
40	4·75	4·77	4·80	4·82	4·85	4·87	4·90	4·92	4·95	4·97
50	5·00	5·03	5·05	5·08	5·10	5·13	5·15	5·18	5·20	5·23
60	5·25	5·28	5·31	5·33	5·36	5·39	5·41	5·44	5·47	5·50
70	5·52	5·55	5·58	5·61	5·64	5·67	5·71	5·74	5·77	5·81
80	5·84	5·88	5·92	5·95	5·99	6·04	6·08	6·13	6·18	6·23
90	6·28	6·34	6·41	6·48	6·55	6·64	6·75	6·88	7·05	7·33
	0·0	0·1	0·2	0·3	0·4	0·5	0·6	0·7	0·8	0·9
99	7·33	7·37	7·41	7·46	7·51	7·58	7·65	7·75	7·88	8·09

TABLE 6

Weighting factors corresponding to probits

Probits	Factors									
	0·0	0·1	0·2	0·3	0·4	0·5	0·6	0·7	0·8	0·9
1	0·001	0·001	0·001	0·002	0·002	0·003	0·005	0·006	0·008	0·011
2	0·015	0·019	0·025	0·031	0·040	0·050	0·062	0·076	0·092	0·110
3	0·131	0·154	0·180	0·208	0·238	0·269	0·302	0·336	0·370	0·405
4	0·439	0·471	0·503	0·532	0·558	0·581	0·601	0·616	0·627	0·634
5	0·637	0·634	0·627	0·616	0·601	0·581	0·558	0·532	0·503	0·471
6	0·439	0·405	0·370	0·336	0·302	0·269	0·238	0·208	0·180	0·154
7	0·131	0·110	0·092	0·076	0·062	0·050	0·040	0·031	0·025	0·019
8	0·015	0·011	0·008	0·006	0·005	0·003	0·002	0·002	0·001	0·001

FOUR-FIGURE LOGARITHMS

	0	1	2	3	4	5	6	7	8	9	1	2	3	4	5	6	7	8	9
10	0000	0043	0086	0128	0170	0212	0253	0294	0334	0374	4	8	12	17	21	25	29	33	37
11	0414	0453	0492	0531	0569	0607	0645	0682	0719	0755	4	8	11	15	19	23	26	30	34
12	0792	0828	0864	0899	0934	0969	1004	1038	1072	1106	3	7	10	14	17	21	24	28	31
13	1139	1173	1206	1239	1271	1303	1335	1367	1399	1430	3	6	10	13	16	19	23	26	29
14	1461	1492	1523	1553	1584	1614	1644	1673	1703	1732	3	6	9	12	15	18	21	24	27
15	1761	1790	1818	1847	1875	1903	1931	1959	1987	2014	3	6	8	11	14	17	20	22	25
16	2041	2068	2095	2122	2148	2175	2201	2227	2253	2279	3	5	8	11	13	16	18	21	24
17	2304	2330	2355	2380	2405	2430	2455	2480	2504	2529	2	5	7	10	12	15	17	20	22
18	2553	2577	2601	2625	2648	2672	2695	2718	2742	2765	2	5	7	9	12	14	16	19	21
19	2788	2810	2833	2856	2878	2900	2923	2945	2967	2989	2	4	7	9	11	13	16	18	20
20	3010	3032	3054	3075	3096	3118	3139	3160	3181	3201	2	4	6	8	11	13	15	17	19
21	3222	3243	3263	3284	3304	3324	3345	3365	3385	3404	2	4	6	8	10	12	14	16	18
22	3424	3444	3464	3483	3502	3522	3541	3560	3579	3598	2	4	6	8	10	12	14	15	17
23	3617	3636	3655	3674	3692	3711	3729	3747	3766	3784	2	4	6	7	9	11	13	15	17
24	3802	3820	3838	3856	3874	3892	3909	3927	3945	3962	2	4	5	7	9	11	12	14	16
25	3979	3997	4014	4031	4048	4065	4082	4099	4116	4133	2	3	5	7	9	10	12	14	15
26	4150	4166	4183	4200	4216	4232	4249	4265	4281	4298	2	3	5	7	8	10	11	13	15
27	4314	4330	4346	4362	4378	4393	4409	4425	4440	4456	2	3	5	6	8	9	11	13	14
28	4472	4487	4502	4518	4533	4548	4564	4579	4594	4609	2	3	5	6	8	9	11	12	14
29	4624	4639	4654	4669	4683	4698	4713	4728	4742	4757	1	3	4	6	7	9	10	12	13
30	4771	4786	4800	4814	4829	4843	4857	4871	4886	4900	1	3	4	6	7	9	10	11	13
31	4914	4928	4942	4955	4969	4983	4997	5011	5024	5038	1	3	4	6	7	8	10	11	12
32	5051	5065	5079	5092	5105	5119	5132	5145	5159	5172	1	3	4	5	7	8	9	11	12
33	5185	5198	5211	5224	5237	5250	5263	5276	5289	5302	1	3	4	5	6	8	9	10	12
34	5315	5328	5340	5353	5366	5378	5391	5403	5416	5428	1	3	4	5	6	8	9	10	11
35	5441	5453	5465	5478	5490	5502	5514	5527	5539	5551	1	2	4	5	6	7	9	10	11
36	5563	5575	5587	5599	5611	5623	5635	5647	5658	5670	1	2	4	5	6	7	8	10	11
37	5682	5694	5705	5717	5729	5740	5752	5763	5775	5786	1	2	3	5	6	7	8	9	10
38	5798	5809	5821	5832	5843	5855	5866	5877	5888	5899	1	2	3	5	6	7	8	9	10
39	5911	5922	5933	5944	5955	5966	5977	5988	5999	6010	1	2	3	4	5	7	8	9	10
40	6021	6031	6042	6053	6064	6075	6085	6096	6107	6117	1	2	3	4	5	6	8	9	10
41	6128	6138	6149	6160	6170	6180	6191	6201	6212	6222	1	2	3	4	5	6	7	8	9
42	6232	6243	6253	6263	6274	6284	6294	6304	6314	6325	1	2	3	4	5	6	7	8	9
43	6335	6345	6355	6365	6375	6385	6395	6405	6415	6425	1	2	3	4	5	6	7	8	9
44	6435	6444	6454	6464	6474	6484	6493	6503	6513	6522	1	2	3	4	5	6	7	8	9
45	6532	6542	6551	6561	6571	6580	6590	6599	6609	6618	1	2	3	4	5	6	7	8	9
46	6628	6637	6646	6656	6665	6675	6684	6693	6702	6712	1	2	3	4	5	6	7	7	8
47	6721	6730	6739	6749	6758	6767	6776	6785	6794	6803	1	2	3	4	5	5	6	7	8
48	6812	6821	6830	6839	6848	6857	6866	6875	6884	6893	1	2	3	4	4	5	6	7	8
49	6902	6911	6920	6928	6937	6946	6955	6964	6972	6981	1	2	3	4	4	5	6	7	8
50	6990	6998	7007	7016	7024	7033	7042	7050	7059	7067	1	2	3	3	4	5	6	7	8
51	7076	7084	7093	7101	7110	7118	7126	7135	7143	7152	1	2	3	3	4	5	6	7	8
52	7160	7168	7177	7185	7193	7202	7210	7218	7226	7235	1	2	2	3	4	5	6	7	7
53	7243	7251	7259	7267	7275	7284	7292	7300	7308	7316	1	2	2	3	4	5	6	6	7
54	7324	7332	7340	7348	7356	7364	7372	7380	7388	7396	1	2	2	3	4	5	6	6	7

FOUR-FIGURE LOGARITHMS

	0	1	2	3	4	5	6	7	8	9	1	2	3	4	5	6	7	8	9
55	7404	7412	7419	7427	7435	7443	7451	7459	7466	7474	1	2	2	3	4	5	5	6	7
56	7482	7490	7497	7505	7513	7520	7528	7536	7543	7551	1	2	2	3	4	5	5	6	7
57	7559	7566	7574	7582	7589	7597	7604	7612	7619	7627	1	2	2	3	4	5	5	6	7
58	7634	7642	7649	7657	7664	7672	7679	7686	7694	7701	1	1	2	3	4	4	5	6	7
59	7709	7716	7723	7731	7738	7745	7752	7760	7767	7774	1	1	2	3	4	4	5	6	7
60	7782	7789	7796	7803	7810	7818	7825	7832	7839	7846	1	1	2	3	4	4	5	6	6
61	7853	7860	7868	7875	7882	7889	7896	7903	7910	7917	1	1	2	3	4	4	5	6	6
62	7924	7931	7938	7945	7952	7959	7966	7973	7980	7987	1	1	2	3	3	4	5	6	6
63	7993	8000	8007	8014	8021	8028	8035	8041	8048	8055	1	1	2	3	3	4	5	5	6
64	8062	8069	8075	8082	8089	8096	8102	8109	8116	8122	1	1	2	3	3	4	5	5	6
65	8129	8136	8142	8149	8156	8162	8169	8176	8182	8189	1	1	2	3	3	4	5	5	6
66	8195	8202	8209	8215	8222	8228	8235	8241	8248	8254	1	1	2	3	3	4	5	5	6
67	8261	8267	8274	8280	8287	8293	8299	8306	8312	8319	1	1	2	3	3	4	5	5	6
68	8325	8331	8338	8344	8351	8357	8363	8370	8376	8382	1	1	2	3	3	4	4	5	6
69	8388	8395	8401	8407	8414	8420	8426	8432	8439	8445	1	1	2	2	3	4	4	5	6
70	8451	8457	8463	8470	8476	8482	8488	8494	8500	8506	1	1	2	2	3	4	4	5	6
71	8513	8519	8525	8531	8537	8543	8549	8555	8561	8567	1	1	2	2	3	4	4	5	5
72	8573	8579	8585	8591	8597	8603	8609	8615	8621	8627	1	1	2	2	3	4	4	5	5
73	8633	8639	8645	8651	8657	8663	8669	8675	8681	8686	1	1	2	2	3	4	4	5	5
74	8692	8698	8704	8710	8716	8722	8727	8733	8739	8745	1	1	2	2	3	4	4	5	5
75	8751	8756	8762	8768	8774	8779	8785	8791	8797	8802	1	1	2	2	3	3	4	5	5
76	8808	8814	8820	8825	8831	8837	8842	8848	8854	8859	1	1	2	2	3	3	4	5	5
77	8865	8871	8876	8882	8887	8893	8899	8904	8910	8915	1	1	2	2	3	3	4	4	5
78	8921	8927	8932	8938	8943	8949	8954	8960	8965	8971	1	1	2	2	3	3	4	4	5
79	8976	8982	8987	8993	8998	9004	9009	9015	9020	9025	1	1	2	2	3	3	4	4	5
80	9031	9036	9042	9047	9053	9058	9063	9069	9074	9079	1	1	2	2	3	3	4	4	5
81	9085	9090	9096	9101	9106	9112	9117	9122	9128	9133	1	1	2	2	3	3	4	4	5
82	9138	9143	9149	9154	9159	9165	9170	9175	9180	9186	1	1	2	2	3	3	4	4	5
83	9191	9196	9201	9206	9212	9217	9222	9227	9232	9238	1	1	2	2	3	3	4	4	5
84	9243	9248	9253	9258	9263	9269	9274	9279	9284	9289	1	1	2	2	3	3	4	4	5
85	9294	9299	9304	9309	9315	9320	9325	9330	9335	9340	1	1	2	2	3	3	4	4	5
86	9345	9350	9355	9360	9365	9370	9375	9380	9385	9390	1	1	2	2	3	3	4	4	5
87	9395	9400	9405	9410	9415	9420	9425	9430	9435	9440	0	1	1	2	2	3	3	4	4
88	9445	9450	9455	9460	9465	9469	9474	9479	9484	9489	0	1	1	2	2	3	3	4	4
89	9494	9499	9504	9509	9513	9518	9523	9528	9533	9538	0	1	1	2	2	3	3	4	4
90	9542	9547	9552	9557	9562	9566	9571	9576	9581	9586	0	1	1	2	2	3	3	4	4
91	9590	9595	9600	9605	9609	9614	9619	9624	9628	9633	0	1	1	2	2	3	3	4	4
92	9638	9643	9647	9652	9657	9661	9666	9671	9675	9680	0	1	1	2	2	3	3	4	4
93	9685	9689	9694	9699	9703	9708	9713	9717	9722	9727	0	1	1	2	2	3	3	4	4
94	9731	9736	9741	9745	9750	9754	9759	9763	9768	9773	0	1	1	2	2	3	3	4	4
95	9777	9782	9786	9791	9795	9800	9805	9809	9814	9818	0	1	1	2	2	3	3	4	4
96	9823	9827	9832	9836	9841	9845	9850	9854	9859	9863	0	1	1	2	2	3	3	4	4
97	9868	9872	9877	9881	9886	9890	9894	9899	9903	9908	0	1	1	2	2	3	3	4	4
98	9912	9917	9921	9926	9930	9934	9939	9943	9948	9952	0	1	1	2	2	3	3	4	4
99	9956	9961	9965	9969	9974	9978	9983	9987	9991	9996	0	1	1	2	2	3	3	3	4

Mean differences span columns 1–9 on the right.

FOUR-FIGURE ANTILOGARITHMS

	0	1	2	3	4	5	6	7	8	9	1	2	3	4	5	6	7	8	9
·00	1000	1002	1005	1007	1009	1012	1014	1016	1019	1021	0	0	1	1	1	1	2	2	2
·01	1023	1026	1028	1030	1033	1035	1038	1040	1042	1045	0	0	1	1	1	1	2	2	2
·02	1047	1050	1052	1054	1057	1059	1062	1064	1067	1069	0	0	1	1	1	1	2	2	2
·03	1072	1074	1076	1079	1081	1084	1086	1089	1091	1094	0	0	1	1	1	1	2	2	2
·04	1096	1099	1102	1104	1107	1109	1112	1114	1117	1119	0	1	1	1	1	2	2	2	2
·05	1122	1125	1127	1130	1132	1135	1138	1140	1143	1146	0	1	1	1	1	2	2	2	2
·06	1148	1151	1153	1156	1159	1161	1164	1167	1169	1172	0	1	1	1	1	2	2	2	2
·07	1175	1178	1180	1183	1186	1189	1191	1194	1197	1199	0	1	1	1	1	2	2	2	2
·08	1202	1205	1208	1211	1213	1216	1219	1222	1225	1227	0	1	1	1	1	2	2	2	3
·09	1230	1233	1236	1239	1242	1245	1247	1250	1253	1256	0	1	1	1	1	2	2	2	3
·10	1259	1262	1265	1268	1271	1274	1276	1279	1282	1285	0	1	1	1	1	2	2	2	3
·11	1288	1291	1294	1297	1300	1303	1306	1309	1312	1315	0	1	1	1	2	2	2	2	3
·12	1318	1321	1324	1327	1330	1334	1337	1340	1343	1346	0	1	1	1	2	2	2	2	3
·13	1349	1352	1355	1358	1361	1365	1368	1371	1374	1377	0	1	1	1	2	2	2	3	3
·14	1380	1384	1387	1390	1393	1396	1400	1403	1406	1409	0	1	1	1	2	2	2	3	3
·15	1413	1416	1419	1422	1426	1429	1432	1435	1439	1442	0	1	1	1	2	2	2	3	3
·16	1445	1449	1452	1455	1459	1462	1466	1469	1472	1476	0	1	1	1	2	2	2	3	3
·17	1479	1483	1486	1489	1493	1496	1500	1503	1507	1510	0	1	1	1	2	2	2	3	3
·18	1514	1517	1521	1524	1528	1531	1535	1538	1542	1545	0	1	1	1	2	2	2	3	3
·19	1549	1552	1556	1560	1563	1567	1570	1574	1578	1581	0	1	1	1	2	2	3	3	3
·20	1585	1589	1592	1596	1600	1603	1607	1611	1614	1618	0	1	1	1	2	2	3	3	3
·21	1622	1626	1629	1633	1637	1641	1644	1648	1652	1656	0	1	1	2	2	2	3	3	3
·22	1660	1663	1667	1671	1675	1679	1683	1687	1690	1694	0	1	1	2	2	2	3	3	3
·23	1698	1702	1706	1710	1714	1718	1722	1726	1730	1734	0	1	1	2	2	2	3	3	4
·24	1738	1742	1746	1750	1754	1758	1762	1766	1770	1774	0	1	1	2	2	2	3	3	4
·25	1778	1782	1786	1791	1795	1799	1803	1807	1811	1816	0	1	1	2	2	2	3	3	4
·26	1820	1824	1828	1832	1837	1841	1845	1849	1854	1858	0	1	1	2	2	3	3	3	4
·27	1862	1866	1871	1875	1879	1884	1888	1892	1897	1901	0	1	1	2	2	3	3	3	4
·28	1905	1910	1914	1919	1923	1928	1932	1936	1941	1945	0	1	1	2	2	3	3	4	4
·29	1950	1954	1959	1963	1968	1972	1977	1982	1986	1991	0	1	1	2	2	3	3	4	4
·30	1995	2000	2004	2009	2014	2018	2023	2028	2032	2037	0	1	1	2	2	3	3	4	4
·31	2042	2046	2051	2056	2061	2065	2070	2075	2080	2084	0	1	1	2	2	3	3	4	4
·32	2089	2094	2099	2104	2109	2113	2118	2123	2128	2133	0	1	1	2	2	3	3	4	4
·33	2138	2143	2148	2153	2158	2163	2168	2173	2178	2183	0	1	1	2	2	3	3	4	4
·34	2188	2193	2198	2203	2208	2213	2218	2223	2228	2234	1	1	2	2	3	3	4	4	5
·35	2239	2244	2249	2254	2259	2265	2270	2275	2280	2286	1	1	2	2	3	3	4	4	5
·36	2291	2296	2301	2307	2312	2317	2323	2328	2333	2339	1	1	2	2	3	3	4	4	5
·37	2344	2350	2355	2360	2366	2371	2377	2382	2388	2393	1	1	2	2	3	3	4	4	5
·38	2399	2404	2410	2415	2421	2427	2432	2438	2443	2449	1	1	2	2	3	3	4	4	5
·39	2455	2460	2466	2472	2477	2483	2489	2495	2500	2506	1	1	2	2	3	3	4	5	5
·40	2512	2518	2523	2529	2535	2541	2547	2553	2559	2564	1	1	2	2	3	4	4	5	5
·41	2570	2576	2582	2588	2594	2600	2606	2612	2618	2624	1	1	2	2	3	4	4	5	5
·42	2630	2636	2642	2649	2655	2661	2667	2673	2679	2685	1	1	2	2	3	4	4	5	6
·43	2692	2698	2704	2710	2716	2723	2729	2735	2742	2748	1	1	2	3	3	4	4	5	6
·44	2754	2761	2767	2773	2780	2786	2793	2799	2805	2812	1	1	2	3	3	4	4	5	6
·45	2818	2825	2831	2838	2844	2851	2858	2864	2871	2877	1	1	2	3	3	4	5	5	6
·46	2884	2891	2897	2904	2911	2917	2924	2931	2938	2944	1	1	2	3	3	4	5	5	6
·47	2951	2958	2965	2972	2979	2985	2992	2999	3006	3013	1	1	2	3	3	4	5	5	6
·48	3020	3027	3034	3041	3048	3055	3062	3069	3076	3083	1	1	2	3	4	4	5	6	6
·49	3090	3097	3105	3112	3119	3126	3133	3141	3148	3155	1	1	2	3	4	4	5	6	6

Mean differences: columns 1–9

FOUR-FIGURE ANTILOGARITHMS

	0	1	2	3	4	5	6	7	8	9	\multicolumn{9}{c}{Mean differences}
											1	2	3	4	5	6	7	8	9
·50	3162	3170	3177	3184	3192	3199	3206	3214	3221	3228	1	1	2	3	4	4	5	6	7
·51	3236	3243	3251	3258	3266	3273	3281	3289	3296	3304	1	2	2	3	4	5	5	6	7
·52	3311	3319	3327	3334	3342	3350	3357	3365	3373	3381	1	2	2	3	4	5	5	6	7
·53	3388	3396	3404	3412	3420	3428	3436	3443	3451	3459	1	2	2	3	4	5	6	6	7
·54	3467	3475	3483	3491	3499	3508	3516	3524	3532	3540	1	2	2	3	4	5	6	6	7
·55	3548	3556	3565	3573	3581	3589	3597	3606	3614	3622	1	2	2	3	4	5	6	7	7
·56	3631	3639	3648	3656	3664	3673	3681	3690	3698	3707	1	2	3	3	4	5	6	7	8
·57	3715	3724	3733	3741	3750	3758	3767	3776	3784	3793	1	2	3	3	4	5	6	7	8
·58	3802	3811	3819	3828	3837	3846	3855	3864	3873	3882	1	2	3	4	4	5	6	7	8
·59	3890	3899	3908	3917	3926	3936	3945	3954	3963	3972	1	2	3	4	5	5	6	7	8
·60	3981	3990	3999	4009	4018	4027	4036	4046	4055	4064	1	2	3	4	5	6	6	7	8
·61	4074	4083	4093	4102	4111	4121	4130	4140	4150	4159	1	2	3	4	5	6	7	8	9
·62	4169	4178	4188	4198	4207	4217	4227	4236	4246	4256	1	2	3	4	5	6	7	8	9
·63	4266	4276	4285	4295	4305	4315	4325	4335	4345	4355	1	2	3	4	5	6	7	8	9
·64	4365	4375	4385	4395	4406	4416	4426	4436	4446	4457	1	2	3	4	5	6	7	8	9
·65	4467	4477	4487	4498	4508	4519	4529	4539	4550	4560	1	2	3	4	5	6	7	8	9
·66	4571	4581	4592	4603	4613	4624	4634	4645	4656	4667	1	2	3	4	5	6	7	9	10
·67	4677	4688	4699	4710	4721	4732	4742	4753	4764	4775	1	2	3	4	5	7	8	9	10
·68	4786	4797	4808	4819	4831	4842	4853	4864	4875	4887	1	2	3	4	6	7	8	9	10
·69	4898	4909	4920	4932	4943	4955	4966	4977	4989	5000	1	2	3	5	6	7	8	9	10
·70	5012	5023	5035	5047	5058	5070	5082	5093	5105	5117	1	2	4	5	6	7	8	9	11
·71	5129	5140	5152	5164	5176	5188	5200	5212	5224	5236	1	2	4	5	6	7	8	10	11
·72	5248	5260	5272	5284	5297	5309	5321	5333	5346	5358	1	2	4	5	6	7	9	10	11
·73	5370	5383	5395	5408	5420	5433	5445	5458	5470	5483	1	3	4	5	6	8	9	10	11
·74	5495	5508	5521	5534	5546	5559	5572	5585	5598	5610	1	3	4	5	6	8	9	10	12
·75	5623	5636	5649	5662	5675	5689	5702	5715	5728	5741	1	3	4	5	7	8	9	10	12
·76	5754	5768	5781	5794	5808	5821	5834	5848	5861	5875	1	3	4	5	7	8	9	11	12
·77	5888	5902	5916	5929	5943	5957	5970	5984	5998	6012	1	3	4	5	7	8	10	11	12
·78	6026	6039	6053	6067	6081	6095	6109	6124	6138	6152	1	3	4	6	7	8	10	11	13
·79	6166	6180	6194	6209	6223	6237	6252	6266	6281	6295	1	3	4	6	7	9	10	11	13
·80	6310	6324	6339	6353	6368	6383	6397	6412	6427	6442	1	3	4	6	7	9	10	12	13
·81	6457	6471	6486	6501	6516	6531	6546	6561	6577	6592	2	3	5	6	8	9	11	12	14
·82	6607	6622	6637	6653	6668	6683	6699	6714	6730	6745	2	3	5	6	8	9	11	12	14
·83	6761	6776	6792	6808	6823	6839	6855	6871	6887	6902	2	3	5	6	8	9	11	13	14
·84	6918	6934	6950	6966	6982	6998	7015	7031	7047	7063	2	3	5	6	8	10	11	13	15
·85	7079	7096	7112	7129	7145	7161	7178	7194	7211	7228	2	3	5	7	8	10	12	13	15
·86	7244	7261	7278	7295	7311	7328	7345	7362	7379	7396	2	3	5	7	8	10	12	13	15
·87	7413	7430	7447	7464	7482	7499	7516	7534	7551	7568	2	3	5	7	9	10	12	14	16
·88	7586	7603	7621	7638	7656	7674	7691	7709	7727	7745	2	4	5	7	9	11	12	14	16
·89	7762	7780	7798	7816	7834	7852	7870	7889	7907	7925	2	4	5	7	9	11	13	14	16
·90	7943	7962	7980	7998	8017	8035	8054	8072	8091	8110	2	4	6	7	9	11	13	15	17
·91	8128	8147	8166	8185	8204	8222	8241	8260	8279	8299	2	4	6	8	9	11	13	15	17
·92	8318	8337	8356	8375	8395	8414	8433	8453	8472	8492	2	4	6	8	10	12	14	15	17
·93	8511	8531	8551	8570	8590	8610	8630	8650	8670	8690	2	4	6	8	10	12	14	16	18
·94	8710	8730	8750	8770	8790	8810	8831	8851	8872	8892	2	4	6	8	10	12	14	16	18
·95	8913	8933	8954	8974	8995	9016	9036	9057	9078	9099	2	4	6	8	10	12	15	17	19
·96	9120	9141	9162	9183	9204	9226	9247	9268	9290	9311	2	4	6	8	11	13	15	17	19
·97	9333	9354	9376	9397	9419	9441	9462	9484	9506	9528	2	4	7	9	11	13	15	17	20
·98	9550	9572	9594	9616	9638	9661	9683	9705	9727	9750	2	4	7	9	11	13	16	18	20
·99	9772	9795	9817	9840	9863	9886	9908	9931	9954	9977	2	5	7	9	11	14	16	18	20

Answers to problems

Chapter 1
1. (i) 2^{12} (ii) 2^6 (iii) 2^{12}
   (iv) $2^{3.5}$ (v) $2^{-3.7}$ (vi) $2^{2.7}$

2. (i) 3.73×10^{-3} (ii) 2.756×10^6 (iii) 1.02×10^{-3}
   (iv) 1.324×10^3 (v) 1.07×10^{-6} (vi) 1.57394×10

3. 0.867 mg

4. 2.65×10^{17}

5. (i) 3.010 (ii) 0.1806 (iii) $\bar{2}.5050$ (iv) 0.1070
   (v) 1.8060 (vi) 1.903 (vii) 4.6990 (viii) 2.796

6. (i) 0.6112 (ii) 26.95 (iii) 1547
   (iv) 0.610 (v) 1.498×10^8 (vi) 1.239×10^{-6}
   (vii) 8.996 (viii) 6.418

7. (i) 9.14 (ii) 3.75 (iii) 4.87
   (iv) 2.87 (v) 9.92

8. $v = k\eta/rd$

9.   DIMENSION X(24)
   (input) FORMAT (12F6.0)
   (output) FORMAT (6F10.3)

Chapter 2
1. $b = a^{\log_a b}$, $a = b^{\log_b a} = (a^{\log_a b})^{\log_b a}$

2. $a = 0, b = 4, c = 7$ and $d = 3$

3. (i) $x = 3$ or 4 (ii) $x = 0.6$ or -2

4. (i) $x = -0.5 \pm \frac{1}{2}i\sqrt{11}$ (ii) $x = 0.33 \pm \frac{1}{3}i\sqrt{11}$
   (iii) $x = 1.175$ or -0.425 (iv) $x = -0.298$ or -6.701

5. (i) $x = 2, y = 1$ or $x = 1, y = 2$
   (ii) $x = 8, y = -12$

6. $x = 1, 3$ or -2

7. $\begin{pmatrix} 1 & 1 & 2 \\ 2 & -1 & 3 \\ 4 & 2 & -1 \end{pmatrix} \begin{pmatrix} x \\ y \\ z \end{pmatrix} = \begin{pmatrix} 7 \\ 16 \\ -4 \end{pmatrix}$
   $x = 1, y = -2, z = 4$

8. Estimated break-even, PPM = 1030; accurate value, BEPPM = 1034.55

Programme for calculation:

```
C     COMPARISON OF TWO PROCESSES
      PPM = 0.
      WRITE(6,1)
    1 FORMAT(30H1    PPM        PROCA        PROCB    )
      DO 3 I = 1,30
      PPM = PPM + 50.
      PROCA = 1.062 + 0.104*PPM/550.
      PROCB = 0.009 + 0.667*PPM/550.
      WRITE(6,2) PPM,PROCA,PROCB
    2 FORMAT(1H0,3F10.4)
    3 CONTINUE
      BEPPM = 1.059*550./0.563
      WRITE(6,4) BEPPM
    4 FORMAT(9HOBEPPM = ,F10.4)
      STOP
      END
```

9. SOLUTIONS
   C1 = ·7412190504
   C2 = ·9177717301
   C3 = 1·0004232023
   C4 = 1·2032464534
   KS = 0

Chapter 3

1. (i) $x = 1$ or 2 (ii) $x = 1, 2,$ or -1
   (iii) $x = 1·32$
2. $x = 1·77, y = 2·77$
3. $y = 3x - 6$

Chapter 4

1. 153
2. 441
3. 1·0510, 1·0212, 1·0150, 0·9615, 1·0007
4. 495. In such a forecast, the order in which the winning home teams are chosen will usually be unimportant, and so the answer is $^{12}C_8$ rather than $^{12}P_8$
5. 10,626
6. Expand both functions by the exponential theorem, neglecting all but the first two terms of each
7. 1·1052
8. 0·0953
9. 120

Chapter 5

1. (i) $15x^4 + 8x - 6$ (ii) $18x^2 + 8x - 3$
   (iii) $-2x/(x^2+1)^2$ (iv) $(2-4x)/(x+1)^4$
   (v) $-26/(4x-7)^2$ (vi) $6ax^2 + 2(ad+2b)x + bd + 2c$
   (vii) $2bx^{(2b-1)} - 3bx^2$ (viii) $18x^2 + 26x + 9$
   (ix) $(3x)^{-2/3} + (2x)^{-1/2} - 27/x^4$

ANSWERS TO PROBLEMS

2. 5, 7, 9 and 17
3. (i) $6x^2(x^3-2)$ (ii) $54x^3(2-x^3)(4-5x^3)$
   (iii) $2 \exp x$ (iv) $7/x$
   (v) $2x/(x^2+4)$ (vi) $x^3(4+x) \exp x$
   (vii) $2/(3-x)^2$ (viii) $(x^2+4x+1)/(x+2)^2$
   (ix) $(\ln x+1)/2 \cdot 303$ (x) $(6x^2+3) \exp(2x^3+3x+7)$
   (xi) $-x/y$ (xii) $(5y-3x^2)/(3y^2-5x)$
   (xiii) $(5y-4x+3)/(2y-5x)$ (xiv) $-(x/y)^{2n-1}$
4. $\bar{v} = 0.1$; $v = 0.160, 0.130, 0.103$ and 0.081
5. $\Delta y/\Delta x = 28.0, 27.3$ and 27.1; $dy/dx = 27.1$ (algebraic value 27.0)
6. Fortran programme and solution:

```
C      COMP 3
C      NEWTON SOLUTION OF EQUATION  FT=0
       ITER = 1
       X = 2.0
1      FT = X**5 - 10.* X**2 - 4.
       DFT = 5. * X**4 - 20. * X
       TREV = X - FT/DFT
       WRITE(6,2) ITER,TREV
2      FORMAT(1H0,I5,15X,F10.6)
       IF (ABS(X-TREV) .LT. 1.E-6 .OR. ITER .GE. 20) STOP
       ITER = ITER + 1
       X = TREV
       GO TO 1
       END
```

1	2.300000
2	2.220535
3	2.211733
4	2.211631
5	2.211631

Chapter 6

1. (i) $-1/(2 \cdot 3x^2)$ (ii) $12x^2+6$
   (iii) $2(2x^2+1) \exp(x^2)$ (iv) $4/x^3$
2. (i) $y = 1.75$ (minimum)   (ii) $y = 0$ (minimum)
   (iii) $y = 28$ (maximum),   (iv) $y = 92.26$ (maximum),
        $y = -4$ (minimum) $y = 1.75$ (minimum)
   (v) $y = 3$ (horizontal inflection)   (vi) $y = 1.15$ (maximum),
                  $y = 1$ (minimum)
3. $4.09°$
4. $h/g = 1$ gives a minimum value of $(h+g)$
5. (i) $5x^4+12x^2y^2+y^4$ (ii) $2x \exp(-y^2)$
   (iii) $3x^2/(x^3+3)$ (iv) $4xy^3+y^2$
   (v) $2(x+y)$ (vi) $-1/(x+y)^2$
6. Rearrange equation to $p = RT/(v-b) - a/v^2$. Evaluate $(\partial p/\partial v)_T$ and $(\partial^2 p/\partial v^2)_T$. At the critical point, both these functions are equal to zero; substituting $v = v_c$, $T = T_c$ in the two equations obtained gives the results $v_c = 3b$ and $T_c = 8a/27Rb$. From the original equation it is found that $p_c = a/27b^2$.

Chapter 7

1. 2.33 (approximately)

296   ANSWERS TO PROBLEMS

2. (i) $\frac{1}{4}x^4+x^3+8x+C$ (ii) $\ln(9+x^2)+C$
   (iii) $\frac{1}{6}\ln\{(x-3)/(x+3)\}+C$ (iv) $-2\exp(-\frac{1}{2}x)+C$
   (v) $(x-1)\exp x+C$ (vi) $\frac{1}{2}\exp(x^2)+C$
   (vii) $-1\cdot6$ (viii) $e-1$
   (ix) $2\cdot303$ (x) $6\cdot998$
   (xi) $\ln(4/3) = 0\cdot287$ (xii) 5

3. (i) 0·3167 (ii) 0·1831 (iii) 0·16665

4.
```
C       COMP 4
C       INTEGRAL BY SIMPSONS RULE
        DIMENSION Y(21)
        X = 0.
        H = 1.0
        Y(1) = 0.
        DO 1 I = 2,21
        X = X + H
1       Y(I) = X**2 * ALOG(X + 3.)
        YE = 0.
        YO = 0.
        DO 2 I = 2,20,2
2       YE = YE + Y(I)
        DO 3 I = 3,19,2
3       YO = YO + Y(I)
        TG = (H/3.) *(Y(1) + 4.*YE + 2.*YO + Y(21))
        WRITE(6,4) TG
4       FORMAT(11H1INTEGRAL =,F10.5)
        STOP
        END
              INTEGRAL = 7630.75119
```

5.
```
C       COMP 4
C       PROBABILITY INTEGRAL BY SIMPSONS RULE
        DIMENSION Y(21)
        U = -0.5
        H = 0.05
        Y(1) = EXP(-0.5)
        DO 1 I = 2,21
        U = U + H
        USQ = -(U**2)/2.
1       Y(I) = EXP(USQ)
        YE = 0.
        YO = 0.
        DO 2 I = 2,20,2
2       YE = YE + Y(I)
        DO 3 I = 3,19,2
3       YO = YO + Y(I)
        TG = (H/3.) *(Y(1) + 4.*YE + 2.*YO + Y(21))
        PROB = TG/SQRT(2.*3.14159)
        WRITE(6,4) PROB
4       FORMAT(14H1PROBABILITY =,F10.5)
        STOP
        END
              PROBABILITY =      .38109
```

ANSWERS TO PROBLEMS

6. INTEGRAL = 17·0523

Chapter 8

1. $\frac{1}{2}$ $\frac{1}{2}\sqrt{3}$ $1/\sqrt{3}$ $\frac{1}{2}$
   $-\frac{1}{2}$ $-\frac{1}{2}$ $-1/\sqrt{3}$

2. $\pi/6$ $5\pi/6$ $13\pi/6$ $17\pi/6$

3. 0·7666

4. (i) $\frac{1}{2}\pi$ (ii) $\frac{1}{2}$ (iii) $\frac{1}{6}\tan^{-1}(\frac{2}{3})$
   (iv) $\frac{1}{2}\ln(2) = 0.347$ (v) $\frac{1}{8}$

5. (a) $2/\pi$, 0, 1/2, 1/2
   (b) 0, 0, 1/2, 1/2

Chapter 9

1. From (9.10), $dc/dy = -c_0 \exp(-y^2)/\sqrt{\pi}$ for all conditions, i.e. $(\partial c/\partial y)_x$ and $(\partial c/\partial y)_t$ will both have this value. If x is constant

$$\frac{\partial c}{\partial t} = \frac{\partial c}{\partial y} \cdot \frac{\partial y}{\partial t};$$

$\partial c/\partial y$ is given above, and $\partial y/\partial t$ can be found by differentiating the equation defining y, treating x as a constant In a similar way $\partial c/\partial x$ is found and then differentiated, keeping t constant, to give $(\partial^2 c/\partial x^2)_t$

2. $k = 0.205$

3. $x = \sin(2 \times 10^3 \times \pi t)$

4. (i) $x^2y^2 + x^2 - y^2 = C$ (ii) $x^2 + y^2 = C$
   (iii) $x^2 y = \frac{1}{3}\exp(3x) + C$ (iv) $y = 3x^2/2 + Ax + B$
   (v) $\ln y = 4x^3/3 + 3x^5/5 + C$

5. $\lambda = 0.0277$ days^{-1}

Chapter 10

1. $a = 10^{-6}$ $b = 2 \times 10^{-3}$ $c = 6.80$

2. $pV = 22.5 - 1/V$

3. $x/m = 0.33\ p^{1.34}$

4. ENERGY MINIMUM = -91.0039
   PARAMETER AT MINIMUM = 147·16

Chapter 11

1. 0·104

2. 0·772 and 0·656

3. 15

4. $m = 2, V = 1.33$

5.

X (observed frequency)	O	E (frequency calculated from Poisson series)	O−E	(O−E)²/E
0	24	29	−5	0·862
1	47	44	+3	0·205
2	36	33	+3	0·273
3	18	17	+1	0·059
4	4*	6		
5	2*	2		

$$\chi^2 = 1·399$$

The theoretical value of χ^2 for $P = 0·95$ and $\phi = 3$ is 7·8 (obtained from Table 3, Appendix XII). The observed frequencies can therefore be considered to form a Poisson distribution.

* The last two values are not used in the calculation, as they contain less than six observations.

Chapter 12

1. $1·078 \pm 0·048$ ($P = 0·95$). The difference between the potencies is likely to be significant, since both the upper and lower limiting values of the potency ratio are greater than one.

2. $m = 4·00 \quad s = 0·27$

3. $m = 34·94$ and $s = 5·4$; graphically, $m = 34·4$ and $s = 5·61$. For $P = 0·95$, the weight of a further single rat will lie between 24 g. and 46 g. For $P = 0·99$, the limits would be 21 g. and 49 g. These results assume a normal distribution of weights about their mean, the factor t being used to calculate them. To avoid this assumption, the Camp–Meidell expression can be used, which gives limits ($P = 0·95$) of $2·98s = \pm 16·1$, i.e. limiting values of the weight of 19 g. and 51 g.

Chapter 13

1. $r = 0·997$, and therefore there is a significant correlation.
$b = 7·73$.
Equation of the regression line is $y = 7·73x - 0·22$.
$X = 4·58 \pm 0·67$ ($P = 0·95$).

2. $u = 1·25$; theoretical value ($P' = 0·05$) is 1·96, and therefore the improvement is not significant.

3. $m_A = 8·040$, $s_A = 0·0332$; $m_B = 8·103$, $s_B = 0·0673$.
$t = 2·65$; theoretical value ($P' = 0·05$) is 2·10, and therefore the difference between means is significant.
$F = 4·08$; theoretical value ($P' = 0·05$) is 3·2, and therefore the variance ratio is significant.
These results indicate that meter B has an unsatisfactorily large random error. One of the meters is giving a systematic error in its reading.

4. $\chi^2 = 0·65$; theoretical value ($P' = 0·05$) is 3·84, and therefore the modification has not produced a significant effect.

5. $t = 1·49$; theoretical value ($P' = 0·05$) is 2·10, and therefore the effect of the preparation is not significant.

ANSWERS TO PROBLEMS

6. NA = 10
AM = 4·534000
SSQA = ·155440
NB = 10
BM = 4·679000
SSQB = ·344490
VALUE OF T = 1·945515

7. Add to programme statements for F in Chapter 13, § 3 (page 197):
NA = 10
AM = 8·040000
SSQA = ·010000
NB = 10
BM = 8·103000
SSQB = ·040810
VALUE OF T = 2·651473
VALUE OF F = 4·081000

8.
```
REGRESSION - ONE INDEPENDENT VARIABLE

XMEAN.....                  4.00000
YMEAN.....                 30.70000

INTERCEPT (A VALUE)...       -.21429
REG. COEFFICIENT......       7.72857
STD. ERROR OF REG. COEF.      .36427

CORRELATION COEF......        .99449
```

	ANALYSIS OF VARIANCE FOR SIMPLE LINEAR REGRESSION			
SOURCE OF VARIATION	DEGREE OF FREEDOM	SUM OF SQUARES	MEAN SQUARE	F VALUE
DUE TO REGRESSION............	1	1672.46286	1672.46286	450.13996
DEVIATION ABOUT REGRESSION....	5	18.57714	3.71543	
TOTAL....	6	1691.04000		

$$r = 0 \cdot 99449 \qquad Y = 7 \cdot 72857 X - 0 \cdot 21429$$

9. Data are read in with t of Problem 4, Chapter 5 (page 92) as abscissa, x as ordinate (5 1·09 10 2·06 15 2·92...). A special transgeneration card is included which performs three operations on x: (i) multiplies by -1, (ii) adds the result to 26·92, (iii) takes the logarithm to base ten of $26 \cdot 92 - x$. The regression is then carried out on values of log $(26 \cdot 92 - x)/\text{time}$.

The equation of the regression line is

$$\log (a-x) = 1 \cdot 41377 - 0 \cdot 00223\, t$$

the half-life is the time taken for log $(26 \cdot 92 - x)$ to become log $(13 \cdot 46)$

$$1 \cdot 1290 = 1 \cdot 41377 - 0 \cdot 00223\, t_{1/2}$$

$$t_{1/2} = 127 \cdot 7 \text{ minutes.}$$

PROBLEM NO.   9
SAMPLE SIZE   12
TRANSGENERATION CARD

CODE	CONSTANT
9	−1.00000
8	26.92000
3	−0.00000

REGRESSION − ONE INDEPENDENT VARIABLE

XMEAN.....	32.50000
YMEAN.....	1.34133
INTERCEPT (A VALUE)...	1.41377
REG. COEFFICIENT......	−.00223
STD. ERROR OF REG. COEF.	.00009
CORRELATION COEF......	−.99107

ANALYSIS OF VARIANCE FOR SIMPLE LINEAR REGRESSION

SOURCE OF VARIATION	DEGREE OF FREEDOM	SUM OF SQUARES	MEAN SQUARE	F VALUE
DUE TO REGRESSION.............	1	.01776	.01776	552.31567
DEVIATION ABOUT REGRESSION....	10	.00032	.00003	
TOTAL....	11	.01809		

TABLE OF RESIDUALS

NO.	X VALUE	Y VALUE	Y PREDICTED	RESIDUAL
1	5.0000	1.4121	1.4026	.0095
2	10.0000	1.3955	1.3915	.0040
3	15.0000	1.3802	1.3803	−.0001
4	20.0000	1.3660	1.3692	−.0031
5	25.0000	1.3530	1.3580	−.0051
6	30.0000	1.3410	1.3469	−.0059
7	35.0000	1.3302	1.3358	−.0055
8	40.0000	1.3201	1.3246	−.0045
9	45.0000	1.3111	1.3135	−.0023
10	50.0000	1.3028	1.3023	.0004
11	55.0000	1.2953	1.2912	.0042
12	60.0000	1.2885	1.2800	.0084

Index

A—ampere, 2, 5
a—atto-, 3
abscissae, 51
absolute zero, 257
acceleration, 4
  due to gravity, 257
accuracy, 174
active ingredient in tablets, limits for, 243–248
algebra, 27–50
  in Fortran, 34–36
algebraic definitions, 27
alternate hypothesis, 190
ampere, 2, 5
analysis, dimensional, 5–6
analysis of variance, 197, 226–232, 253–256
analytical geometry, 261–264
ångström, 3, 5
angular measure, 123
answers to problems, 293–300
antilogarithms, 11
  table of, 290–291
A.P.—arithmetic progression, 66
approximations, 257
area, 4
  plane, 262
area integral, 264
area
  of circle, 260
  of ellipse, 260
  of parallelogram, 260
  of rectangle, 260
  of trapezium, 107, 260
  of triangle, 106, 259, 260
  under chromatographic peak, 119
  under curve, 108, 119
areas of solids, 264
argument, algebraic, 28
arithmetic and computing, 1–26
arithmetic mean—*see* mean
arithmetic progression, 66
Arrhenius equation, 141
asymptotes, 59, 152
asymptotic curve, 152
asymptotic graphs, 152
atmosphere, 3
atto-, 3
Avogadro number, 6, 7, 21, 257
axes, 51
  of ellipse, 58

bacterial growth, 138
bacteriological counts, 232–235
bacteriology, applications of statistics to, 210–235
bar, 3, 4
batch sampling, 236–238
binary numbers, 17
binomial distribution, 162–165
  mean and variance of, 269
binomial probability, 160–162
binomial theorem, 72–74
  for any index, 282
biological assay, 162, 164, 179
  applications of statistics to, 210–235
biomedical computer package, 206
BMD, 206
Boltzmann's constant, 257
borderline samples, 245
Boyle's law, 59
brackets in Fortran, 35
byte, 17

χ^2 for data in two classes, 275
χ^2 test, 190, 197–199, 233
χ^2, values of, 285
c—centi-, 3
cal—calorie, 3, 15
calculus, 80–92
calorie, 3, 15
Camp-Meidell expression, 185, 237
candela, 3, 5
Cartesian co-ordinates, 63
cd—candela, 3, 5
cell, computer, 17
Celsius (degree), 3
centi-, 3
centigrade, 3
cgs—centimetre-gramme-second, 13
change of origin, 55
change of variable, 86
changing the base of logarithms, 30
changing the variable, 112
characteristic, 10
charge on electron, 257
chi-squared—*see* χ^2, above
chromatographic peak, area under, 119
circle, 57, 58
  area of, 260
  circumference of, 260

circumference of ellipse, 260
Clausius-Clapeyron equation, 140
coefficients, algebraic, 28
combination of errors, 244
combinations, 70–71
common logarithms, 10, 30
comparison of data by statistical methods, 190–209
compiler, computer, 17
complete differentials, 100–101
complex numbers, 32
complex series, 153
compound probability, 159
computation, integration by, 116–120
computer cells, 17
computer compilers, 17
computer, curve fitting by, 155–157
computer input, 18–20
computer language, vii, 17
computer programmes—*see* page vi
computer units, 16
computer words, 17
computers, digital, 16–18
computing, arithmetic and, 1–26
concentration, 4
concentration gradient, 4
constants, fundamental, 257
convergency, tests for, 68
convergent series, 67
conversion factors, 257–258
co-ordinates,
    Cartesian, 63
    polar, 63
    rectangular, 51
    three-dimensional, 63
correlation coefficient, 203
correlation, significant linear, 203
cosecant, 122
cosine, 122
    derivatives of, 129
    difference of, 128
    exponential form of, 131
    integration of, 132
    power series for, 130
    sum of, 128
cosine
    of difference of two angles, 126
    of sum of two angles, 126
    variation with angle, 124–126
cotangent, 122
counts, bacteriological, 232–235
counts, particle, 169, 170
covariance, 201–203
Cramer's rule, 45
critical point of gas, 100
cube root, 2
cubic decimetre, 3, 4
cubic equations, 33
cumulative frequency, 185–187, 222
    percentage, 187
cumulative frequency curve, 187, 222
cumulative frequency integral, 187

curve, area under, 108, 119
curve fitting by computer, 155–157
curved graphs, 149
curved lengths, 261
curved regression, computer programme for, 206
curves,
    asymptotic, 152
    exponential, 60
    hyperbolic, 60, 152
    logarithmic, 60
    parabolic, 60
    reciprocal, 59
    sigmoid, 186

d—deci-, 3
da—deka-, 3
deci-, 3
decimal system, 1
decimals, 2
definite integrals, 110
    of odd and even functions, 133–134
definitions, algebraic, 27
degree Celsius, 3
degree centigrade, 3
degree kelvin, 2, 5, 16
degree of equations, 31
degrees of freedom, 172
deka-, 3
denominator, 2
density, 4
dependent variables, 52
derivatives—*see* differential coefficients
derived functions—*see* differential coefficients
derived SI units, 4
determinants, 43–45
deviate, normal, 176
    test, 190
    values of, 284
deviation, standard, 174, 175, 187
difference of sines and cosines, 128
difference of variates, 273
differential calculus, 77–92
    applications of, 88
differential coefficients, 79, 80, 93, 100–103
differential equations, 136–147
    with separable variables, 136
differentials, 80
    exact, 145–147
    partial and complete, 100–101
differentials of variables, 136
differentiation,
    partial, 100
    rules of, 81–86
differentiation of a function of a function, 86
diffusion coefficient, 4

INDEX

diffusion equation, 143
   solution of, 279–281
diffusion, Fick's laws of, 279
digital computers, 16–18
dimensional analysis, 5–6
dimensions, 5
   of derived SI units, 4
directrix of parabola, 56
dispensing errors, 248–253
dispensing of liquids, 248–252
dispensing of solids, 252–253
dissociation constants, 12
distribution,
   binomial, 162–165
      mean and variance of, 269
   normal, 166–168, 176–178, 270–272
   Poisson, 75, 168–170, 232, 276
      mean and variance of, 269
   skewed, 184
distribution equation, 168
division, 1, 7, 11, 14, 29
dose assay, threshold, 210–212
dyne, 3

e, 10, 30, 68, 69, 74–75
EIGEN, 104
electrical energy, 15
electromagnetic unit, 3
electron, charge on, 257
electron orbital, 143
electrostatic unit, 3
ellipse, 57, 58
   area of, 260
   circumference of, 260
empirical equations, 148
end-of-file cards, 18
end-of-record cards, 18
energy, 4, 15–16
entropy and exact differentials, 146
E.O.F. cards, 18
E.O.R. cards, 18
equations
   for experimental measurements, 148–158
   of higher degree, 33
   of the first degree, 31, 52–57
   of the second degree, 31, 55–57
   of the third degree, 33, 57
equations,
   algebraic, 28, 31–34, 38, 42–47
   cubic, 33
   degree of, 31
   differential, 136–147
   distribution, 168
   empirical, 148
   fitting of, to graphs, 148–153
   general solution of, 136
   graphical solution of, 61–63
   linear, graphs of, 53
   linear, homogeneous, 45–47

equations (*continued*),
   Newton method for solution of, 88–91
   parabolic, 149
   partial differential, 143–144
   solutions of, 31
   roots of, 31
erg, 3
error, experimental, 174
error, limits of, 172, 175, 210
   for a single measurement, 185
   in sampling, 236
   of a mean, 178
   probability for, 172, 191
error of prediction, 204
error, random, 172
error, standard, 175
   of potency ratio, 216
error, systematic, 174
errors, combination of, 244
errors in dispensing, 248–253
Euler's criterion, 145
even functions, definite integrals of, 133–134
exact differentials, 145–147
exp, 75
expansion, work of, 15
experimental errors, 174
experimental measurements, 148–158, 168–170
explicit functions, 87
exponent, 60
exponential curves, 60
exponential forms of trigonometric functions, 131
exponential theorem, 74–75
exponents—*see* indices
expressions, algebraic, 28
   in Fortran, 35

f—femto-, 3
factor t, 178
factorials, 69
factors, 1
femto-, 3
Fick's laws of diffusion, 279
first derivative of a function, 93
first-degree equations, 31, 52–57
first-order reaction, 136
floating point numbers, 24
flow, 4
foci of ellipse, 57
foci of hyperbola, 59
focus of parabola, 56
force, 4
Fortran IV, 17–25
Fortran
   card, 18, 19
   equivalents of symbols for mathematical operations, 27
   programmes—*see* **page vi**

INDEX

Fortran,
  algebra in, 34–36
  language of, vii, 17
  numbers, indices, logarithms and square roots in, 20
  operators in, 27, 34
  trigonometric functions in, 135
  variables in, 34
four point assay, 212
Fourier series, 154
Fourier's theorem, 154
fractional cumulative frequency, 185
fractional frequency of occurrence, 163, 165
fractional index, 7
fractions, 2
fractions, partial, integration by, 114
freedom, degrees of, 172
frequency, cumulative, 185–187, 222
  fractional, 185
  percentage, 187
frequency of occurrence, 162, 165, 197
  fractional, 163, 165
frequency of vibration, 143
frequency-distribution histogram, 165–168
functions,
  algebraic, 28
  explicit, 87
  generating, 163
  implicit, 87
  inverse, 87
  moment generating, 268
  periodic, 125
fundamental constants, 257

G—giga-, 3
g—gramme, 258
gas constant, 15, 257
gas, critical point of, 100
Geiger-Müller counter, 169
general solution of equations, 136
generating functions, 163
geometric progression, 67
geometry, analytical, 261–264
giga-, 3
Gothic symbols, 159
G.P.—geometric progression, 66
grain, 258
gramme, 258
gramme-molecule, 5
graph paper,
  logarithmic, 150
  log-log, 151
  probability, 187–189
  rectangular co-ordinate, 51
graphical integration, 114
graphical solution of equations, 61–63
graphical solution of simultaneous equations, 62

graphs, 51–65
  asymptotic, 152
  curved, 149
  fitting equations to, 148–153
  linear, 149
  periodic, 153
graphs of linear equations, 53
graphs with rapidly changing slopes, 149
gravity, acceleration due to, 257

h—hecto-, 3
half-life period, 137
harmonic motion, simple, 141–143
heat absorbed by ideal gas, 146
heat energy, 15
hecto-, 3
hemisphere,
  surface area of, 263
  volume of, 264
higher derivatives and partial differentiation, 93–105
higher partial derivatives, 101–103
higher-degree equations, 33
histograms, 165
homogeneous linear equations, 45–47
horizontal inflection, point of, 98
hydrogen-ion concentration, 12
hyperbola, 59
hyperbolic curves, 60, 152

i, 33
ideal gas,
  heat absorbed by, 146
  molar volume of, 257
identities, algebraic, 27, 28, 30
implicit functions, 87
  differentiation of, 87
inch, 257
increment, infinitesimal, 79
increments, 78
indefinite integrals, 110
independent variables, 52
independent variates, 202
index of significance, 217
indices, 6–9, 28
  in Fortran, 20
inequalities, algebraic, 28
infinite series, 67
infinitesimal increment, 79
infinity, 1
inflection, horizontal, point of, 98
inflection, oblique, point of, 186
input/output formats for real numbers, 21–25
input to the computer, 18–20
integers, 1
  in Fortran, 20, 25

INDEX 305

integral,
 area, 264
 probability, 177, 271
 volume, 264
integral index, 7
integrals,
 definite, 110
  of odd and even functions, 133–134
 indefinite, 110
 limits of, 107
 multiple, 264
 reduction of, to a standard form, 112–114
 standard, 265
integration, 106–120
 graphical, 114
 mean value of a function by, 115
integration
 as a summation, 106–109
 by computation, 116–120
 by partial fractions, 114
 by parts, 113
 by Simpson's rule, 117
 by substitution, 112
 by trapezium rule, 116
 of algebraic functions, 109
  rules for, 109–112
 of trigonometric functions, 132
 with a curve, 108
integration constant, 137
intercept of line, 52
intersection, points of, 54
inverse functions, 87
  relation between derivatives of, 87
inverse trigonometric functions, 132
irrational numbers, 8, 68
isothermals, 94

J—joule, 4, 15

K—kelvin, 2, 5, 16
K values, 12
k—kilo-, 3
kcal—kilocalorie, 15
kelvin, 2, 5, 16
kg—kilogramme, 2, 5, 258
kilo-, 3
kilocalorie, 15
kilogramme, 2, 5, 258
kilogramme-molecule, 5, 6
kilowatt-hour, 15
kinematic viscosity, 4
kmol—kilogramme-molecule, 5, 6
kurtosis, 184
kWh—kilowatt-hour, 15

l—litre, 3, 4, 258
language, computer, vii, 17
language of Fortran IV, vii, 17
Latin square, 213
l-atm—litre-atmosphere, 15
L.E., 210
least squares, method of, 199–201
length, 5
lethal dose, median, 225–226
level of significance, 190, 191, 192
library programmes, 17
lim, 78
limiting values, 78
limits for active ingredient in tablets 243–248
limits of error, 172, 175, 210
 probability for, 172, 191
limits of error
 for a single measurement, 185
 in sampling, 236
 of a mean, 178
limits of integrals, 107
linear correlation, significant, 203
linear equations, graphs of, 53
linear equations, homogeneous, 45–47
linear function of variates, 274
linear graphs, 149
linear regression, 199–206
 computer programme for, 206
litre, 3, 4, 258
litre-atmosphere, 15
ln, 10, 30
log, 10, 30
$\log_e$, 10
log E.D., 222
log effective dose, 222
log potency ratio, 211
logarithmic curves, 60
logarithmic graph paper, 150
logarithmic scales, 12–13
logarithmic series, 75–76
logarithms, 9–13
 changing base of, 30
 common, 10, 30
 natural, 10, 30, 74–76
 table of, 288–289
logarithms in algebra, 29–30
logarithms in Fortran, 20
log-log analysis of extinction time, 235
log-log graph paper, 151

µ—micro-, 3
M—mega-, 3
m—metre, 2, 257
m—milli-, 3
machine language, 17
Maclaurin's theorem, 94–96
mantissa, 10
mass, 5

INDEX

mass of proton, 257
matrices, 39–42
 and simultaneous equations, 42–47
maximum values, 96
mean, 164, 172, 187, 190
 limits of error of, 178
 reliability of, 172
mean
 log potency ratio, 211
 of binomial distribution, 269
 of Poisson distribution, 269
 potency ratio, 210
 single survivor time, 233–235
 value of a function, 115
measurements, experimental, 148–158, 168–170
measurements, repeated,
 large sets of, 182
 statistical analysis of, 172–189
mechanical energy, 15
median lethal dose, 225–226
mega-, 3
method of least squares, 199–201
metre, 2, 257
metre-kilogramme-second, 3
metric system, 2
micro-, 3
milli-, 3
millilitre, 258
millimetre of mercury, 3
minim, 258
minimum from a parabola, 156
minimum values, 97
mks—metre-kilogramme-second, 3
mks units, 3
ml—millilitre, 258
M.L.D.—median lethal dose, 225
mode, 185
mol—mole, 5
molar volume of ideal gas, 257
mole, 5
moment generating functions, 268
multiple integrals, 264
multiplication, 1, 2, 7, 11, 13, 29

N—newton, 4
n—nano-, 3
nanometre, 5
Naperian logarithms, 10
natural logarithms, 10, 30, 74–76
negative angles, 126
negative integral index, 7
negative numbers, 2
newton, 4
Newton method for solution of equations, 88–91
Newton's law of motion, 142
normal deviate, 176
 values of, 284
normal deviate test, 190

normal distribution, 166–168, 176–178, 270–272
nth root, 2
null hypothesis, 190
numbers, 1, 2, 6–8
 binary, 17
 complex, 32
 floating point, 24
 irrational, 8, 68
 rational, 8
 real, 20
numbers in Fortran, 20
numerator, 2

oblate ellipsoid, volume of, 261
oblique inflection, point of, 186
occurrence, fractional frequency of, 163, 165
occurrence, frequency of, 162, 165, 197
odd functions, definite integrals of, 133–134
operators, 27
 in Fortran, 27, 34
optimisation, 39, 103–104
orbital, electron, 143
ordinates, 51
origin, 51
 change of, 55
oscillators, 130, 141–143
ounce, 258
output format for real numbers, 23

π, 68, 69
p—pico-, 3
parabola, 55–56
parabola, minimum from, 156
parabolic curves, 60
parabolic equation, 149
parallel lines, 54
parallelism, test for, 215
parallelogram, area of, 260
partial and complete differentials, 100–101
partial derivatives, higher, 101–103
partial differential equations, 143–144
partial differentiation, 100
partial fractions, integration by, 114
particle counts, 169, 170
parts, integration by, 113
percentage cumulative frequency, 187
percentage response, 222–224
 probit of, 223
percentages, probits corresponding to, 286
periodic functions, 125
periodic graphs, 153
permutations, 69–70
personal errors in dispensing, 248

INDEX

pH, 12
pharmacy, applications of statistics in, 236–256
pico-, 3
picometre, 5
pint, 258
pK, 12
Planck's constant, 21, 257
plane area, 262
plating techniques, 232–233
point of horizontal inflection, 98
point of oblique inflection, 186
points of intersection, 54
Poisson distribution, 75, 168–170, 232, 276
    mean and variance of, 269
Poisson series, 169
polar co-ordinates, 63
positive index, 7
potency ratio, 210
    standard error of, 216
pound, 258
power, 4
power series, 153
    for sine and cosine, 130–131
power series method, 141
powers, 7, 11, 14, 29
precision, 174
prediction, error of, 204
prefixes for SI units, 3
pressure, 4
primes, 1
principal values of inverse trigonometric functions, 132
probability, 159–171
    binomial, 160–162
    compound, 159
    simple, 159
probability
    for limits of error, 172, 191
    for tests of significance, 191
    graph paper, 187–189
    integral, 177, 271
    level, 172, 178
probits, 189, 222–224
    weighting factors corresponding to, 286
probits corresponding to percentages, 286
process velocity, 77–79
product of variates, 273
progression, arithmetic, 66
progression, geometric, 67
prolate ellipsoid, volume of, 261
proton, mass of, 257
Pythagoras' theorem, 259

quadratic equations, 31
quantal response assay, 222–225
quantitative response assay, 212–221

quotient, 1
    of variates, 274

radian, 123
radioactive decay, 137
random error, 172
random sampling, 236
rate processes, 77–79
rational numbers, 8
reaction velocity, 77–79
real numbers, 20
    input/output formats for, 21–25
reciprocal curves, 59
rectangle, area of, 260
rectangular co-ordinates, 51
reduction of integrals to a standard form, 112–114
regression,
    curved, computer programme for, 206
    linear, 199–206
        computer programme for, 206
    sum of squares about, 229
    sum of squares due to, 229
    variance about, 203
regression analysis, 229–232
regression coefficient, 201
    variance of, 277
regression line, 148, 199–208, 277
    error of prediction from, 204
regression variance, 277–278
relations between trigonometric functions, 122
repeated measurements,
    large sets of, 182
    statistical analysis of, 172–189
representative sampling, 236
response, percentage, 222–224
    probit of, 223
roots, 2, 11, 14, 29
    of equation, 31
rules for integration of algebraic functions, 109–112
rules of differentiation, 81–86

sample mean, variance of, 275
samples, 168, 175
    borderline, 245
    qualitative tests on, 243
sampling, 236–238
sampling errors, 236
scatter, 172
scatter diagram, 199–200
scientific subroutine package, 47
secant, 122
second, 2, 5
second derivative of a function, 93
second-degree equations, 31, 55–57

second-order reaction, 138
separable variables, 136
series, 66–76
  complex, 153
  convergent, 67
  Fourier, 154
  infinite, 67
  logarithmic, 75–76
  Poisson, 169
  power, 153
  Tchebychev, 155
series for experimental measurements, 148–158
series to summarise measurements, 153–155
Sheppard's correction, 183
SI units, 2–5
  derived, 4
  prefixes for, 3
sigmoid curve, 186
sign convention for graphical integration, 115
significance, index of, 217
significance, level of, 190, 191, 192
significance, tests of, 190–209
  probability for, 191
significant linear correlation, 203
simple harmonic motion, 141–143
simple probability, 159
Simpson's rule, 117
  integration by, 117
simultaneous equations, 38
  graphical solution of, 62
  matrices and, 42–47
sine, 121
  derivatives of, 129
  difference of, 128
  exponential form of, 131
  integration of, 132
  power series for, 130
  sum of, 128
sine
  of difference of two angles, 126
  of sum of two angles, 126
  variation with angle, 124–126
skewed distributions, 184
skewness, 184
slide rules, 13–15
slope of line, 52, 53
small angles, 126
solid of revolution, surface of, 262
solid of revolution, volume of, 263
solids, areas of, 264
solids, volumes of, 264
solutions of equation, 31
sphere, surface of, 260
sphere, volume of, 261
square root, 2, 14
square roots in Fortran, 20
squares, sum of, 227
  about regression, 229
  due to regression, 229

SSP, 47
standard deviation, 174, 175, 187
standard error, 175
  of potency ratio, 216
standard integrals, 265
statements, algebraic, 28
stationary values, 98–100
statistical analysis of repeated measurements, 172–189
statistical methods, comparison of data by, 190–209
statistical tables, 284–286
statistics,
  applications of, in pharmacy, 236–256
  applications of, to biological assay and bacteriology, 210–235
  theorems in, 275–276
statistics and experimental errors, 174
Stirling's approximation, 257, 266
substitution, integration by, 112
sum and difference of sines and cosines, 128
sum of squares, 227
  about regression, 229
  due to regression, 229
sum of variates, 273
surface of hemisphere, 263
surface of solid of revolution, 262
surface of sphere, 260
surface tension, 4
survivor time, mean single, 233–235
symbols for mathematical operations and their Fortran equivalents, 27
symbols, Gothic, 159
systematic errors, 174
  in dispensing, 248
Système International d'Unités, 2

T—tera-, 3
t, 178
  values of, 284
t test, 190, 192–196
tablets, limits for active ingredient in, 243–248
tablets, uniformity of weight of, 238–243
tangent, 122
  derivatives of, 129
  integration of, 132
  variation with angle, 124–126
Tchebychev series, 155
tera-, 3
test for parallelism, 215
tests for convergency, 68
tests for toxicity, 225
tests of significance, 190–209
  probability for, 191
theorems in statistics, 275–276
third derivative of a function, 93

INDEX

third-degree equations, 33, 57
three-dimensional co-ordinates, 63
three-dimensional oscillator, 143
threshold dose assay, 210–212
time, 5
toxicity tests, 225
trapezium, area of, 107, 260
trapezium rule, integration by, 116
triangle, area of, 106, 259, 260
triangles, 259
trigonometric functions, 121
    derivatives of, 129–130
    exponential forms of, 131
    integration of, 132
    inverse, 132
    power series for, 130–131
    relations between, 122
    sum and difference of, 128
    variation of, with angle, 124–126
trigonometric functions in Fortran, 135
trigonometry, 121–135
turning point, 98

uniformity of weight of tablets, 238–243
universe, 168, 175
universe mean, 190
universe standard deviation, 178, 190

van't Hoff equation, 141
vapour pressure of liquids, 140
variables,
    change of, 86
    changing the, 112
    dependent, 52
    differentials of, 136
    independent, 52
    separable, 136
variables in Fortran, 34
variance, 164, 172
    analysis of, 197, 226–232, 253–256
    regression, 277–278
variance
    about regression, 203
    of a function of variates, 273–274
    of a sample mean, 275
    of binomial distribution, 269
    of Poisson distribution, 269
    of regression coefficient, 277
variance ratio test, 190, 196
variance ratio, values of, 285

variates, 163
    difference of, 273
    independent, 202
    linear function of, 274
    product of, 273
    quotient of, 274
    sum of, 273
    variance of a function of, 273–274
variation, coefficient of, 174
variation of trigonometric functions with angle, 124–126
velocity, 4
velocity constant, 136, 138
velocity gradient, 4
velocity of processes, 77–79
velocity of reactions, 77–79
vertex of parabola, 56
vibration, frequency of, 143
viscosity, 4
    kinematic, 4
volume, 4
    of hemisphere, 264
    of oblate ellipsoid, 261
    of prolate ellipsoid, 261
    of solid of revolution, 263
    of sphere, 261
volume integral, 264
volumes of solids, 264

weight factors, 223
weight of tablets, uniformity of, 238–243
weighting factors corresponding to probits, 286
word, computer, 17
work, 4, 15
work of expansion, 15

x-ray diffraction, 155

Yates' correction, 198

zero, 1
    absolute, 257
    logarithm of, 29